THE OESTRID FLIES

Biology, Host–Parasite Relationships, Impact and Management

Dedication

It is with great pride and not a little sadness that we dedicate this book to two of the most influential and respected oestrid biologists of our time, Paul Catts and Jerry Weintraub. Each has left an indelible mark on our field and their contributions can be seen throughout each of the chapters of this book.

Doug Colwell
Martin Hall
Phil Scholl

THE OESTRID FLIES
Biology, Host–Parasite Relationships, Impact and Management

Edited by

D.D. Colwell

*Agriculture and Agri-Food Canada,
Lethbridge Research Centre, Alberta,
Canada*

M.J.R. Hall

*Natural History Museum, London,
UK*

P.J. Scholl

*USDA–ARS–MLIRU,
Lincoln, Nebraska,
USA*

CABI Publishing

CABI Publishing is a division of CAB International

CABI Publishing	CABI Publishing
CABI International	875 Massachusetts Avenue
Wallingford	7th Floor
Oxfordshire OX10 8DE	Cambridge, MA 02139
UK	USA
Tel: +44 (0)1491 832111	Tel: +1 617 395 4056
Fax: +44 (0) 1491 833508	Fax: +1 617 354 6875
E-mail: cabi@cabi.org	E-mail: cabi-nao@cabi.org
Website: www.cabi-publishing.org	

A catalogue record for this book is available from the British Library, London, UK.

A catalogue record for this book is available from the Library of Congress, Washington, DC.

ISBN   0 85199 6841
          978 0 85199 6844

Typeset by SPI Publisher Services, Pondicherry, India.
Printed and bound in the UK by Biddles Ltd., Kings Lynn.

Contents

Contributors

Dr J.R. Anderson, University of California Berkeley (retired), current address, 1283 NW Trenton, Bend, OR 97701, USA; E-mail: jranderson@bendbroadband.com

Dr C. Boulard, Station de Pathologie Avaire et Parasitologie, INRA, Centre de Tours 37360, Nouzilly, France; E-mail: boulard@tours.inra.fr

Dr D.D. Colwell, Agriculture and Agri-Food Canada, Lethbridge Research Centre, 5403 1st Ave., S. Lethbridge, Alberta, T1J 4B1, Canada; E-mail: colwelld@agr.gc.ca

Dr E. Lello, Rua General Telles 2786, 18 602–120 Botucatu, SP, Brazil; E-mail: dr.montenegro@terra.com.br

Dr P. Dorchies, Interactions Hôtes-Agents Pathogènes, Ecole Nationale Vétérinaire de Toulouse, 23 Chemin des Capelles, F-31076 Toulouse Cedex 3, France; E-mail: p.dorchies@envt.fr

Dr M.J.R. Hall, Department of Entomology, The Natural History Museum, Cromwell Road, London, SW7 5BD, UK; E-mail: m.hall@nhm.ac.uk

Dr H. Hoste, Interactions Hôtes-Agents Pathogènes, Ecole Nationale Vétérinaire de Toulouse, 23 Chemin des Capelles, F-31076 Toulouse Cedex 3, France.

Dr P. Jacquiet, Interactions Hôtes-Agents Pathogènes, Ecole Nationale Vétérinaire de Toulouse, 23 Chemin des Capelles, F-31076 Toulouse Cedex 3, France.

Dr A.C. Nilssen, Zoology Department, Tromso Museum, University of Tromsø, N-9037 Tromsø, Norway; E-mail: arnec@imv.uit.no

Dr D. Otranto, Department of Animal Health and Welfare, Faculty of Veterinary Medicine, University of Bari, Str. Prov. Per Casamassima Km 3, 70010 Valenzano, Bari, Italy; E-mail: d.otranto@veterinaria.uniba.it

Dr T. Pape, Zoological Museum, Universitetsparken 15, DK-2100 Copenhagen, Denmark; E-mail: Tpape@smuc.ku.dk

Dr R. Roncalli Amici, 29 Louise Dr., Milltown, NJ 08850, USA; E-mail: r.ron@att.net

Dr P.J. Scholl, Research Leader, USDA-ARS-MLIRU, 305 Pl Bldg, UNL-EC, Lincoln, NE 68503-0938, USA; E-mail: pscholl@unlserve.unl.edu

Dr J.R. Stevens, Hatherly Laboratory, Department of Biological Sciences, University of Exeter, Prince of Wales Road, Exeter, EX4 4PS, UK; E-mail: J.R.Stevens@exeter.ac.uk

Dr G. Tabouret, UMR 1225, Ecole Nationale Vétérinaire, BP 87614, 23 Chemin des Capelles, F-31076 Toulouse Cedex 3, France; E-mail: g.tabouret@envt.fr

Dr D.M. Wood, ECORC, Agriculture and Agri-Food Canada, K.W. Neatby Bldg. C.E.F., Ottawa, ON K1A 0C6, Canada; E-mail: wooddm@agr.gc.ca

Preface

The oestrid flies are a small but unique group whose larvae cause myiasis in mammals and which have a great impact on the productivity and welfare of domestic animals. While the species important to livestock production have received considerable attention there is much less devoted to the vast majority that parasitize a wide range of other mammals, many of which are now endangered species. Since the last synoptic view of the family, great advances have been made in understanding these fascinating parasites, but the emphasis has remained on species of economic importance. This book is the first in 40 years to give a systematic treatment of the family Oestridae and draws together world-renowned experts to address the advances in systematics, biology, host–parasite relations and history.

Following an introductory chapter that sets the scope of the book there is an all-embracing account of the advancement of knowledge on oestrids from the first recorded references through to modern chemical control practices. Several beautiful plates from old texts illustrate some of the early understanding of life cycles. Two chapters discuss the current understanding of taxonomy and systematics, bringing together both an elegant and a detailed presentation of the family, using cladistics with a thoroughly documented molecular examination. Together these chapters solidify the monophyly of the family.

A brief general overview of the life cycle diversity is followed by detailed chapters on the morphology and biology of the larval and adult stages. Scanning electron microscopy gives a new perspective on the morphological adaptations seen in egg and larval stages. A detailed review of pupariation and the biology of the pupal stage leads into a thorough review of the fascinating behaviour of the adults; no more complete examination of the knowledge of these two life cycle stages can be found in any single place. Many crucial advances have been made in recent years on developing an understanding of the interactions between the invading larvae and its host. Three author groups, active and respected in their fields, summarize the tremendous volume of recent literature for each of three subfamilies: Hypodermatinae, Oestrinae and Cuterebrinae. Knowledge on host–larval interactions in the fourth subfamily,

Gasterophilinae, is scant, despite its importance to the horse industry. The review highlights the large gaps in our knowledge, not only for those larvae that parasitize domestic horses, but also for those that are found in other and perhaps more fascinating hosts. The chapters on host–larval interactions are followed by a thorough treatment of the role of oestrids in zoonoses and their impact on humans. The book continues with a critical analysis of the various approaches for control of oestrids and puts these in the context of integrated pest management. The final section includes a synopsis of the biology and impact of each of the currently recognized genera along with distribution and host lists. This section also provides images of larvae and adults of many poorly known genera. The book contains a large, consolidated list of references cited in each of the chapters which is perhaps the most comprehensive listing available on the Oestridae.

Our hope is that this book will provide new information to a wide range of readers, both the merely curious and the current workers in the field. We also hope that this work will stimulate further study of neglected members of the family. It is highly likely that there is a great deal more diversity in host–parasite relations and adaptations to parasitism than has been observed by the focus on species of veterinary importance.

Douglas D. Colwell
Martin J.R. Hall
Philip J. Scholl

1 Introduction

D.D. COLWELL,[1] M.J.R. HALL[2] AND P.J. SCHOLL[3]

[1]Agriculture and Agri-Food Canada, Canada; [2]Natural History Museum, UK; [3]USDA-ARS-MLIRU, USA

The Oestridae is a family of true flies in the superfamily Oestroidea (Diptera). It is a relatively small family, with only some 151 species in 28 genera (Wood, 1987) compared with two closely related families, the Calliphoridae (blowflies) and the Sarcophagidae (fleshflies), which have approximately 1000+ and 2000+ species, respectively (Shewell, 1987a,b). However, despite low numbers, these flies are, without exception, fascinating for their remarkable lifestyles. As larvae, all oestrids are obligate parasites of living vertebrates, usually mammals, causing a disease condition known as myiasis.

Myiasis – Definition

The infestation of animals with the larvae of insects in general was termed scholechiasis by Kirby and Spence (1818), but Hope (1840) later proposed the term myiasis (Greek, *myia* = fly) to identify those cases caused just by fly larvae. Myiasis has since been defined as: 'the infestation of live vertebrate animals with dipterous larvae, which, at least for a certain period, feed on the host's dead or living tissue, liquid body substances, or ingested food' (Zumpt, 1965). Various aspects of myiasis have been reviewed a number of times in the past decade (Hall and Smith, 1993; Hall, 1995; Hall and Wall, 1995; Guimarães and Papavero, 1999).

Myiasis – Categorization

There are two main systems for categorizing myiasis:

1. The disease agents are classified parasitologically, according to their level of dependence on the host (specific parasites = obligate; semi-specific parasites = facultative (primary, secondary and tertiary)) (Patton, 1922; Zumpt, 1965).

Table 1.1. Oestrid myiasis, categorized according to the anatomical site of infestation. (From Zumpt, 1965.)

Site of myiasis	Description	Subfamilies of Oestridae involved
Dermal/subdermal	Larvae penetrate skin and develop in boil-like swellings (furuncles)	Hypodermatinae, Cuterebrinae
Nasopharyngeal	Larvae deposited in nostrils or mouth and invade sinuses and pharyngeal cavities	Oestrinae
Intestinal	Eggs usually deposited on host and they, or hatched larvae, are ingested in grooming	Gasterophilinae

2. The disease is classified anatomically in relation to the location of the infestation on the host (Bishop in Patton, 1922; James, 1947; Zumpt, 1965).

By the first definition, the Oestridae are all specific or obligate parasites, i.e. they have to develop in or on a living host. They cannot complete their development on dead tissues (e.g. carrion).

Although the second definition might appear somewhat artificial, being based just on a physical site in or on the host, it is actually very appropriate for oestrids, the genera of which target different regions of the host body (Table 1.1). Hence, the anatomical classification of sites of infestation concords with the taxonomic classification of the flies. The oestrids have very different biologies and adaptations for survival at the different sites. The sites infested by oestrids can be divided into three regions: dermal/subdermal, nasopharyngeal and intestinal. Zumpt (1965) proposed two further categories of myiasis based on anatomical locations: sanguinivorous and urinogenital. The sanguinivorous category is of blood-sucking maggots that otherwise do not invade the body. No oestrids adopt this strategy, nor do any oestrids invade the urinogenital tract.

Oestridae – Taxonomy and Evolution

The recent evolution of mammals suggests that oestrid flies are also recently evolved and it is no surprise that they are considered to be 'higher' Diptera. Zumpt (1965) postulated two roots for the evolution of myiasis, the root for the specialized oestrids being the sanguinivorous root, derived from less specialized calliphorids that had ectoparasitic blood-sucking larvae, like present-day *Auchmeromyia* (on mammals) and *Protocalliphora* (on birds). Papavero (1977) speculated on the evolution of Oestrinae (as Oestridae) in relationship to their hosts, i.e. 25 genera in four orders (Marsupialia, Proboscidea, Artiodactyla and Perissodactyla). He noted that the hosts all shared the following characteristics in that they:

- were terrestrial herbivores;
- had a minimum length of 1 m and weight of 20–25 kg;
- were in the geographical and ecological area of the parasite for sufficient time to allow parasite/host adaptation; and
- were gregarious if inhabiting open situations.

Schäfer (1979) included the other oestrid subfamilies in addition to Oestrinae in his discussion of their evolution. He observed that these hosts lived either in herds (because animals in herds are more easy to find than single animals, although this depends on their distribution) or in burrows (because a burrow that does not move and can accumulate attractive odours over many years is more easy to find than a single animal).

The relationship of oestrids with herbivores is very strong. The only species that naturally parasitizes carnivores is *Dermatobia hominis*, although it mainly parasitizes herbivores. Probably the major reason that herbivores are dominant as hosts is that they are more numerous than carnivores and so supply more habitat (Schäfer, 1979). Hence on Isle Royale in Lake Superior, USA, the prey–predator biomass ratio was 452:1 (Jordan *et al.*, 1971) while in five African game reserves it ranged from 94:1 to 301:1 (Schaller, 1972). For the Gasterophilinae, an additional reason might be that the digestive tract of herbivores is a less hazardous environment for larvae than the digestive tract of carnivores.

Although the parasitic lifestyle can expose the parasite to particular hazards, e.g. attack by the hosts' immune system, it also has considerable advantages. There can be few more secure and underexploited environments on the African plain than inside the stomach of a large herbivore such as an elephant or a rhinoceros. There is generally an abundant supply of nutrients and the warm and stable body temperature of mammals is ideal for larval development. It can also be an environment in which the insect can safely overwinter in temperate climates, as has been shown to be the case with the sheep nasal bot fly, *Oestrus ovis*. The length of larval infestation in some species is astonishing, being measured in months rather than days, and it is a testament to the capacity of oestrids to overcome the immune responses of their hosts. However, immune responses do develop and, although they might not prevent a primary infestation, they can act to reduce secondary infestations by the same species (e.g. in *O. ovis*, Rogers and Knapp, 1973).

Oestridae – Adult/Larval Balance

All of the Oestridae have reduced mouthparts in the adult stage and cannot feed, although some may imbibe fluids (*Cephenemyia* and *Cuterebra*; Wood, 1987). Therefore, all of their nutrients for adult activities have to be ingested in the larval stage and they cannot be replenished once the adults emerge. This is in marked contrast to flies in the families Calliphoridae and Sarcophagidae, which include the other major agents of myiasis and which can and do feed as adult flies. While the larval stages of Calliphoridae and Sarcophagidae tend to be shorter in duration than the potential for the adult stages, the larval stages of Oestridae tend to be of longer duration, often much longer than the adult stages. Adult oestrids could be considered as just a sophisticated means of replication and vectoring of the immature stages. Indeed, the females of some species of *Hypoderma* emerge from their puparium with all eggs fully developed and the capacity to mate immediately and oviposit on hosts (Scholl and Weintraub, 1988). Females can be extremely persistent in their efforts to oviposit on hosts, for example, egg-laying females of *Gasterophilus intestinalis* will pursue galloping horses until the horses stop, when oviposition is immediately resumed (Cope and Catts, 1991). The similarly persistent activity of *Hypoderma* can cause a dramatic escape response by cattle, termed 'gadding'.

Oestridae – Host Specificity and Finding

Another striking difference between these families is that whereas calliphorids and sarcophagids tend to be catholic in their selection of hosts, oestrids tend to be highly host-specific. Thus the host associations can be linked to the taxonomic classification of the flies (Table 1.2), as with the sites of infestation (Table 1.1).

Much of the host specificity is a product of the fine balance between parasite and host, and the adaptations made by the parasite to survive in the host. If oestrids are introduced to a host that is widely different from the natural host, then frequently the parasite will not develop properly. Numerous cases of human infestation with the sheep nasal bot fly are reported (Chapter 11), but the larvae do not normally complete development. In one rare case of development to the third instar, the infested human was reported to be HIV-positive and, therefore, immunocompromised (Badia and Lund, 1994).

Because of the parasite–host specificity it is important that female flies are well equipped to locate their hosts, so that the survival of their offspring is optimized. Little is known of the host location strategies of the Oestridae (Hall, 1995). However, there are two dominant forms of oviposition/larvaposition, either on-host or off-host. The vast majority of Cuterebrinae adopt the latter strategy, laying their eggs at sites likely to be visited by hosts such as the entrances of nests. The stimuli that attract females to these sites are not known. They could include olfactory attractants, but Capelle (1970) recorded oviposition of *Cuterebra polita* in the abandoned burrows of pocket gophers, suggesting that visual cues might also be important. The single atypical cuterebrid is *D. hominis*, which lays eggs on to so-called 'porter' flies, zoophilic or anthropophilic flies that it captures, presumably in the vicinity of mammalian hosts. There are few reports of adult *D. hominis* around hosts (Hall, 1995) and it is not known if they show any selective behaviour in approaching hosts. Whatever the actions of these females, once they have attached eggs to the porter flies the final selection of a host is down to the porters, which is the reason for the catholic range of hosts of *Dermatobia*.

Species in genera of Hypodermatinae, Gasterophilinae and Oestrinae usually adopt an on-host strategy for oviposition or larvaposition. Therefore, they require good host location behaviour and this involves both olfactory and visual components. Unlike most myiasis-causing Calliphoridae, which require some predisposing condition of the host to stimulate attraction, e.g. wetting or wounding, oestrids are attracted to normal healthy hosts. Olfactory stimuli are most impor-

Table 1.2.   Relationship of oestrid taxa to host group.

| Oestridae classification | | Host groups |
Subfamily	Genus	
Oestrinae	*Cephenemyia*	Cervidae
	Cephalopina	Cameline
	Gedoelstia	Antelopes
	Kirkioestrus	Antelopes
	Oestrus	Ovine, caprine
	Pharyngobolus	Elephant
	Pharyngomyia	Cervidae, zebras, pigs, giraffe, hippopotamus, springbuck, sheep
	Rhinoestrus	Equine
	Tracheomyia	Kangaroos
Gasterophilinae	*Cobboldia*	Elephants
	Gasterophilus	Equine
	Gyrostigma	Rhinoceros
Hypodermatinae	*Hypoderma*	Bovine
	Oestroderma	Pikas
	Oestromyia	Mice, marmots, pikas
	Pallasiomyia	Saiga antelope
	Pavlovskiata	Goitered antelope
	Portschinskia	Mice and pikas
	Przhevalskiana	Caprine and gazelles
	Strobiloestrus	Bovidae, *Kobus* species
Cuterebrinae	*Cuterebra*	Rodents, lagomorphs, howler monkey
	Dermatobia	Non-specific, larger mammals and birds
Cuterebrinae (probably)	*Neocuterebra*	African elephant
	Ruttenia	African elephant

tant in long-range attraction to hosts and include CO_2, e.g. for *Hypoderma* and *Cephenemyia* (in Hall, 1995). Visual stimuli are more important at close range, especially for those species that do not even land on the host, but are able to direct a spray of larvae towards the site of infestation, e.g. *Rhinoestrus latifrons* and *R. purpureus* (Rastegayev, 1984), *Cephenemyia jellisoni* (Cogley and Anderson, 1981) and *C. trompe* (Anderson and Nilssen, 1990).

The only species of Hypodermatinae, Gasterophilinae and Oestrinae that does not lay its eggs/larvae directly on the host is *G. pecorum*, which instead deposits eggs on grasses where they remain viable for months, until ingested by equids. To counter this rather random method of 'selecting' a host, *G. pecorum* produces more eggs than any other gasterophilid, up to 2000 per female.

Oestridae – Pathology of Infestations

Because the oestrids tend to be very specific to their hosts they also tend to be very well adapted to their hosts. As parasites go, they are largely benign and the majority of them cause little lasting damage to their hosts or even much trauma during an infestation. In economic terms, the impact of oestrid infestations on the host is much more important in respect of the effect on host products harvested by farmers after host death than on any pathological effect on the living host. Thus, the groups of oestrid flies that cause most economic impact are the warble flies (*Hypoderma* species) and the human bot fly, *D. hominis*, a cuterebrid. The furuncles in which these parasites live on the host can seriously degrade the quality of a hide, because of the scarring and small holes produced. Conversely, as long as the infestations are not too severe, the lesions are well tolerated by the living host and, when the larvae exit the host the lesions heal swiftly.

Oestridae – Conservation Perspectives

Many species of Oestridae that parasitize wildlife are rarely encountered as larvae, let alone as adult flies. Most collections are made during culls of wildlife or as a result of other hunting activities. They are also collected from animals recently introduced to zoological gardens. Their high host specificity has important conservation implications. For example, the rhinoceros bot fly, *Gyrostigma rhinocerontis*, is restricted to the black and white rhinoceros of Africa. It is well adapted to finding rhinoceros because the larvae leave the digestive tracts of the hosts, where they have been developing, during defecation. Because rhinoceros, especially white rhinos, tend to defecate in a communal spot (Dorst and Dandelot, 1972), the chances of emerging flies finding a new host are higher than if the larvae were scattered throughout the rhinos' range. However, because these hosts have declined in numbers, largely due to poaching, the numbers of the fly must also have declined. The number of available hosts in Africa was estimated at about 3100 black rhino and 11,670 white rhino in 2001. The intensity of infestation in rhinos is unknown, but related gasterophilids in horses, *G. intestinalis* and *G. nasalis*, have mean levels of infestation of 40–50 larvae per horse (Gawor, 1995; Ribbeck *et al.*, 1998; Lyon *et al.*, 2000). If every rhino in Africa was infested with 50 larvae and the same numbers of the species live outside the host as pupae and adults, then only around 1.5 million individuals of this species would be left in the world today, in any stage from larvae to adults, which are very small numbers for a fly.

The Oestridae – Rationale

In summary, the Oestridae represent a very special group of flies. We hope that this book will stimulate continued and expanded study of these fascinating parasites. They afford a wonderful opportunity for the study of host–parasite relationships. Even if their effects on the host are not often overtly deleterious, they can have a major impact on animals as resources for human food and materials.

Infestations that might not have a serious impact on an individual animal's health can still have an impact on the performance of the animal, which might affect the profitability of milk and meat production, although further work in defining economic impacts is needed (Colebrook and Wall, 2004). Therefore control of many of the oestrid parasites of domesticated animals is usually sought. Livestock owners also owe it to their animals to reduce welfare problems to as low a level as possible. Myiasis can be an important animal welfare problem even if it does not result in death or permanent injury.

It has been almost 40 years since knowledge of the Oestridae was thoroughly summarized (Zumpt, 1965). Tremendous advances have been made since that time in understanding the biology, control and host responses of the economically significant species. While management and reduction of the impact on animal health has advanced significantly there are important challenges. In particular, the need for increased sustainability of agricultural practices drives the search for alternative, integrative approaches for management. We have attempted to bring together a synopsis of the current state of knowledge and place this unique group of flies into the context of the modern understanding of host–parasite relationships, phylogeny and chemotherapy. With this we hope to help frame questions that stimulate further investigation.

In contrast to the advances made in knowledge of the economically important members of the Oestridae, many other members of the group have faded into obscurity. No more is known about their biology and host relationships than when Zumpt (1965) prepared his work. Several of the less well known species are known from rare or endangered hosts and are thus at risk themselves. With this volume we hope to inform readers about these gaps in knowledge (for we are certain that there are many fascinating variations on our current knowledge), to stimulate the readers' curiosity and ultimately encourage new research.

2 Historical Perspectives on the Importance and Impact of Oestrids

R. Roncalli Amici

Milltown, New Jersey, USA

Over the centuries, since the time of the pharaohs, the presence of oestrids and their pathogenic effect on both animals and humans have been reported by a large number of authors. As myiasis agents of veterinary and medical importance the Oestridae have produced great losses to the livestock industry and considerable pathogenicity and, occasionally, death in man. The genera of major economic and clinical importance are *Oestrus, Cephenemyia, Gasterophilus, Hypoderma, Oedemagena* and *Dermatobia.*

The history of myiasis is a most interesting one, including a cast of players ranging from emperors to shepherds, explorers and missionaries. It deals with the suffering of animals and humans and has engaged many scientists in the pursuit of a better understanding of the disease and of its control.

Fossil Oestrids

The oldest representatives of the Oestridae family belong to unidentified specimens collected by Wilmot Hyde Bradley (1899–1979) from the Eocene oil shales of the Green River formation in the Rocky Mountains of Colorado and Utah (Bradley, 1931). There is some debate as to the identification of these specimens despite their superficial resemblance to gasterophilid larvae (see Fig. 3.1).

Oestrids in Antiquity – Egyptian, Greek and Roman periods

The first description of oestrid larvae probably goes back to the Kahum veterinary papyrus which was discovered in Egypt in 1889 by the British archaeologist, Sir Flinders Petrie (1853–1942). This papyrus was written during the reign of Amenemhat III, one of the pharaohs of the XII dynasty (*c.* 1850 BC). According to the interpretation of Griffith (1898) (cited in Penso, 1973) and Oefele (1901) (cited in

Papavero, 1977), two of the several translators of the papyrus, a case of cutaneous myiasis, probably caused by *Hypoderma bovis*, was described in a calf.

The Greeks, who coined the name of Oistros, knew the gadding effects of *Hypoderma* flies and tabanids on cattle. Aristotle (384–322 BC), in the *History of Animals*, reported the presence of what were probably *Cephenemyia* larvae from inside the heads of deer (Papavero, 1977).

The Romans also knew about the presence of the *Hypoderma* fly, which they called *Asilus*. Publius Virgilius Maro (71–19 BC), the Caesarean poet, in his agricultural manual, *Georgics*, reported how winged pests (probably *Hypoderma*) were driving cattle herds through the groves in terror (L. III, verses 146–151). The Roman philosopher Lucius Annaeus Seneca (4 BC–65 AD) mentioned in one of his *Epistles to Lucilius* that the sheep nasal bot called *oistros* by the Greeks was called *asilus* by the Romans. Other Roman writers such as the encyclopaedist Pliny the Elder (Caius Plinius Secundus) (AD 23–79) wrote in his *Historia Naturalis* (L. XI) that 20 vermiculi (*Cephenemyia* larvae) were present in the head of a deer 'beneath the hollow of the tongue'. According to Smithcors (1957), the Romans considered that bots were harmful to horses; he deduced this from Pliny's statement that the powdered hipbones of swine 'conveighed downe by a horne into the throat of horses and such like beasts, will cure the wringing torments of the bots that fret and gnaw them in the bellies'.

In the 5th century AD, Publius Vegetus Renatus (450–510) wrote the first veterinary treatise *Artis veterinariae sive mulomedicinae* ('The Veterinary Art') in the Christian era. In Chapter XLIV (L. I) of this work, which was first printed in 1528, Vegetius reported the presence of parasites in the anus of horses, probably larvae of *Gasterophilus haemorrhoidalis*, which he describes as being similar to cooked beans (*in ano fabae coctae similes*).

Byzantine and Medieval Times

During the medieval period, accounts on oestrids were scanty; there are, however, some reports of interest from Byzantine and Arabic sources. Alexander Trallianus (525–605), one of the most famous Byzantine physicians, reported in his book *De Arte Medica*, published in the 6th century, that the oracle of Delphi recommended the use of larvae (*Oestrus ovis*) expelled through sneezing from the nostrils of goats for the treatment of epilepsy (Papavero, 1977).

In the 10th century, under the reign of the Emperor Constantine VII Porphyrogenitus (905–959), *Geoponika*, an agricultural encyclopaedia, was written by Cassianus Bassus or possibly by the emperor himself. It reported that *myopes* (probably *Hypoderma* spp.) terrorize cattle. Prevention could be achieved by scattering infusions of bay fragments on pastures or they could be cured by the use of plaster (Papavero, 1977). An Hippiatrika compiled by an anonymous author during the same period noted the presence in horses of σκαλόυες or τερηδόνς, interpreted as larvae of *Gasterophilus* spp. (Penso, 1973).

In the 11th century, Raschi (Rabbi Schlomo Jizchaki) (1030–1105), a Talmud commentator born in the Champagne region, reported 'Wrümer des Fleisches' or 'Grabeleisch' (i.e. *Hypoderma* spp.) located 'zwischen der haut und dem fleisch' (between the skin and the flesh) of cattle (Bodenheimer, 1928).

The Arabic physician, Abu-Merwan ibn Zuhr (1113–1162), also known as Avenzoar, described a case of human myiasis, caused by *Hypoderma* spp. He noted that a worm sometimes breeds between the skin and flesh (*inter cutem et carnem*); this infestation is called 'cattle disease' since it affects cattle. He also reported that the best method to kill the worm is to burn it with a red-hot iron. The same condition, i.e. worms between skin and flesh, was also described by Pietro d'Abano (1250–1316), one of the most illustrious physicians of the late Middle Ages and, also, a follower of Arabic medicine (Penso, 1973).

Damiri, the common name of Kamal ijd-Din Muhammad ibn Musa ad-Damiri (1344–1405), a well-known Arabic writer on natural history as well as on canon law, described in his 'Hayat al-Hayawan' (Life of Animals), under 'Nagal', worms located in the nose of camels (*Cephalopina titillator*) and sheep (*O. ovis*) (Bodenheimer, 1928).

The 1500s and the Hippiatrics

During the 16th century, *Gasterophilus* spp. were observed and reported by a number of hippiatrics (the Greek word describing equine veterinarians, also known as ueterinarii or mulomedici by the Romans). The value of horses, for war, pleasure and work, made observations on their parasites of great importance. *Gasterophilus* larvae in horses were described by Agostino Columbre (1518), a practising veterinarian. He referred to their resemblance to 'cooked beans', repeating a comment made by the Roman author Vegetius.

Antonio Brasavola Musa (1500–1555), physician to the Farnese family in Ferrara, Italy, found, at necropsy, 'red worms' shaped like pumpkin seeds 'cucurbitae' in the stomach of horses. *Gasterophilus* larvae were also reported in horses by other skilled hippiatrics such as Pasquale Caracciolo and Filippo Scacco (Roncalli, 2001).

Carlo Ruini (1598), an accomplished veterinary anatomist from Bologna, Italy observed, at necropsy, the presence of *Gasterophilus* larvae in the stomach of horses. Shortly thereafter, Ulisse Aldrovandi (1602) reviewed the Greek and Latin literature on oestrids, describing their knowledge of infestations in cattle. In that period, the first depiction of *Gasterophilus* spp. appeared in a Japanese pictorial scroll (Momoyama Jidai Kaibo zu) of the Momoyama period (1582–1602); in spite of the ravages of wars and earthquakes the scroll still exists in Tokyo. In a book (Mouffet, 1634) published after his death, the Elizabethan physician and entomologist Thomas Mouffet (1553–1604), reported that a fly called 'oestrum' caused cattle 'to run mad'.

The Conquistadores Bring Oestrids to America; but Find New Ones Locally

Christopher Columbus (1451–1506) in his second voyage to America (1493) brought with him cattle, sheep, pigs, horses and donkeys. Other live animals were carried to the American shores from Europe by the 'Conquistadores' during successive landings. It is conceivable that a number of oestrids, e.g. *O. ovis*, *Hypoderma* spp., *Gasterophilus* spp., came along. On the other hand, the Europeans found in the New World previously unknown oestrids, i.e. *Dermatobia hominis*, and *Cuterebra* spp. For example, Father Bernardino de Sahagún (1500–1590), the Spanish missionary and anthropologist who wrote about the history of the Aztec people in Mexico, reported in his 'Historia general de las cosas de Nueva España', written between 1576 and 1577, on the presence of *Cuterebra* larvae in rabbits and rats (Papavero, 1977). In 1626, Franciscan Father Pedro Simón reported on *Dermatobia* infestations among Spanish 'Conquistadores' travelling in 1569 along the Rio Magdalena in Colombia (Guimãraes and Papavero, 1966). In 1745, the French explorer and mathematician Charles Marie de la Condamine (1701–1774) reported on *Dermatobia* infestations among humans and animals along the Amazon River. According to this author, the Mayan Indians were familiar with *Dermatobia* and called it 'saglacuru'. The larvae caused 'une douleur insupportable' (an unbearable pain) in humans. La Condamine made a drawing of a *Dermatobia* larva and carried with him the original sample, kept in alcohol. He believed that the real cause of *Dermatobia* infestations was not yet known. Similar cases were reported among people travelling along the Rio Tietê in Brazil in 1769 by the 'Sargento-mor' Theotõnio José Juzarte in his 'Diário da navegação' (1769–1770).

The 17th Century and the Beginning of Scientific Research

In the 1600s, the study of parasitology gained great impetus with the work of Franceso Redi (1626–1697), a court physician of the Medici family in Florence. He contributed to rational thought through debunking the theory of spontaneous generation of insects, by demonstrating that flies develop on putrefied flesh only when flies deposit eggs. He also published some lucid engravings of insects, including the first depiction of a larva of *Cephenemyia*, the nasal bot of deer (Redi, 1668).

Antonio Vallisnieri (1661–1730), a physician at the University of Padua, published several books containing his findings on human and animal parasites. In two of them, 'Saggio de dialoghi sopra la curiosa origine di molti insetti' (Essay on Dialogs on the Curious Origin of Many Insects) (1696) and 'Esperienze, ed osservazioni intorno all'origine, sviluppi e costume di varj insetti' (Experiences and Observations on the Origin, Development and Customs of Various Insects) (1713a) he reported the presence of worms in the superior part of nasal passages and in the frontal sinuses of sheep and goats and in the skin of cattle. He also published fine drawings of larvae and adults of *O. ovis* and *Hypoderma* spp. (Figs 2.1 and 2.2). In a letter to Senator Garzoni of Venice, a well-known hippiatric, Vallisnieri (1713b) described the life cycle of *Gasterophilus* spp. He was the first to rear adult *O. ovis* from mature third instars and to describe its life cycle in sheep. Vallisnieri also reported on the trephining of a frontal sinus of a peasant with a recovery of a mature larva of *O. ovis* by a physician who was also Vallisnieri's relative (1733). This was one of the first records of oestrosis in humans and, possibly, the first attempt ever to treat this myiasis surgically either in man or in sheep. Another treatment, for bots in horses, was proposed by Thomas De Gray (1684), a British hippiatric, who suggested that red mercury mixed with butterballs was effective.

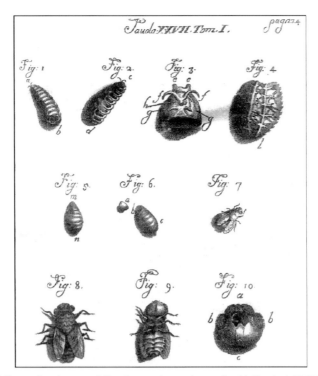

Fig. 2.1.   Life cycle stages of *Oestrus ovis* as drawn by Vallisnieri (1733).

The 18th Century: the Issuance of *Systema Naturae* and Appearance of Veterinary Schools

One of the most brilliant researchers of the 18th century was the French natural-ist René-Antoine Ferchault de Réaumur (1683–1757). He published several illus-trated reports (1734–1740) on the life histories of Oestridae in sheep (mouche du ver du nez des moutons) (Vol. IV, Plate 35) (Réaumur, 1734), deer (Vol. V, Plate 9) (Réaumur, 1740), cattle (Vol. IV, Plate 36) (Réaumur, 1738) and horse stom-ach bots (Vol. IV, Plate 35) (Réaumur, 1738).

At the same time, Carl von Linné (Linnaeus) (1707–1787), a Swedish physi-cian as well as a naturalist, studied the cutaneous parasites of reindeer in Lapland (Linnaeus, 1758). He also observed the nasal bot in sheep and was responsible for coining and introducing its name in the scientific literature, first in 1746 as *Oestrus sinus frontis* and, later, in 1761 under its now valid name *O. ovis* (Brauer, 1863). Linnaeus also named the species *Oestrus hominis* (= *D. hominis*) in a letter addressed to and published by Peter Simon Pallas in 1781 (Guimarães and Papavero, 1966). As reported by Papavero (1977, p. 18), Linnaeus, in his 10th edition, fixed defi-nitely the concept of *Oestrus*, creating that genus for those flies with reduced mouthparts whose larvae were parasitic on the skin, head or digestive tract of some mammals. His genus *Oestrus* included the species *bovis, tarandi, nasalis, haemorrhoidalis* and *ovis*, parasitic on the ox, reindeer, horse and sheep.

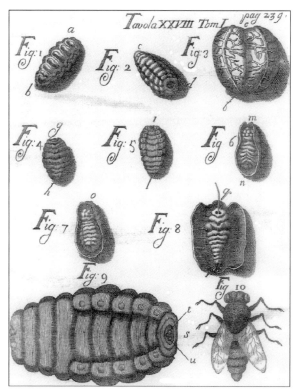

Fig. 2.2. Life cycle stages of *Hypoderma* sp. as drawn by Vallisnieri (1733).

The German naturalist Peter Simon Pallas (1741–1811), a follower of Linnaeus, who became professor at the Academy of Sciences in St. Petersburg (1768), described in *Spicilegia zoological* (1776, 1777) (Papavero, 1977) a number of oestrid larvae (some for the first time) found in mammals during his 6-year expedition to Siberia. Other entomologists of note who worked in the area of systematic dipterology were the Swede Carl Baron de Geer (1720–1778) and the Dane Johann Christian Fabricius (1745–1808), a pupil of Linnaeus, who emphasized the mouthparts in his classification of insects, in contrast to Linnaeus, who used wings as the primary characteristic for his classification.

With the creation of veterinary schools, in France and in other European nations, during the second half of the 18th century more attention was paid to the clinical aspects of myiasis in livestock and to methods of treatment. A French hippiatric, Françe Alexandre Pierre de Garsault (1692–1778), reported on the presence of bots in the stomach of horses, and included drawings of *Gasterophilus* (Garsault 1770, Plate V). Philippe Etienne La Fosse (1738–1820), in his famous treatise *Cours d'hippiatrique* (La Fosse, 1772), described the presence of bots in the stomach of horses and featured illustrations of larvae *in situ* as well as seen under a microscope (Plate L). Louis Jean Marie Daubenton (1716–1799), a renowned French naturalist, published several editions of a book for shepherds (first edition –1782) in which there was a good description of lesions caused by oestrids in sheep. Daubenton also studied bots in the stomach of horses and published depictions of them in *Histoire naturelle* by George Louis Leclerc Buffon (1707–1788) in 1753. Giovanni Brugnone (1741–1818) of the Veterinary School of Turin in his treatise on horse breeds (Brugnone, 1781) described in great detail *Gasterophilus* infestations in horses; he disputed Vallisnieri's belief that bots had a definite clinical effect on horses; also, he provided advice on how to prevent and treat bots in horses.

Of interest is a dissertation on oestrids in sheep and cattle, enriched by several plates, published in 1787 at the University of Leipzig by Bernhard Nathanael Gottlob Schreger (1766–1825), who later became a prominent surgeon, and by Johann Leonhard Fischer (1760–1833), an anatomist and parasitologist in his own right, who in cooperation with Paul Christian Friederich Werner (1751–1785) had in 1782 issued a work illustrating larvae of *Oestrus haemorrhoidalis* (Penso, 1973). In 1782, Philip Chabert (1737–1814), Director of the Veterinary School of Maison-Alfort in France, reviewed conditions caused by oestrids in horses, cattle and sheep. For the treatment of nasal bots in sheep he recommended the use of 'huile empyreumatique', a concoction containing oil, horn powder and turpentine – developed by Chabert himself, and which, according to him, was requested in all the French provinces as well as abroad. The cost of this product, sold at Alfort, was 3 francs/pint. 'Huile empyreumatique' was administered via intranasal injection or better by trephining the frontal sinuses of sheep; he also illustrated the trephination sites in Plate (I), which appeared in his first edition of *Traité des maladies vermineuses dans les animaux* (1782).

The 19th Century: Development of Systematics; Reporting of New Oestridae; Recognition of Losses Produced by Oestrids in Livestock

In 1811, Henry Alexander Tessier (1741–1837) of the Veterinary School of Maison-Alfort reported that the use of trephining for the removal of bots from the nasal cavities of sheep was often successful.

In 1815, Bracy Clark Esq. (1771–1860), a British veterinarian and Fellow of the Linnean Society, published *An essay on the bots of horses and other animals*, a first-class work on his parasitological observations on local oestrids in England and also on some received from abroad. The publication also contains a good historical review of myiasis; the illustrations that accompany the text are excellent and some of them well convey to the reader the suffering of animals affected with myiasis (Fig. 2.3). Of interest, is his description of *Cuterebra* from specimens sent from Savannah, Georgia (USA).

The studies of Clark stimulated a number of researchers, especially in Holland where Alexander Numan (1780–1852) (Numan, 1834) published a book on the pathogenicity of bots in horses. Later, another Dutchman, Jacobus Ludovicus Conradus Schroeder van der Kolk (1797–1862), issued a publication ('*Die Ostraciden*') on the anatomy and physiology of *Gastrus equi*. Also inspired by the work of Clark was a treatise published by Konrad Ludwig Schwab (1780–1859) in Munich in 1840; which provided insight on oestrids at that time. In 1818, William Elford Leach (1790–1836), a British diptera taxonomist, divided oestrids

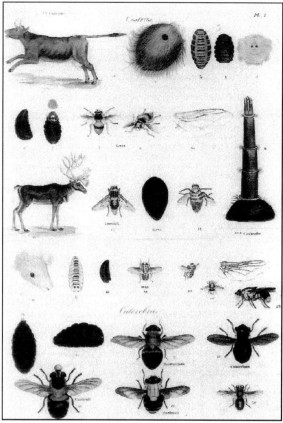

Fig. 2.3. Representative specimens and hosts of Oestridae. (From Clark, 1815.)

into two groups: *Oestrus* and *Gasterophilus*; to these, the French zoologist Pierre André Latreille (1762–1833) added the genus *Hypoderma* in 1825 (Penso, 1973).

One of the greatest entomologists of the 19th century was the Austrian Friederich Moritz Brauer (1832–1904). He published a monumental text (Brauer, 1863) that summarized all the knowledge on oestrids at that time and provided the basis for the classification of the Oestroidea.

In 1872, the Russian Colonel Nikolaj Michajlowvicz Przhevalsky (1839–1888), a geographer, botanist and naturalist, while on an expedition in Tibet noted large larvae under the skin of chiru that he had shot. K.Y. Grunin, while examining untanned skins of chiru at the Zoological Institute of Leningrad, found a number of *Przhevalskiana orongonis*; a year later, Grunin erected the genus *Przhevalskiana* (Zumpt, 1965).

A French scientific journalist, Louis Figuier, with the cooperation of eminent parasitologists such as Raphael Blanchard (1875–1919), also a well-known historian, produced *Les insects* (Figuier, 1867), a popular account of the life of insects, including oestrids. Figuier used the services of Émile Bayard (1837–1891), one of the best French engravers of the 19th century, to depict oestrid-produced 'gadding' scenes for sheep, cattle and horses, which, over the years, were reproduced in several textbooks.

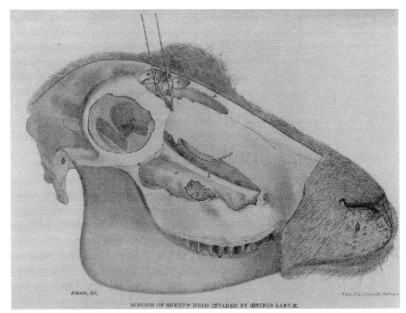

SINUSES OF SHEEP'S HEAD INVADED BY ŒSTRUS LARVÆ.

Fig. 2.4. Diagrammatic section of a sheep head showing the technique for removal of *Oestrus ovis* larvae. Dashed lines indicate the location and orientation of the trephine. (From Curtice, 1890.)

In the middle of the 19th century, with the expansion of the livestock industry worldwide, there was a parallel search for the improvement of health conditions of animals. In the USA, with the creation of the Bureau of Animal Industry in 1884, particular attention was directed to the study and treatment of parasitic diseases (including *O. ovis*, *Hypoderma* spp.). Cooper Curtice (1856–1939) placed *O. ovis* (Curtice, 1890) as the first of a series of nine most destructive parasites in sheep. Curtice (1890) dedicated nine pages and four beautiful plates to *O. ovis*. He also discussed the use of preventive (smearing noses with tar and grease), medicinal (nasal injection with turpentine and linseed oil) and surgical (trephining) treatments (Fig. 2.4). The British economic entomologist Eleanor Ormerod (1828–1901) conducted a series of investigations on cattle grubs and on the economic losses produced by this pest; she also suggested that the larvae of *Hypoderma* penetrate through the skin of cattle (Wallace and Ormerod, 1904; Bishopp *et al.*, 1926).

The 20th Century: Elucidation of the Biology and Life Cycle of Important Oestrids

During the first part of the 20th century, a number of researchers in Europe, North and South America and Africa studied the biology and the life cycle of a number of oestrids. *Hypoderma* spp. and *Dermatobia* were considered serious pests for the cattle industry, with deleterious consequences, especially with regard to the tanning industry; therefore, many efforts were made to better understand their biology and to develop a suitable treatment for their control.

In 1912, the American parasitologist, Brayton Howard Ransom (1879–1925) (USDA), while reading a paper to the National Association of Tanners, proposed an eradication programme to eliminate cattle grubs in the USA. A few years later (1926), Fred Corry Bishopp (1884–1970) *et al.* (USDA) published a good monograph on the cattle grubs, their biologies and suggestions for control. In Europe, noteworthy were the publications in several languages by August W.E. Gansser (1923–1956), a Swiss scientist; these publications, some of them sponsored by 'The Hide and Allied Trades Improvement Society' in England, dealt with the biology and control of warble flies. It would take, however, many decades to see the eradication of *Hypoderma* spp. in cattle in a number of European countries and a drastic reduction in its frequency in cattle in the USA.

Edmond Sergent (1876–1969), a French parasitologist, who spent some 50 years in Algeria working for the Institute Pasteur, wrote, along with his brother Etienne (1878–1948), a pioneering paper on 'Thimnii', a local word for opthalmomyiasis by *O. ovis*, a condition at that time frequently observed among Algerian shepherds.

In 1917, in Brazil, Arturo Neiva (1880–1943) and Florencio Joâo Gomes (1886–1919) described the biology of *D. hominis* in all its phases. In the same year, 1917, Seymour André Hadwen (1887–1947) a grub specialist in Canada, published on the life cycle of *H. bovis* and *Hypoderma lineatum*. In Ireland, Geoffrey Douglas Hale Carpenter (1862–1953) and his associates worked for many years (1908–1922) on the infestation of *Hypoderma* spp. in cattle; in their conclusions it was stated: 'no further doubt is possible as to the entrance of young *Hypoderma* larvae into the host through the skin' (Bishopp *et al.*, 1926).

Also, at this time, there was an intensification in the search for new medicaments for the treatment of myiasis. Early in the 20th century, the use of tobacco water (injected into the rectum) was recommended for the treatment of bots in horses. This system was later abandoned following the report (1917) by Maurice C. Hall (1881–1938), the eclectic American parasitologist, that carbon disulphide, orally, was very effective for the treatment of infestations with *Gasterophilus* spp. in horses, a therapy that was then used for many decades. With regard to cattle oestrids, until the development of the chlorinated hydrocarbon insecticides, their control, as was the case with *D. hominis*, was conducted in the old-fashioned way, i.e. manual extraction and application of tobacco leaves, pork fat, gum petroleum jelly, kerosene, turpentine, rotenone or garlic at the warbles' openings (Roncalli, 1984a).

In the second half of the 20th century medical and veterinary research focused on the development of new drugs with the goal of reducing the impact of parasitic diseases in humans and livestock. A great breakthrough took place with the introduction, in 1984, of ivermectin, the first in a group of compounds known as macrocyclic lactones which were collectively referred to as 'endectocides'. Ivermectin showed excellent efficacy against oestrids of domestic animals (*Gasterophilus* spp., *O. ovis*, *D. hominis*, *Hypoderma* spp. and *Cuterebra* spp.).

The taxonomic literature published on oestrids in the 20th century is voluminous. Of note is the *Manual of Myiology* in 12 parts (which deals in depth with the muscoid flies (1934–1942), which was published in Brazil by Charles Henry Tyler Townsend (1863–1944)). Konstantin Yakovlevich Grunin (1911–1981) of the Zoological Museum of Moscow produced numerous publications on taxonomy; two of them – one on Hypodermatidae (1964) and another on Palaearctic Oestridae (1957 in Russian; 1966 in German) – are outstanding. As stated by Papavero (1977): '[T]he inclusion of valuable data on the biology, morphology, anatomy, distribution, control, etc. of the Oestridae, makes Grunin's works fundamental for the study of this group.'

In the 20th century, a number of researchers distinguished themselves in the field of dipterology and in the study of myiasis: the Belgians, Louis M. Gedoelst (1861–1927) and Alphonse Hubert Jérome Rodhain (1876–1956); the Frenchmen, Emile Roubaud (1882–1962) and Eugène Séguy (1890–1985), also an excellent artist; the Briton, Walter Scott Patton (1867–1960) and the Italian, Mario Bezzi (1868–1927).

A great contribution to the understanding of oestrids was the publication of a massive monograph, *Myiasis in Man and Animals of the Old World* (1965), resulting from a painstaking collection of data and assemblage of his own findings in Africa by Fritz Konrad Ernst Zumpt (1905–1985), a German entomologist transplanted to South Africa.

In 1977, Nelson Papavero, a Brazilian entomologist-historian, published a tentative classification of the Oestridae based on his reinterpretation of the systems proposed by Grunin (1957, 1966) and Zumpt (1965). Ten years later, in 1987, Donald Montgomery Wood, a Canadian entomologist, described the Oestridae to be a monophyletic family composed of four subfamilies: Cuterebrinae (restricted to the New World), Gasterophilinae, Hypodermatinae, and Oestrinae. Two distinguished American entomologists – Elmer Paul Catts (1930–1996) in 1982 and Curtis William Sabrosky (1910–1997) in 1986 – examined the biology and taxonomy of the North American rodent and lagomorph bot flies. In 1999, the Brazilian José Enrique Guimarães and Nelson Papavero reviewed the taxonomy and biology of myiasis and pseudomyiasis producing flies in the neotropical region. In 2001, Thomas Pape, a Swedish taxonomist, analysed the phylogeny of Oestridae at the generic level using 118 characters from all developmental stages (including morphology, ontogeny, physiology and behaviour). He gave 'subfamilial rank with the philogenetic relationship (Cuterebrinae (Gasterophilinae (Hypodermatinae + Oestrinae)))'.

During the last 30 years great advances have been made in the field of electron microscopy with the use of scanning and transmission electron microscopes which permit a more complete study of dipterans (e.g. Guitton and Dorchies, 1993; Colwell *et al.*, 1999). Also, increasing numbers of studies on Oestridae are being conducted by the use of the new molecular techniques. Of interest in this regard is the review by Otranto and Stevens (2002) on the molecular approaches to the studies of myiasis-causing larvae and the work by Otranto *et al.* (2003) on the molecular differentiation of *Hypoderma bovis* by polymerase chain reaction–restriction fragment length polymorphism (PCR–RFLP); this assay could offer additional diagnostic and epidemiological opportunities for the study of cattle grub infestations.

Over the centuries explorers, missionaries, physicians, entomologists and the like have been reporting on human myiasis. Oestridae, in fact, can also attack human beings producing lesions and pain; this problem was well documented by Maurice Theodore James (USDA) (1905–1982) in his excellent monograph in 1947. A number of human myiases are occupational; in Sardinia, for example, shepherds have been and still are subject to attack from *O. ovis* especially at milking time. A relatively recent phenomenon, which must be taken into consideration, is that of globalization that includes the widespread movement of human beings and animals; this offers opportunities for the spread of diseases, including myiasis. Another phenomenon is the increase in tourism and specific activities (bird-watching, archaeology, etc.) in particular areas where myiasis-producing flies are present. In spite of the advances in the treatment of animal myiasis [the 'reservoir for human myiasis'], numerous cases have been reported in recent years in humans in the USA, Europe, China and Japan. These published cases, however, do not reflect the true situation, since a number of human cases are not reported. Oestrids have also left their footprint on the turmoil of the 20th century.

The history of oestrid myiasis, known since Pharaonic times, is a most interesting one; it includes a vast microcosm involving emperors and shepherds, explorers and missionaries; it deals with the suffering of animals and humans; it has engaged many scientists in the pursuit of a better understanding of the disease and of its control, it is a fascinating history and will continue to be so in the years to come.

3   Phylogeny and Evolution of Bot Flies

T. PAPE

Zoological Museum, Denmark

Introduction

Bot fly monophyly is well established (Wood, 1987; Pape, 1992, 2001a). Only Pollock (1999) still maintains that *Gyrostigma* and *Gasterophilus* of the stomach bot flies cannot be part of the Oestridae, yet his evolutionary scenario is neither founded on parsimony- nor on likelihood-based phylogeny and is as such outside formal scientific testing. There is growing evidence in support of the hypothesis that the Oestridae have a phylogenetic position nested within the non-rhiniine Calliphoridae (Rognes, 1997; Pape and Arnaud, 2001). Having Oestridae and Calliphoridae equally ranked, yet with the former included within the latter, is incompatible with traditional Linnaean classification and will require either lumping or splitting of families. This, however, is beyond the scope of the present chapter.

Bot flies are mammal parasites, and as will be discussed below, they probably appeared during the mammal radiation following the K–T boundary. With only some 150 extant species, the bot flies can hardly be considered successful in terms of sheer numbers. Animal parasitism may in itself contribute little to evolutionary diversification (Wiegman *et al.*, 1993), and the strict association with mammals may have been a constraint to bot fly radiation by providing restricted niche opportunities. A low species count, however, does not necessarily mean evolutionary stasis, and bot fly evolution most probably was spurred by the considerable morphological and biological diversification and the extensive migrations taking place in the mammalian host pool.

Methods

The bot fly genus-level cladogram presented by Pape (2001a) was used as a phylogenetic framework in a slightly improved form as given in Fig. 3.1 and further discussed in the following section and in Appendix 3.1. Evolutionary scenarios were

©CAB International 2006. *The Oestrid Flies: Biology, Host–Parasite Relationships, Impact and Management* (eds D.D. Colwell, M.J.R. Hall and P.J. Scholl)

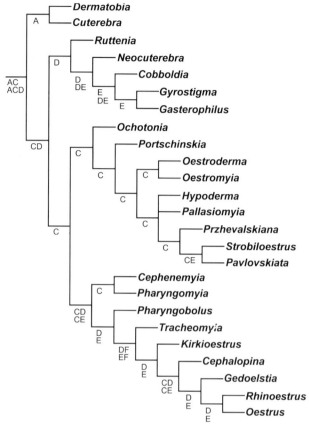

Fig. 3.1. Cladogram of bot fly genera with possible ancestral distributions as obtained from DIVA of the data in Table 3.1 indicated under each branch. Distributional abbreviations: A = Neotropical, C = Palaearctic, D = W + C Africa, E = S + E Africa, F = Australia. Autapomorphic character states for individual branches are given in Appendix 3.1.

developed by mapping ecological and biological traits on to the cladogram and tracing ancestral states and state transformations in MacClade ver. 3.0.1 (Maddison and Maddison, 1992). Biogeographical hypotheses were produced with the use of dispersal–vicariance analysis (DIVA) (Ronquist, 1996), which takes its theoretical background from Ronquist (1994, 1997).

Phylogeny

Our key to bot fly evolution is phylogeny, and the generic cladogram presented by Pape (2001a) is here adopted as the currently most elaborate and best-corroborated working hypothesis. Pape (2001a) used a suite of 118 morphological as well as non-morphological characters drawn from immatures and adults in a parsimony-based phylogenetic analysis at the generic level. Larval and adult characters were largely in agreement, and overall rescaled consistency index was high (RI = 0.69), indicating little homoplasy among the characters. Branch support, calculated as Bremer support and bootstrap percentiles, was in general weak to modest, yet much of this uncertainty could be ascribed to the unsettled position of the biologically and morphologically little-known species *Neocuterebra squamosa* and *Ruttenia loxodontis*. For support values of particular branches Pape (2001a) can be referred to. Several biological features were used as phylogenetically informative characters (i.e. genetically based inheritable traits), which raises the issue of circularity when optimizing such characters on the cladogram (e.g. de Queiroz, 1996; Zrzavy, 1997), but I fully agree with Grandcolas *et al.* (2001) that there is no justification for omitting phylogenetically informative characters from a phylogenetic analysis. Incidentally, pruning these biological characters individually from the analysis will not change the topology of the cladogram, but will at most collapse the branch carrying the hypothetical character transformation, indicating that for each of these characters the relevant part of the topology is not crucially dependent on the inclusion of the particular character or at least that the inclusion of the character in question causes no conflict with the pruned analysis.

The currently favoured generic phylogeny (Fig. 3.1) shows a basal split between the New World Cuterebrinae and the remaining three subfamilies, and the Gasterophilinae as the sister group to Hypodermatinae + Oestrinae. This topology, however, is far from strongly supported. The monophyly of each of the bot fly subfamilies is reasonably well corroborated, except for the still somewhat tentative position of *Ruttenia* and *Neocuterebra* as basal lineages within the Gasterophilinae.

The subfamily Cuterebrinae, here divided into only two genera, *Cuterebra* and *Dermatobia*, is defined largely from a suite of larval characters. Yet even from the possession of a setose suprasquamal ridge in the adult fly, which is a unique feature in the Oestridae, the monophyly of the subfamily as such seems well corroborated. Zumpt (1965) referred *Ruttenia* and *Neocuterebra* to the Gasterophilinae, yet found no morphological evidence in favour of this. Wood (1986, 1987) noted that *N. squamosa*, like all Cuterebrinae, everts the anterior spiracular felt chamber during pupariation, and that similarities in larval spine armature between species of *Ruttenia*, *Neocuterebra* and *Cuterebra* could possibly indicate phylogenetic relationship. Pape (2001a) grouped *Ruttenia* and *Neocuterebra* with the stomach parasites and argued for this through the fused condition of tergite and sternite in the female abdominal segment 7. The relative position of *Ruttenia* and *Neocuterebra* near the base of the Gasterophilinae is still an open question. The stomach parasites form a very well corroborated group, with exceedingly strong evidence for a sister-group relationship between *Gyrostigma* and *Gasterophilus*.

The monophyly of the Hypodermatinae may seem reasonably well corroborated, although still based entirely on the morphology of immature stages. Even the internal phylogenetic topology, with *Ochotonia* at the base, *Portschinskia* branching off next and *Oestroderma* + *Oestromyia* being the sister group of the ungulate-parasitizing species, has good morphological and biological support. The clade of the ungulate-parasitizing species, however, has not been satisfactorily resolved into well-defined genera. Part of this problem stems from the monotypic genus *Pallasiomyia*, which is still unknown from adult specimens, and from the lack of synapomorphic characters for species of *Przhevalskiana*. Pape (2001a) coded *Pavlovskiata*, *Przhevalskiana* and *Strobiloestrus* for a checkerboard-like abdominal pattern that changes with the incidence of light, yet with *Pallasiomyia* left uncoded the level of synapomorphy could not be established unambiguously by the parsimony-based analysis. Species of these three checkerboard-patterned genera also share a characteristic deep black colour of the first flagellomere, and perhaps even a reduction in the size of the antennal pedicel, and their inclusion in a separate clade – with or without *Pallasiomyia* – seems likely (see further discussion in Appendix 3.1). The monotypic *Pavlovskiata* shares some probably apomorphic larval features with the Afrotropical *Strobiloestrus*, and this possible sister-group relationship may be considered justification for maintaining the monotypic genus. The lack of corroboration of the monophyly of *Przhevalskiana*, however, means that this genus in its current definition could be paraphyletic with regard to *Pavlovskiata* + *Strobiloestrus*, in which case a lumping would be justified and the latter name will take precedence.

The Oestrinae or nasal bot fly is a biologically and morphologically very well defined group. Within this group, the branching pattern is still only moderately supported, yet evidence so far is supportive of a basal position of the sister genera *Pharyngomyia* and *Cephenemyia*, and with *Pharyngobolus*, *Tracheomyia*, *Kirkioestrus*, *Cephalopina*, *Gedoelstia*, *Rhinoestrus* and *Oestrus* following in a phylogenetic sequence. First-instar morphology remains unknown for *Pharyngobolus*, *Tracheomyia* and *Kirkioestrus*, and that of *Gedoelstia* needs restudy. Even adult morphology is still superficially screened for phylogenetically informative characters, as adults of especially *Pharyngobolus*, *Tracheomyia* and *Kirkioestrus* are rare in collections and often of substandard quality.

Appendix 3.1 provides an explicit list of character states supporting individual bot fly clades, with a few additional characters and minor improvements in a few character state codings relative to Pape (2001a).

Molecular studies on bot fly phylogeny are currently in their early stages. Nirmala *et al.* (2000) included one to two species from each of the bot fly subfamilies in their study of calyptrate phylogeny based on 18S and 16S ribosomal DNA. Surprisingly, bot fly monophyly could not be established. Also, although the hypodermatine taxa *Hypoderma diana* and *Oestromyia leporina* emerged as sister groups, the long-held notion of a sister-group relationship between the Hypodermatinae and the Oestrinae (the latter taxon represented by *Cephenemyia stimulator*) was not supported by the data. Otranto and Stevens (2002, see also chapter 4), in a taxonomically much more inclusive study on myiasis-causing Oestridae and Calliphoridae, presented a molecular cladogram based on COI sequences in which the Oestridae as well as its constituent subfamilies emerge as monophyletic groups. The only conflict with the non-molecular cladogram presented by Pape (2001a) is that Otranto and Stevens (2002) have the Gasterophilinae emerging as the sister group of the Hypodermatinae rather than of the combined Hypodermatinae + Oestrinae.

Temporal Origin of Bot Flies

Bot fly larvae have no impact on the skeletal anatomy of their hosts, and the adult life of most bot flies is an ephemeral burst of mating and oviposition well away from anoxic fossiliferous sites like stagnant dystrophic waters and sticky mud flats. Also, the size and forceful flight makes adult bot flies little prone to entombment in amber-producing resin exudates. Accordingly, bot fly fossil evidence is sparse (Evenhuis, 1994) and most records need a critical reassessment (Pape, 2001a, and unpublished data). Townsend (1917), for example, boldly stressed the 'undoubted oestrid affinities' of *Musca ascarides* from the American Eocene, yet he found the species so 'markedly distinct from *Hypoderma*' that he erected a new nominal genus *Lithohypoderma* to accommodate it. Townsend saw no problems in assuming an Eocene bovine host, and he explained the remarkably high number of larvae, even puparia, in several freshwater deposits as due to mature larvae leaving hosts (?accidentally) at water hole congregation sites, a scenario left unsupported by comparable observations from the recent fauna of bot flies and their hosts. Similarly, Townsend (1938), in an almost self-contradictory allegation, stated that the Upper Miocene *Adipterites obovatus* 'Evidently [is of] cuterebrid or hypoder-matid stock, but unlike any form hitherto known'. The discovery of a spiracular plate of a *Cobboldia* larva in the stomach of a woolly mammoth preserved in the Siberian permafrost is truly outstanding (Grunin, 1973), yet this subfossil is no more than some 100,000 years old, possibly even less. It is noteworthy that fossils of other calyptrate families are equally sparse and go back no longer than the lower Eocene (Evenhuis, 1994; Michelsen, 2000; Amorim and Silva, 2002), with an interesting exception being the Upper Cretaceous (?calliphorid) puparia reported by McAlpine (1970) and described under the name *Cretaphormia fowleri*. The calliphorid and possible chrysomyine affinities of this fossil, however, were argued from rather circumstantial evidence. Papavero (1977, p. 210) interpreted the basal split of subfamily Oestrinae (his Oestridae) as a Laurasia–Gondwana vicariance event, dating the origin of the nasopharyngeal bot flies (Oestrinae) to 'between the late Jurassic and early Cretaceous'. Papavero's combination of phylogenetic and distributional information may have more subtle interpretations, and Gondwana distributions have not yet been documented within the Calyptratae. One very solid piece of information bearing on the minimum age of bot fly ancestry is the presence in Australia of the nasal bot fly *Tracheomyia macropi*. With a host-borne immigration from South America via Antarctica into Australia being most likely, as suggested by Wood (1987) and discussed further below, the *Tracheomyia–Kirkioestrus–Cephalopina–Gedoelstia–Rhinoestrus–Oestrus* clade should have a minimum age of 38 million years ago (mya), after which time South America became separated from Antarctica (Raven and Axelrod, 1972, 1974). If the basal New World–Old World dichotomy of the Oestridae is considered a vicariance event, as supported by the biogeographic analysis given below, the minimum age of the bot flies would be pushed further back and into the late Cretaceous, i.e. to the separation of South America and Africa at some 100–90 mya (Raven and Axelrod, 1974). It is important, however, that New World–Old World disjunctions do not have to be considered as western Gondwanan vicariance events. Biotic interchange between the New World and the Eurasian continent has been possible at various times throughout the Tertiary (Davis *et al.*, 2002).

Little information is currently available from other calyptrate families. Michelsen (1991) suggested that the Neotropical species of the endemic New World genera of Anthomyiidae, *Coenosopsia* and *Phaonantho*, could be derived from ancestors migrating south in the early Tertiary. Carvalho (1999) and Carvalho *et al.* (2003) suggested that the Muscidae originated in the early to middle Cretaceous, yet the arguments presented are open to alternative and equally likely interpretations. Due to the absence of well-documented Gondwana distributions in the Calyptratae, it seems unlikely that this taxon, and therefore also the bot flies, are older than the late Cretaceous. Even a late Cretaceous origin is difficult to reconcile with the striking absence of fossil calyptrates as well as acalyptrates in pre-K–T and early post-K–T deposits (pending a better corroborated decision on *Cretaphormia*; see Grimaldi and Cumming (1999) for an unidentified cyclorrhaphan larva from the Upper Cretaceous of eastern North America, and Amorim and Silva (2002) for an unconfirmed record of an acalyptrate from the Lower Cretaceous of Australia). The oldest known fossil, which certainly is a member of the Calyptratae, is the anthomyiid *Protanthomyia minuta* described by Michelsen (2000). With Baltic amber dated as 40+ million years old, and Anthomyiidae a possible sister group of the Muscidae (Michelsen, 1991, 2000), it seems inescapable that the Oestroidea were present at least in the mid Eocene.

Summarizing this fragmented and still inconclusive evidence, the bot flies appear to have originated and diversified in the early Tertiary.

Evolution of Vertebrate Parasitism (Myiasis)

Zumpt (1965, p. 1) boldly stated that 'it is not too difficult to reconstruct the evolution of myiasis', and he hypothesized that 'the highly specialized oestroid flies probably originated from calliphorid ancestors which found their way into the host's skin via an ectoparasitic blood-sucking stage' (p. 2). With the recent documentation of probable calliphorid non-monophyly (Rognes, 1997; Pape and Arnaud, 2001), Zumpt's hypothesis of bot fly 'calliphorid ancestry' has gained some support. Also, a possible phylogenetic position of the bot flies close to the Bengaliinae (including Auchmeromyiinae of Rognes, 1997) is interesting, as this taxon contains myiasis-producing species, some of which are sanguinivorous (Zumpt, 1965; Ferrar, 1987). Lack of phylogenetic resolution within the Bengaliinae, however, leaves the question of vertebrate myiasis as a possible ancestral bengaliine trait largely unanswered. Unless the Bengaliinae are shown to be the sister taxon of the bot flies *and* to have the production of myiasis as a ground plan character state, an independent evolution of bot fly versus bengaliine vertebrate myiasis would seem most likely.

Optimizing life habit on the most recent cladograms of the Tachinidae family-group (Rognes, 1997; Pape and Arnaud, 2001) is not very informative due to presence of deep polytomies, low support for several branches, unique ground plan life habits for several terminal taxa (e.g. wood-louse-parasitizing Rhinophoridae, insect-parasitizing Tachinidae, earthworm-predatory Polleniinae, snail-scavenging Helicoboscinae, etc.) and limited biological information for important terminal taxa like the Rhiniinae. Therefore, the plesiomorphic life habit at the very root of the oestrid branch and from which the vertebrate parasitism developed remains obscure.

Ancestral Bot Fly Hosts

Apart from accidental bird and reptile hosts in the genus *Cuterebra* (Catts, 1982; Sabrosky, 1986) and occasional bird hosts in *Dermatobia hominis* (which according to Guimarães and Papavero (1999) probably relate to misidentifications of the muscid genus *Philornis*), all bot fly hosts are mammals. Accordingly, the most recent ancestor of extant bot flies probably was a mammal parasite. More specific hypotheses of ancestral bot fly host taxa within the mammals may be pursued by treating host data as character states and optimizing these on the cladogram. Host data are conventionally, and most easily, defined taxonomically, but coding host taxa as phylogenetic units introduces *a priori* hypotheses of homology relating to the taxonomic inclusiveness of the host coding, as discussed by Pape (2001a, p. 154). Generalized taxonomic host coding is particularly prone to character state homology problems for the ground plan coding approach applied by Pape (2001a) and followed here. Coding host taxa at the level of the component host species increases resolution but often the intended homology information is lost, with no information left in cases where parasites share no host species. *A priori* hypothetical transformation series between taxonomically defined host character states would be dubious, which in itself documents that such character states are mere substitutes for the more elusive, 'real' character states: the olfactory and visual stimuli involved in host choice. Parasites simply do not target taxa as such at any rank, for which reason taxonomically defined host groups may be poor character states, yet host taxon coding is our shortcut to the much more complex information on host-seeking behaviour and physiological adaptations that is built into the genetic code.

Treating host–parasite associations like distributional data with a host switch equal to dispersion coupled with local extinction in the ancestral area, and parallel speciation equal to vicariance, as suggested by Ronquist (1995), may be intrinsically coherent and consistent, yet assessing evolutionary costs for host switches versus parallel speciation may not be feasible for real examples. Also, such costs should be expected to differ from one event to the next, reducing the explanatory power of models based on generalized costs. With these reservations in mind, optimizing taxonomic host data on the cladogram may still provide valuable insight if interpreted cautiously.

The Cuterebrinae have a multitude of hosts partly due to the low specificity of *D. hominis*, but the unique carrier- or vector-mediated host infection used by *D. hominis* is most likely a specialization, for which reason *Dermatobia* is a poor model for the ancestral cuterebrine. Even the genus *Cuterebra* covers a broad range of hosts, encompassing the four eutherian taxa Marsupialia, Primates, Lagomorpha and Rodentia (Guimarães and Papavero, 1999; the single record from Everard and Aitken (1972) of a *Cuterebra* sp. from an introduced Indian mongoose (Carnivora) is here considered accidental and could relate to an infected rodent falling prey to the mongoose). Primates are hosts only to a single species, *Cuterebra baeri*, and this unique association probably is autapomorphic. The extensively arboreal hosts of *C. baeri* (mainly howler monkeys) would intuitively indicate an oviposition site somewhere in the canopy, which in itself would be unique within *Cuterebra*. Six species of *Cuterebra* have been recorded from marsupial hosts that apparently are at least partly terrestrial species (cf. Emmons, 1997). It is noteworthy that of these six bot fly species, three are also recorded as bred from a rodent host (Guimarães and Papavero, 1999; note that the list of natural hosts, p. 234, does not include all records given under the particular species).

Notwithstanding that lagomorph diversity in the Neotropics is very low, and probably of more recent origin (Emmons, 1997), it is still noteworthy that all cuterebrine lagomorph parasites are recorded from North America, even if one Central American species, *Cuterebra maculosa*, is suspected to have a lagomorph host (Sabrosky, 1986; Guimarães and Papavero, 1999). The known (for *C. maculosa* suspected) lagomorph parasites are the nine morphologically similar species of Sabrosky's (1986) *buccata* group plus the three members of his *cuniculi* group. These species together possibly constitute a monophyletic group defined by the presence of red eye spots (at least present in those six species of the *buccata* group and one from the *cuniculi* group where sufficiently fresh individuals have been observed). With the lagomorph parasites possibly being a monophyletic group embedded within the mainly North American clade of species with a facial carina, a lagomorph would not be the most likely ancestral host for the genus *Cuterebra*. It is furthermore interesting that in the New World only lagomorphs of the family Leporidae (hares, rabbits) are regular hosts for *Cuterebra*. The few records from Ochotonidae (pikas) were considered by Baird and Smith (1979) to be accidental. In the Old World the situation is the reverse as only the Ochotonidae have been recorded as lagomorph hosts for the lagomorph–rodent-parasitizing hypodermatine bot flies (Grunin, 1965).

The majority of Neotropical rodent host records are drawn from the subfamily Sigmodontinae (= Cricetidae of Guimarães and Papavero, 1999), which is noteworthy considering that this subfamily did not appear in South America until at the earliest the late Miocene, some 10–14 mya ago (D'Elía, 2000). Within the Hystricomorpha (Hystricognatha, or caviomorphs, i.e. the diverse New World rodent clade containing the porcupines, chinchillas, guinea pigs, acouchis, agoutis, nutria, spiny rats, tree rats, etc.), only a handful of species have been recorded as cuterebrid hosts, and the only two species of *Cuterebra* recorded as hystricomorph parasites also parasitize other non-hystricomorph rodents or even marsupials. (Note that Guimarães and Papavero (1999) mention a specimen of *C. funebris* with the label 'ex *Loncheres guianae*', which here is considered a host record from the spiny mouse *Neacomys guianae* (Muridae), yet this is not given in their host–parasite list.) The hystricomorphs are interesting in this context as they have a much longer Neotropical history dating back to at least the Oligocene–Eocene transition (Wyss *et al.*, 1993), and because they have been suspected on molecular grounds to have a phylogenetic position outside the remaining rodents (e.g. Luckett and Hartenberger, 1993; D'Erchia *et al.*, 1996; but see Novacek, 1992; Philippe, 1997). Recent molecular data, however, are in favour of rodent monophyly and indicate that the hystricomorphs may have a position at or near the base of the rodent clade (Huchon *et al.*, 2002; DeBry, 2003).

With the number of Neotropical host records of *Cuterebra* spp. still fragmentary and grossly insufficient for reliable assessment of host–parasite associations, a tentative informed guess at the ancestral cuterebrine host would point to a rodent. The broad host range of *Dermatobia* is difficult to interpret and provides limited information in this context if the low specificity is considered autapomorphic. The apparent absence of native non-hystricomorph rodents on the *D. hominis* menu is noteworthy (Guimarães and Papavero, 1999) and contrasts with the situation in *Cuterebra*. Perhaps even more noteworthy is the lack of reliable published records of *Dermatobia* larvae from larger native mammals, e.g. tapirs and white-tailed deer, which through their status as much-favoured game animals should be reasonably well known with regard to macroparasite load. Hogue (1993) enumerated both tapir and deer as hosts for *D. hominis*, yet Janzen (1983) explicitly noted the absence of *D. hominis* larvae on a corraled Baird's tapir 'despite large numbers of these fly larvae in cattle in nearby pastures'. Janzen (1976) listed deer and peccaries, among other animals, as potential hosts for *D. hominis* in the riparian evergreen forest of Guanacaste, Costa Rica, but he gave no documented records. Guimarães and Papavero (1999) noted that 'little is known of the native mammalian hosts of *Dermatobia*', and they listed 'monkeys, jaguars, pumas, agoutis, grisons, armadillos, toucans and ant birds', yet rejected the avian hosts as cases involving *Philornis* (Muscidae). I have not been able to find reliable records of bot fly larvae from monkeys that could not be ascribed to the Central American *C. baeri*. Also, a carnivore would not seem likely to be the primary host (see discussion of carnivore hosts for *Cuterebra* spp. by Catts, 1982). The extensive use of species of calyptrate flies, especially Muscidae but even Fanniidae, Sarcophagidae, and Calliphoridae, as egg-carriers (Guimarães and Papavero, 1999) would point to a regular use of larger mammals as hosts because of the much stronger appeal these have to the rather non-specific zoophilous calyptrates. It cannot be ruled out that one or more important hosts may have been found among the now extinct Neotropical megafauna. While numerous large mammals could have served as hosts (e.g. Simpson, 1980), the most distinctive megafauna element of the humid Neotropical forest was perhaps the various species of proboscideans, which would have attracted large numbers of mosquitoes as well as zoophilous calyptrates. The gomphotheres, i.e. the proboscidean relatives of modern elephants, were relatively recent immigrants to the Neotropics (Shoshani and Tassy, 1996), but the phylogenetically basal position of *Dermatobia* within the Cuterebrinae does not in itself rule out that the carrier-mediated host seeking could have evolved long after the basal split of the Cuterebrinae took place.

It is difficult to escape the conclusion that the ancestral host of the extant Gasterophilinae was an elephant (Proboscidea), and this optimization will hold under all three possible positions of *Ruttenia* and *Neocuterebra* at the base of the subfamily cladogram (Fig. 3.2). Rubtzov (1939) provided an early hypothesis of parallel evolution between the stomach bot flies and their hosts (Fig. 3.3), noting an apparent match between host and parasite phylogenies. This, however, was caused partly from Rubtzov's exclusion of hyraxes, tapirs and sea cows (currently considered extant members of the Proboscidea–Sirenia–Hyracoidea–Perissodactyla clade), none of which are known to be hosts for stomach bot flies. Parallel host–parasite evolution would imply independent extinctions of stomach bot flies in all these groups (Fig. 3.4), although sea cows arguably may be excluded from such considerations as their aquatic life would seem to provide *a priori* evidence of a barrier to bot fly parasitism. Whether hyraxes are the sister group of the elephant–sea cow clade or of the perissodactyls (Fischer, 1989; Novacek, 1992) does

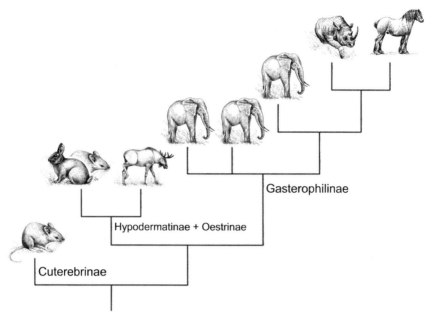

Fig. 3.2. Ancestral hosts for bot fly clades with emphasis on the Gasterophilinae. See text for further discussion of host switch scenario. (Drawings by E. Binkiewicz.)

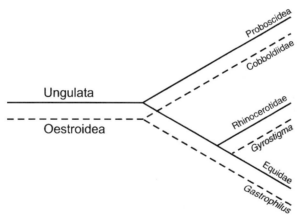

Fig. 3.3. An early hypothesis of parallel evolution of the gasterophiline stomach parasites (modified from Rubtzov, 1939).

not influence the argument. The more independent the extinctions that are implied by the phylogenetic data, the less likely is the scenario of parallel evolution, and the more likely the host switching. Without relative costs for host switches versus parallel speciation, however, a rational choice between the two scenarios is not possible. If the relative values of 1 (parallel evolution) and 3 (host switch), as used in the example given by Ronquist (1995), are applied, the most probable host of the immediate ancestor of all Gasterophilinae as well as of the stomach bot flies will indeed be shown to be an elephant (J. Bonet, unpublished data).

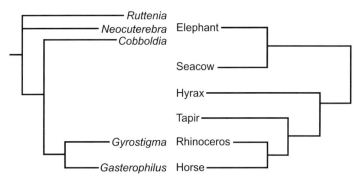

Fig. 3.4. Topological match between parasite and host cladograms for the Gasterophilinae.

Under a one-host-per-parasite assumption, calculating the host of the most recent ancestor of *Gasterophilus* and *Gyrostigma* would give equal probability for an elephant, a rhinoceros and an equid, as well as for the rhino + equid ancestor. In other words, under the given costs, host switching and parallel evolution would be equally likely scenarios for the evolution of host choice in these two genera. The size of the stomach parasites, however, may provide indirect evidence against parallel evolution. Bot flies are well known for their inability to produce undersized individuals, and all extant bot flies are rather bulky. Stomach bot fly larvae grow to a size as large as that of other bot flies or even larger (*Gyrostigma*), and large larvae may seem incompatible with the physical constraints provided by small hosts. Hyraxes (with the possible exception of extinct giant forms, e.g. Court and Hartenberger, 1992) as well as the fox-sized equid ancestor *Hyracotherium* (or *Eohippus*) may not have had a size sufficient for serving as hosts for a stomach bot fly. Egg deposition may give further indirect evidence. Species of *Cobboldia* and *Gyrostigma* glue their eggs to bare surfaces of their hosts (tusks, skin), and pending further information from *Ruttenia* and *Neocuterebra*, it is here considered most likely that the ancestor of the stomach bot flies glued eggs to similar bare surfaces. The ancestors at the various nodes of the elephant–hyrax–rhino–equid cladogram were all rather small and therefore probably fur-clad species that would present neither the intestinal capacity nor the bare surfaces suitable for egg-laying. Parallel evolution finds no support, and the only real match in phylogeny – *Gasterophilus*–*Gyrostigma* versus rhinos–equids – may as well be explained by a reasonable match in host ecology and distribution. It is tentatively suggested that the host shift was from elephants to rhinoceroses and later to equids. Gluing eggs to elephant tusks (*Cobboldia*) may be more similar to gluing eggs to rhino skin (*Gyrostigma*) than to grass blades (*Gasterophilus pecorum*) or horse hairs (other *Gasterophilus* spp.).

An approximate date for the host shifts may be hypothesized by considering the host evolution. During the Eocene (~50 mya) the proboscidean clade evolved on the African continent (Shoshani and Tassy, 1996), and equids and rhinoceroses arose soon after, diversifying in North America and Asia (Prothero and Schoch, 1989). When the African plate collided with the Eurasian plate some 17 mya ago

(Dewey *et al.*, 1973), elephants, rhinoceroses and equids migrated extensively between the continents (Prothero and Schoch, 1989), and similarities in ecology and alimentary physiology, combined with a suitable size, may have facilitated the host shifts hypothesized above. Note that the fossil evidence for an early Eocene New World presence of Gasterophilinae (Rognes, 1997) cannot be upheld (Pape, 2001a).

In the Hypodermatinae, *Ochotonia* is known to parasitize lagomorphs of the family Ochotonidae, while *Portschinskia* as well as the *Oestromyia–Oestroderma* clade include parasites of ochotonid lagomorphs as well as rodents, some species parasitizing both host taxa. The most probable host for the ancestral Hypodermatinae would, by simple optimization, be ochotonid lagomorphs, yet with *Ochotonia* known in only a single third-instar larva, further host records are needed and a combined lagomorph–rodent ancestral host cannot be excluded.

Hosts of the exclusively ovo-larviparous Oestrinae range widely, from kangaroos and various artiodactyls to elephants, horses, hippopotamuses and giraffes. The most likely ancestral host among these is not obvious, and the slight dominance of artiodactyl taxa on the oestrine branch can hardly be more than a hint at an ancestral artiodactyl host. It is noteworthy that species of the *Cephenemyia–Pharyngomyia* clade parasitize mainly Cervidae (*Pharyngomyia dzerena*, however, is a parasite of the bovid *Procapra gutturosa*), while the evidence is slightly in favour of a bovid host for the ancestor of the clade *Kirkioestrus–Cephalopina–Gedoelstia–Rhinoestrus–Oestrus*, with camel parasitism in *Cephalopina* possibly due to a host switch.

An overall picture emerges with all four bot fly subfamilies possibly having different hypothetical ancestral hosts and the initial question left largely unanswered, although rodent (or rodent–lagomorph) parasitism has the edge. Even looking at host choice from an entirely different angle – that of sheer body size – will not provide an unambiguous answer. The ancestors of subfamilies Gasterophilinae and Oestrinae probably had large hosts the size of sheep or larger, while that of the Hypodermatinae most likely had small hosts (rabbit-sized or smaller). The issue of the Cuterebrinae is difficult because of the wide host range and vector-mediated larval infection of *D. hominis*, yet there is a noteworthy lack of records of *Dermatobia* from mouse- and rat-sized native rodents and marsupials (Guimarães and Papavero, 1999). If *Dermatobia* is disregarded because of its probably derived egg-deposition strategy, and with *C. baeri* being considered a howler monkey specialist, all other *Cuterebra* have small hosts, which indicates that the ancestral host of the Cuterebrinae was a small mammal. Optimizing host size on the bot fly cladogram does not provide an unambiguous ancestral state.

A bold attempt was made to use DIVA for the reconstruction of the ancestral host by coding hosts rather than areas for the terminal taxa, although the assumption of one-host-per-parasite does not hold throughout the Oestridae. Again, *D. hominis* provides difficulties because of its broad host range, but it was here coded for rodents, monkeys, carnivores and armadillos (note that the addition of unique hosts may be considered as 'dispersion' into new host space and as such will have no influence on the optimization of the ancestral bot fly host). The host codings used in the analysis are given in Table 3.1 and represent my best estimates of the ancestral host taxon of the terminal taxon in question. The large

Table 3.1. Codings for ancestral hosts and ancestral areas for clades analysed with DIVA. Note that codings are estimates of ancestral conditions and for some taxa they differ from present distribution and currently known host range as further explained in the text.

Bot fly taxon	Ancestral host	Ancestral distribution
Cuterebra	Rodent	Neotropical
Dermatobia	Rodent, jaguar, monkey, armadillo	Neotropical
Ruttenia	Elephant	W + C Africa
Neocuterebra	Elephant	W + C Africa
Cobboldia	Elephant	W + C Africa, S + E Africa, Oriental
Gyrostigma	Rhinoceros	S + E Africa, Oriental
Gasterophilus	Equid	Palaearctic, S + E Africa
Ochotonia	Lagomorph (Ochotonidae)	Palaearctic
Portschinskia	Lagomorph (Ochotonidae), rodent	Palaearctic
Oestroderma	Lagomorph (Ochotonidae)	Palaearctic
Oestromyia	Lagomorph (Ochotonidae), rodent	Palaearctic
Hypoderma	Cervidae, Bovidae	Palaearctic
Przhevalskiana	Bovidae	Palaearctic
Pallasiomyia	Bovidae	Palaearctic
Pavlovskiata	Bovidae	Palaearctic
Strobiloestrus	Bovidae	Afrotropical
Cephenemyia	Cervidae	Palaearctic, Nearctic
Pharyngomyia	Cervidae, Bovidae	Palaearctic
Pharyngobolus	Elephant	W + C Africa, S + E Africa
Tracheomyia	Kangaroo	Australia
Kirkioestrus	Bovidae	S + E Africa
Cephalopina	Camel	Palaearctic
Gedoelstia	Bovidae	W + C Africa, S + E Africa
Rhinoestrus	Bovidae	Palaearctic, W + C Africa, S + E Africa
Oestrus	Bovidae	Palaearctic, W + C Africa, S + E Africa

majority of bot flies are restricted to one or a few host species, and parallel host–parasite evolution is not apparent, as discussed above. Assuming that bot fly ancestors were restricted to one or a few hosts, as are their extant descendants, DIVA was constrained to give host distributions of at most two taxa (the minimum value). This resulted in a combined group, rodents + elephants being the ancestral host. An identical result is obtained if the ancestral host of *D. hominis* is coded as rodents only. If the probable sister taxa lagomorphs and rodents are coded as the same taxon (i.e. as Glires), thereby assuming that new hosts are most easily recruited among those phylogenetically closest to the original host, the ancestral bot fly host is optimized as either Glires alone or the combined group Glires + elephants. If all artiodactyl hosts are coded identically, the ancestral bot fly host should be found in the group rodents + elephants, rodents + lagomorphs or rodents + artiodactyls. If artiodactyls are given an identical coding and rodents and lagomorphs are coded as Glires, the ancestral host will be found in Glires alone. The quest for the ancestral host seems to focus on the rodent or rodent–lagomorph clade. Interestingly, molecular dating puts the split between Rodentia and Lagomorpha around the K–T boundary and early rodent diversification at the Palaeocene–Eocene transition (Huchon *et al.*, 2002).

With the Oestridae probably emerging in the midst of an extensive mammal radiation, pressing the issue of the ancestral host any further probably leads to pure speculation, yet it should be noted that the ancestral host by no means has to be found among the extant mammal orders. The very rodent-like Multituberculata, for example, roamed the northern hemisphere from the late Jurassic until the early Eocene, dwindling into extinction after the rodents and lagomorphs appeared on the scene (Romer, 1966).

Ovi- and Larviposition Habits

I have previously (Pape, 1992) discussed the deposition of progeny (eggs, larvae) in relation to the host or food source, and to cover the entire Tachinidae family group (Oestroidea). I chose a coding focusing on deposition contact, i.e. deposition on or away from the food source, and with no distinction between a living host and a dead or decomposing food source. This may possibly be stretching the concept of primary homology too far, and alternative codings may exist that are equally informative. But even a distinction between oviposition directly on a host, versus oviposition on non-host media, may appear crude. Evaluating the ancestral oviposition habit of the bot flies is obscured partly by the lack of relevant information for the genera *Neocuterebra*, *Ruttenia*, *Ochotonia* and *Portschinskia* and partly by the uncertainty regarding the sister-group relationship of the Oestridae. In the Cuterebrinae, no species are known to oviposit on their host, and several *Cuterebra* spp. glue eggs on grass blades or stones (Beamer, 1950; Catts, 1982; Ferrar, 1987). While one may suspect at least *Ruttenia loxodonta*, with larvae being scattered on the buttocks, flanks and chest of the elephant, to oviposit directly on the skin of its host, it cannot be excluded that *N. squamosa* will glue eggs to objects on elephant trails in the forest, as suggested by larvae being found exclusively in the elephant sole. Eggs dissected from females of *Portschinskia* were reported by Grunin (1965) to lack an attachment organ, yet their shape with a concave ventral part is reminiscent of eggs of *Cuterebra* spp., and gluing of eggs to either host or non-host surfaces should be expected. Nothing is known regarding eggs and oviposition habit of *Ochotonia*.

Females of Oestrinae incubate their eggs, and larvae hatch when forcefully expelled from the female ovo-larvipositor. The close host contact during deposition of progeny may arguably be more similar to oviposition directly on the host than to oviposition on the ground. While no non-cuterebrine bot fly apart from *G. pecorum* has been documented as ovipositing on a non-host surface, it is indeed noteworthy that such oviposition remains a possibility for the basal genera of both the Gasterophilinae (pending better data on the exact phylogenetic position of *Neocuterebra*) and the Hypodermatinae. With this important information still lacking, the possibility exists that the cuterebrine way of depositing eggs on non-host surfaces could be the ancestral oviposition habit. Formal optimization on the cladogram also depends on homology assessment and outgroup coding. Pape (2001a) used *Calliphora* as the nearest outgroup, coded for oviposition directly on the food source, which resulted in the cuterebrine way being optimized as apomorphic, as indicated in Appendix 3.1. A note on *D. hominis* is pertinent here, as this species certainly does not oviposit on hosts, and the unique habit of usurping host-seeking vectors has given rise to evolutionary speculations. Disney (1974) built up an evolutionary scenario on the bold assumption 'that the ancestral *Dermatobia* laid its eggs direct onto the vertebrate host . . .' (p. 71). Catts (1982, p. 331), on the other hand, stated: 'it is unlikely that *D. hominis* ever used oviposition on the host as an egg dispersal tactic', which was argued from the shared occurrence in all Cuterebrinae of egg-hatching stimulated by a sudden increase in temperature, which would not work if eggs were deposited on hosts. Hall and Wall (1995) pointed out that the best egg vectors are those flies that are attracted to potential hosts and that such flies would most easily be encountered on the host, possibly indicating egg-laying on the host as an ancestral trait of the *Dermatobia* lineage.

Within the oviparous Oestrinae, *Cephenemyia* spp. may possess a host-infecting mechanism similar to that of the hypothetical oestrine ancestor. Gravid females hover in front of and somewhat below the head of suitable hosts and perform rapid darts towards the host head, forcefully ejecting incubated larvae through the ovipositor in one droplet of viscous fluid, or a spray of watery, non-sticky droplets (Anderson, 1989, 2001; Anderson and Nilssen, 1990). The adult female will aim either at the nostrils *per se* or at the upper lip and the lower part of the muzzle. In the last case, larvae migrate down into the mouth and from there up into the pharyngeal area where they find their final position. An informed guess would be that similar conditions hold for the remaining oestrine genera, yet particularly for *Pharyngobolus*, a nasal larval entry would seem contradicted by the very anatomy of the elephant host, and one would suspect the female to aim at the mouth.

The literature gives *Oestrus ovis* as larvipositing into the nostrils of its host (e.g. Grunin, 1966; Harwood and James, 1979; Kettle, 1990), and Zayed (1998, p. 66) states that larvae of *Cephalopina titillator* 'are deposited in the nostrils of camels'. Deposition into the nasal openings makes perfect sense considering the more sheltered position and the shorter migration distance, yet observations are not explicitly documented and unambiguous evidence is not presented that deposition could not be on the muzzle or upper lip with larvae migrating downwards towards the mouth rather than upwards and into the nostrils. The fact that females of *O. ovis*, *Rhinoestrus purpureus* and *Pharyngomyia picta* may occasionally larviposit into the human eye (Sauter and Huber, 1988; Hall and Wall, 1995) could perhaps be taken to indicate that females of these species aim at large, dark openings, i.e. at the nostrils. A single record exists of *C. stimulator* larvipositing 'on the skin of man' (Minár and Breev, 1982, p. 91), yet no further information was provided. A noteworthy specialization is that of *Gedoelstia* spp., with females apparently injecting larvae into the eye of the host, and larvae either entering the bloodstream or migrating along nerve tracts (Basson, 1962a,b, 1966; Horak, 1987). This specialization may have its origin from such occasional ocular myiasis, which in itself could be a side effect of a shift in larviposition target from muzzle to nostril. Incidentally, species of *Gedoelstia* are only very rarely reported as a cause of human ocular myiasis (Zumpt, 1965; Bisley, 1972).

First-instar Tissue Migration

Evaluation of the evolution of first-instar tissue migration in the bot flies suffers to some extent from lack of information from the same genera within the Gasterophilinae and Hypodermatinae, for which we have no information on oviposition. Available evidence from *N. squamosa* and *R. loxodontis* on larval position in the host may be taken to indicate absence of migration. In *Gasterophilus*, newly hatched first-instars migrate in the connective tissue of the skin (creeping myiasis) or in the mucosa of the tongue or gingiva before entering the intestinal tract (Grunin, 1969; Cogley *et al.*, 1982). How first-instars of *Cobboldia* and *Gyrostigma* pass from where the eggs are glued on to the host and into the intestinal tract of elephants and rhinoceroses still remains undocumented.

Grunin (1965) considered the position of larval boils in hosts of *Portschinskia* spp. to indicate absence of migration. No migration is seen to take place in the *Oestromyia–Oestroderma* clade. No information is available for *Ochotonia*.

All Oestrinae are larviparous and forcefully inject their progeny on to the host. If conditions in *Cephenemyia* can be considered representative for the subfamily, the first-instars will migrate after deposition on the host, creeping on the mucosa of the host's nasal or oral cavities. Such physical displacement to reach a point suitable for attachment is not here considered a (primary) homology to the often extensive non-feeding tissue migration seen in *Cuterebra* spp. and in the Hypodermatini. Optimizing migration on the cladogram will provide tentative support for the multiple evolution of first-instar tissue migration in the bot flies: in the lineage leading to *Cuterebra*, in the lineage leading to the ungulate-parasitizing Hypodermatini and somewhere along the gasterophiline lineage, the exact position depending on the presence/absence of migratory behaviour in first-instars of *Cobboldia* and *Gyrostigma*.

Main Position of Larvae

The main position of a larva is here defined as the position in the host where most of the growth takes place. With all Cuterebrinae, all Hypodermatinae and *Ruttenia* and *Neocuterebra* of the Gasterophilinae being subcutaneous parasites, there is little doubt that larvae of the bot fly ancestor produced boils in the skin of its host. From this ancestral life habit developed the intestinal parasitism in the lineage leading to *Cobboldia–Gyrostigma–Gasterophilus*, and the nasopharyngeal parasitism in the lineage leading to the Oestrinae.

The *Cobboldia–Gyrostigma–Gasterophilus* ancestor probably was an intestinal parasite developing in the ventricle and probably with larvae attached to the stomach wall. The condition in *Cobboldia*, where larvae were stated by Zumpt (1963) to crawl around freely in the stomach between the wall and the contents, needs further investigation as such observations could relate to serious stress following death of the host. All species of *Cobboldia* and *Gyrostigma*, and *Gasterophilus haemorrhoidalis*, *Gasterophilus intestinalis*, *Gasterophilus meridionalis*, *G. pecorum* and *Gasterophilus ternicinctus* show the plesiomorphic ventricular larval position, while larvae of *Gasterophilus nasalis* and *Gasterophilus nigricornis* have their main development in the duodenum and those of *Gasterophilus inermis* in the lower part of the colon (Zumpt, 1965; Grunin, 1969). Further resolution of species-level phylogeny within *Gasterophilus* is needed to evaluate this possible evolution of niche partitioning.

In the Oestrinae, there is a shift from a pharyngeal position in the basal genera *Cephenemyia*, *Pharyngomyia*, *Pharyngobolus* and *Tracheomyia* to a position in the nasal passages and frontal sinuses in the clade *Kirkioestrus–Cephalopina–Gedoelstia–Rhinoestrus–Oestrus* (Zumpt, 1965; Grunin, 1966). Zayed (1998), however, found that the large majority of third-instar larvae of *Cephalopina* extracted from Egyptian camels (92% of 691 larvae) were located in the pharyngeal area. This is contrary to information provided by Zumpt (1965) and Grunin (1965) that larvae lodge in the head cavities and nasal sinuses. For *Tracheomyia*, the reported occurrence of larvae in the lower part of the trachea and even in bronchia and bronchioles (Mykytowycz, 1964) may be suspected to be a post-mortem artefact, as nasal bot fly larvae may migrate extensively when stressed by oxygen deficiency after the death of their host (Dudzinski, 1970).

Bot Fly Biogeography

Reconstructing ancestral areas involves the definition of unit areas. The New World falls naturally into Nearctic and Neotropical regions. Central America may be difficult to place unambiguously, yet this may not pose serious problems as I consider Central America as having little potential as an ancestral area in the present context of bot fly higher phylogeny. The Afrotropical mainland is here divided into 'West + Central Africa' containing the humid forests, at present mainly southern Cameroon and the Congo Basin, and the much drier remaining 'South + East Africa'. This may be unnecessarily crude, yet there is a point in coding the area covered in humid forest separately, as will be discussed below. In the present context, splitting Europe from non-tropical Asia seems unwarranted, and I have lumped all of non-tropical Asia with Europe and the northern parts of the African continent into what largely corresponds to the Palaearctic region. Treating tropical Asia separately is considered warranted to stress the remarkably low bot fly diversity, and Australia is coded because of the presence of *T. macropi*.

A synthesis of present-day Old World distributions appears in the regional Diptera catalogues (Pont, 1977a,b, 1980a,b; Soós and Minár, 1986a,b,c), and of New World distributions in Sabrosky (1986), Wood (1987) and Guimarães and Papavero (1999). The ancestral area of *Cuterebra* is coded as Neotropical because the North American species are considered likely to be phylogenetically nested within the South American species. Sabrosky (1986) split the non-dermatobiine Cuterebrinae into six genera: the Neotropical *Alouattamyia*, *Andinocuterebra*, *Metacuterebra*, *Pseudogametes* and *Rogenhofera*, and the largely Nearctic *Cuterebra*. Guimarães and Papavero (1999) shared this view except that they treated *Alouattamyia* as a synonym of *Metacuterebra*, while Pape (2001a) lumped all non-dermatobiine Cuterebrinae into one genus *Cuterebra*. Sabrosky (1986) argued for his generic classification partly from the morphology of the facial plate, whether a facial carina is present (*Cuterebra*) or absent (*Metacuterebra*), and partly from the male terminalia, whether the cercal bases are separated in the midline by membrane (*Cuterebra*) or contiguous (*Metacuterebra*). The character states diagnosing the genus *Metacuterebra* are here considered plesiomorphic at that level and as such they imply no corroboration of monophyly (but note that even *Dermatobia* may have the male cerci separated by membrane, as illustrated by Guimarães and Papavero, 1999: Fig. 107). The morphological evidence may be considered slightly in favour of a hypothesis that the North American Cuterebrinae form a monophyletic group possibly close to the similarly 'keeled' Neotropical species separated as *Rogenhofera* by Guimarães and Papavero (1999). The most parsimonious hypothesis is that the Cuterebrinae originated in the Neotropics and not in North America, which goes counter to Sabrosky's (1986, p. 26) statement that ' . . . it seems clear that the cuterebras arose in the New World, probably in North America, . . . '.

Cobboldia has extant representatives in the Afrotropical as well as in the Oriental region (Zumpt, 1965), and an extinct Pleistocene representative in the Palaearctic region (Grunin, 1973). Rather than including all three regions in the hypothetical ancestral area, I have coded *Cobboldia* for an ancestral Afrotropical distribution, which may (tentatively) be argued from the Afrotropical origin of the elephant hosts. The ungulate-parasitizing Hypodermatinae are mainly Palaearctic, with only *Strobiloestrus* being strictly Afrotropical. A few species of *Przhevalskiana* extend south into East Africa (Kenya, Tanzania), and *Hypoderma tarandi* shares a Holarctic distribution with its reindeer host, yet these cases are considered as range extensions within originally Palaearctic taxa, and their ancestral distribution has been coded accordingly. For the Oestrinae, *Cephenemyia* is represented in the Neotropical region with a single species (*Cephenemyia jellisoni* in Costa Rica; T. Pape unpublished data), yet this is obviously a southern outlier of a mainly Nearctic range, and the genus *Cephenemyia* has been coded as Nearctic as well as Palaearctic. *Cephalopina titillator* is considered a Palaearctic species that has extended its range into at least eastern and southern Africa, even Australia, following the domesticated host (Barker, 1964; Haeselbart *et al.*, 1966). The codings of hypothetical ancestral distributions are summarized in Table 3.1. The calliphorid subfamily Bengaliinae was chosen as outgroup, for which the ancestral distribution was coded as the Afrotropical and Oriental regions.

Running the data in DIVA results in the most likely distribution for the bot fly ancestor being either Neotropical region + West Africa + Palaearctic region or Neotropical region + Palaearctic region (Fig. 3.3). A restriction to the non-contiguous Neotropical + Palaearctic regions has obvious problems as a hypothesis of an ancestral area. Alternatively, a close connection between the Neotropical region and the West African area would indicate a basal vicariance event within the Oestridae, splitting the New World Cuterebrinae from the Old

World bot fly lineage, leading to the three other subfamilies. The conflict between a probably early Tertiary origin of the bot flies and a western Gondwanan vicariance event has already been discussed.

The split between an originally (western) Afrotropical Gasterophilinae and a Palaearctic ancestor of Hypodermatinae + Oestrinae may likewise be interpreted as a vicariance event. The Afrotropical origin of the Gasterophilinae is straightforward, yet the interpretation of the current Oriental occurrence of *Cobboldia* will depend on the phylogeny of the three included species. In the present case, *Cobboldia* has been considered to be of Afrotropical origin, yet coding *Cobboldia* as two terminal taxa with an Afrotropical and an Oriental distribution, respectively, opens the possibility of an early dispersion into the Oriental region of the ancestor of the stomach parasites.

The extant Hypodermatinae are almost exclusively Palaearctic. The only 'exotic' element is the three Afrotropical species of *Strobiloestrus* forming a distal branch phylogenetically emdedded within the Hypodermatini (Fig. 3.1). The Oestrinae are more complex, with DIVA suggesting an Afrotropical–Palaearctic ancestor, which through vicariance split into a Palaearctic ancestor of the *Cephenemyia–Pharyngomyia* lineage and an Afrotropical ancestor of the remaining oestrines.

An Old World ancestor of the clade Hypodermatinae + Oestrinae would require at least three dispersions into the New World: once by *H. tarandi*, and at least once in the genus *Cephenemyia* and once by the ancestor of the *Tracheomyia–Kirkioestrus–Cephalopina–Gedoelstia–Rhinoestrus–Oestrus* clade. Note that the DIVA optimization in Fig. 3.3 would seem to imply that the ancestor of the latter clade should have a distribution stretching from Africa through the Palaearctic region and via the New World into Australia. Dispersal into Australia from the Palaearctic region via South-east Asia and the New Guinean terrain seems ruled out by the combination of an absence of oestrine bot flies in host-rich South-east Asia and the apparent lack of dispersal opportunities for mammals large enough to be hosts to nasopharyngeal bot flies. It is here considered most likely that a host-borne dispersal into Australia happened via the New World, as already suggested by Wood (1987). A dispersal of the *Tracheomyia* ancestor via South America and Antarctica into Australia would imply massive extinction of the *Tracheomyia* lineage (or any of its descendants) on several continents, leaving the extant Australian *T. macropi* as the sole survivor.

Ancestral Habitat

With the risk of moving on to thin ice, it is interesting to look at the ecological requirements of the various bot fly groups, even if only sparse information is available. *Dermatobia hominis* has a broad host menu, although apparently mainly non-native, yet in spite of potential hosts being available practically everywhere, the species is restricted to humid, or at least evergreen, tropical forests and does not occur north of Mexico (except as introduced larvae). The species appears to require high relative humidity for eggs to survive, and moist soil for the pupa to develop (Sancho, 1988, but see Janzen, 1976). In *Cuterebra*, the North American species have distributions largely matching that of their hosts (Sabrosky, 1986), while very little information is available for the Neotropical species. Still, with *Cuterebra* probably having evolved in the Neotropics, the ancestral cuterebrine habitat may well have been humid tropical forests.

In the Gasterophilinae, *Neocuterebra* and *Ruttenia* are restricted to West and Central African rainforests (Zumpt, 1965). Unless both bot fly species are physiologically constrained to the forest elephant (which may be a separate species, *Loxodonta cyclotis*, versus the savannah elephant, *Loxodonta africana*, see Roca *et al.*, 2001), the absence of records from savannah elephants could indicate ecological requirements not unlike those of *Dermatobia*. In tropical Asia, *Gyrostigma sumatrensis* is associated with the rainforest-dwelling Sumatran rhino, but other members of the Gasterophilinae as well as species of Oestrinae appear ecologically more versatile and are generally found wherever suitable hosts occur. *Cobboldia russanovi* even thrived under subarctic conditions in the Pleistocene (but it did not survive the demise of its host, the woolly mammoth, and as such may represent the first anthropogenic extinction of a bot fly species). The basal lineages of Hypodermatinae are unique in having adapted to cool montane conditions.

This information certainly is insufficient for an unambiguous optimization on the cladogram, but it is interesting that species ecologically constrained to the humid tropics have a basal phylogenetic position in the Cuterebrinae and Gasterophilinae, suggesting that the ancestor of the bot flies could have been a species of the tropical rainforest. More ecological studies and a better phylogenetic resolution within the genus *Cuterebra* would seem a potentially rewarding research topic.

Conclusions

Evolutionary scenarios are no stronger than their underlying data. Much remains to be discovered, documented and analysed in the Oestridae, yet a rough picture is emerging. The bot flies may have originated in the early Tertiary, possibly during the early radiation of the rodents some 55 mya, and possibly in the Afrotropical region. The ancestral bot fly is likely to have been an inhabitant of humid tropical forests, and it may have been a dermal parasite on one of the early rodents, or perhaps less specifically on the still very similar rodents and lagomorphs. Eggs could have been glued to leaves or stones in places frequented by suitable hosts, and larvae clinging to passing hosts probably developed in subcutaneous boils formed at the site of entry.

An early bot fly lineage may have followed the ancestor of the hystricomorph rodents when this, perhaps during a late Eocene thermal maximum with humid tropical forests extending into the boreal zone, dispersed from Africa via Eurasia and North America into South America, leaving no survivors of either parasites or hosts outside South America when temperatures declined and tropical forests retreated.

Another early bot fly lineage had extended its distribution into the Palaearctic region, with subsequent vicariant speciation creating an Afrotropical lineage and a Palaearctic lineage. The Afrotropical lineage evolved into the Gasterophilinae, with a significant host switch from rodents to proboscideans, and later even a switch from a subcutaneous to an intestinal position of the maturing larvae. When grasslands and savannas appeared during the Eocene, stomach bot flies followed their proboscidean hosts into this major new habitat. Not until the Miocene, when the colliding Eurasian and African plates brought extensive migrations of large herbivores between the continents, did the stomach bot flies broaden their host range by parasitizing first rhinoceroses and later even equids.

The Palaearctic lineage divided, with one branch developing into a group of species, coping successfully with the still cooler climate by adapting to rodent and lagomorph hosts living at medium to high altitudes around and above the tree-line. Within this boreo-montane clade, one species successfully made a host switch to the much larger-sized ungulates, diversifying and spreading into North America and Africa. The other branch of the Palaearctic lineage evolved ovo-larvipary, enabling gravid females to squirt progeny directly on to the muzzle of selected hosts. Larvae were initially developing in the pharyngeal area, but the first-instar larva of one sublineage migrated further into the nasal and frontal sinuses. The entirely new niche facilitated a much broader host range, and the very mobile hosts brought extensive dispersion into all continents. A single upper Eocene lineage managed to follow its host across Antarctica and into Australia before the latter continent started its slow northward drift, creating the Tasmanian Gateway and making space for the Antarctic Circumpolar Current, which through extensive global cooling has had a profound effect on mammal faunas, and thereby on their bot fly parasites.

Acknowledgements

I am grateful to Dr D.M. Wood, Ottawa, for inspiring discussions on bot fly evolution. Dr H. Shima, Fukuoka, generously facilitated studies of bot fly specimens in Kyushu University, Fukuoka, Canadian National Collection of Insects, Ottawa, and the US National Museum of Natural History, Washington, DC, with travel support from his Grant-in-Aid for Scientific Research (B) from the Japan Society for the Promotion of Science (No. 12440241).

Appendix 3.1

Clades of the Oestridae and their character support in terms of autapomorphies. Adapted from Pape (2001a) with minor corrections and additional evidence as discussed for the relevant branches. A re-analysis was not considered necessary from the few changes made, but affected local as well as overall support is expected to show a minor increase. The generic cladogram is shown in Fig. 3.1. Monotypic genera are not diagnosed as monophyly is considered as default from type-species validity.

Family Oestridae

- Egg glued to substrate.
- Egg-hatching by discarding an anterodorsal cap.

- First-instar larva with bands of thorn-like spines encircling several segments.
- Second- and third-instars with ecdysial scar of spiracular plates near 3 and 9 o'clock (left and right spiracle, respectively).
- Puparium hatching by splitting off only the dorsal valve.
- Postocular setae reduced.
- Postcranium concave.
- Adult mouthparts reduced and never used for feeding purposes.
- Adult with most setae hair-like.
- Anatergite bare (i.e. infrasquamal setulae absent).
- Subcostal vein running parallel to costa before joining this (i.e. without curved or sinuous middle part).
- First radial vein without a knob or hump at level of subcostal break.
- Abdominal sternite 2 (even most of the following sternites) freely exposed and widely separated from tergal margins by more or less folded membrane.
- Male sternite 5 with posterior margin simple or with shallow emargination.
- Male tergite 6 fused to syntergosternite 7 + 8.

Subfamily Cuterebrinae

- Larva III anterior spiracle wide and tube-shaped.
- Larva III anterior spiracle everting its felt chamber at pupariation.
- Larva III capable of retracting the posterior spiraclular plates into the preceding segments.
- Suprasquamal ridge setose.
- Eggs deposited on non-host surfaces, i.e. not directly on the host.
- Phallic dorsolateral processes fused to wall of distiphallus throughout.

As discussed above, oviposition habits are unknown for *Neocuterebra*, *Ruttenia*, *Ochotonia* and *Portschinskia*, and the possibility remains that the deposition of eggs on non-host surfaces is apomorphic at a more inclusive (i.e. more basal) level.

Genus *Dermatobia* Brauer, 1861
Monotypic: *D. hominis* (Linnaeus Jr, 1781)

Genus *Cuterebra* Clark, 1815

- Larva I with terminal adhesive sack.
- Larva I migrating in host tissue.
- Larva III with spines scattered on entire surface.
- Larva III with posterior spiracular slits tortuous and intertwining.
- Wing fumose.
- Tarsi large and flattened.
- Alula broad distally, almost square.
- Femora ventrally with large, shining, bare area.

With this broad definition of genus *Cuterebra* there is little doubt as to its monophyly. The adhesive sack of the first-instar larva and the ontogenetic transformation of the posterior spiracular slits into a unique, tortuous pattern are found nowhere else in the Oestridae. The awkward lack of unique adult autapomorphies when larvae are unknown for most species (Pape, 2001a) is here improved by the inclusion of three character states shared by all *Cuterebra* spp. examined and not found in *Dermatobia* or in any other Oestridae. Splitting off of cuterebrine species with strongly deviating biology, like the howler monkey parasitizing *C. baeri* (in *Alouattamyia*) or morphological 'outliers' like the bee-like *C. fassleri* (in *Andinocuterebra*) in monotypic genera, finds no justification from phylogeny.

Gasterophilinae + Hypodermatinae + Oestrinae

* Arista bare.
* One spermatheca of reduced size.
* Oviposition within 2–3 days after female eclosion.

The first split within the Oestridae has no convincing support. The bare arista has at least one homoplasious occurrence within the Cuterebrinae, i.e. in species sometimes separated as *Rogenhofera* Brauer, and possibly in *C. fassleri* (Guimarães), the arista of which was given as bare by Sabrosky (1986). The last two character states need documentation from *Neocuterebra*, *Ruttenia* and *Ochotonia*.

Subfamily Gasterophilinae

* Female abdominal segment 7 with tergite and sternite fused along the margins.

Only one character state has been found in support of the Gasterophilinae in its broadest sense, as first applied by Zumpt (1957), yet this author had no arguments for the inclusion of *Ruttenia* and *Neocuterebra*. The unique structure of the female abdominal segment 7 is unparalleled in the remaining bot flies.

The basal cladistic events within the Gasterophilinae are still uncertain. According to Pape (2001a), parsimony is narrowly in favour of a basal position of *Ruttenia*, with the monophyly of *Neocuterebra* plus the remaining gasterophiline genera supported by a fumose wing and the extension of wing vein $A_1 + CuA_2$ to or almost to wing margin.

Genus *Neocuterebra* Grünberg, 1906
Monotypic: *N. squamosa* Grünberg, 1906

Genus *Ruttenia* Rodhain, 1924
Monotypic: *R. loxodontis* Rodhain, 1924

Cobboldia + *Gyrostigma* + *Gasterophilus*

* Larva II–III with posterior spiracular plates fused.
* Larva III with posterior spiracular slits vertical.
* Larva III with main position in the digestive tract of its host.
* Tarsomere 5 apicoventrally with one pair of bristly setae.
* Head predominantly orange.

The stomach parasites form a biologically very well-defined group. Larval adaptations to the near anoxic intestinal environment, like haemoglobin, tracheal cells, and the capability of closing off the posterior spiracular cavity, may be autapomorphies for this clade (Pape, 2001a), yet the distribution of these features within the Gasterophilinae needs to be documented. I have here added the orange colour of the head, which is shared by all species of the three genera, while both *R. loxodontis* and *N. squamosa* have a brownish or blackish head.

Cobboldia Brauer, 1887

- Larva I with mandible laterally tripartite.
- Adult body flattened.
- Postpronotal lobe swollen.
- Metathoracic spiracle with opercular posterior lappet.
- Wing cell r_{4+5} closed.
- Male cerci fused.
- (Anterior end of egg slits open along preformed hatching sutures.)

A well-defined genus of elephant stomach parasites. Note that intercalary segments present in the second and third instars are found also in species of *Gyrostigma*. This feature may therefore either be synapomorphic for species of *Cobboldia*, in which case they have developed convergently in *Gyrostigma*, or be synapomorphic for all stomach bot flies, with an apomorphic loss in *Gasterophilus*. The last character state, given in parentheses, may be a further synapomorphy of *Cobboldia* spp., yet optimization is ambiguous as explained further under Hypodermatinae below.

Gyrostigma + *Gasterophilus*

- Larva I with labrum elongate.
- Larva I posterior spiracles situated on tubular projections.
- Larva II–III with intermandibular sclerites.
- Larva III antennomaxillary lobe sclerotized at least dorsally.
- Larva III mandible ventrally with tile-like surface structure.
- Larva III with posterior spiracular slits curved.
- Subalar greater ampulla small.
- Metathoracic spiracular fringes slanting or horizontal.
- Apical scutellar setae directed upwards.
- Wing vein M straight.
- Lower calypter very short and with practically no free median margin.
- Lower calypter bordered by dense fringe.
- Female segment 7 with dorsal setae perpendicular to surface.
- Female sternite 8 keel-shaped.
- Epiphallus reduced.

The sister-group relationship between *Gasterophilus* and *Gyrostigma* is remarkably well corroborated and needs no further comments.

Gasterophilus Leach, 1817

- Larva I with mandible almost straight.
- Larva I with mandible directed dorsally.
- Larva I with discal ventrolateral setae.
- Larva II–III with two or three groups of small spines between mandibles and antennomaxillary lobes.
- Larva III with antennomaxillary lobe sclerotized all around.
- Metathoracic spiracle with posterior lappet reduced and much smaller than anterior lappet.

- (Body at least partly with yellow clothing setae.)
- (Phallus short.)
- (One spermatheca completely lost.)

Species of *Gasterophilus* are stomach parasites in equids. The last three character states have been put in parentheses as they have not yet been convincingly shown to be synapomorphic for species of this genus. Colour characters are often difficult to code unambiguously when dealing with numerous, and especially non-monotypic, terminal taxa, but at least within the Gasterophilinae, the yellow clothing setae of *Gasterophilus* spp. appear as a probable autapomorphy. Both *R. loxodontis* and *Gyrostigma conjungens* Enderlein, however, have yellow or yellowish clothing setae, indicating that colour of clothing setae is prone to homoplasy. Incidentally, the black clothing setae on the female abdomen of *G. pecorum* may possibly be plesiomorphic.

Being short is an entirely relative term, but I have at this time found no satisfying parameter for quantifying the relative length of the gasterophiline phallus. Still, by browsing existing figures (e.g. Patton, 1937; Grunin, 1969; Pollock, 1999), it appears that the distiphallus is very short in *Gasterophilus*. *G. pecorum* has the longest phallus within the genus, which most likely is further evidence of its basal phylogenetic position.

Determining the level at which the spermathecal character state is apomorphic will have to await information from *Gyrostigma* to be properly settled.

Gyrostigma Brauer, 1885

- Larva II–III with posterior spiracular slits convoluted.
- Antennal pedicel greatly enlarged and covering flagellomere 1.
- Antennal pedicel apically with long and robust setae.
- Ocelli absent.
- Anatergite conical.
- Metathoracic spiracular fringes horizontal.
- Tarsomere 5 with an irregular row of apicoventral bristly setae.

This genus is biologically characterized as rhino parasites. The monophyly is well corroborated from one larval and several adult autapomorphies, even if the adult character states cannot be confirmed for *G. sumatrensis*, adults of which are still unknown. The two African species of *Gyrostigma* are large, long-legged species with a superficial similarity with pompilid wasps in appearance as well as flight behaviour (C. Dewhurst, unpublished data) (see Fig.14.78).

Hypodermatinae + Oestrinae

- Posterior spiracular slits shaped as a porous plate.
- Phallus with short epiphallus.

Even if formal support values are low (Pape, 2001a), the sister-group relationship between the Hypodermatinae and Oestrinae is generally accepted.

Subfamily Hypodermatinae

- Larva I–III with mandible fused to remaining cephalopharyngeal skeleton.
- Larva I–III with Keilin's organ absent.

- (Anterior end of egg slits open along preformed hatching sutures.)
- (Vein M with a bend at or almost at cross-vein r–m.)

The subfamily seems reasonably well defined by the unique fusion of the larval cephalopharyngeal skeleton into one piece and the absence of the pair of triple setae (Keilin's organ) on each larval thoracic segment. Synapomorphies for members of Hypodermatinae can only be drawn from the third-instar larva as long as *Ochotonia* is known from this stage only, yet the last two character states, given in parentheses, may be autapomorphic either for the entire subfamily or for its non-*Ochotonia* members. Eggs of all Cuterebrinae, *Neocuterebra*, *Gyrostigma* and *Gasterophilus* are equipped with a small, round cap that detaches entirely when hatching. *Cobboldia* spp. and at least the non-*Ochotonia* Hypodermatinae produce eggs that open along anterior slits. Pape (2001a) cautiously coded the latter hatching mechanism like the probably plesiomorphic condition in *Calliphora*, resulting in an ambiguous optimization on the cladogram. Further morphological studies are needed on eggs without cap, as well as detailed information on the eggs of *Ruttenia* and *Ochotonia*.

Ochotonia Grunin, 1968
Monotypic: *Ochotonia lindneri* Grunin, 1968

Portschinskia + *Oestroderma* + *Oestromyia* + *Hypoderma* + *Pallasiomyia* + *Przhevalskiana* + *Pavlovskiata* + *Strobiloestrus*

- Larva III with reduced antennomaxillary lobe.

Portschinskia Semenov, 1902

- Head proportionately small.
- Antennal pedicel enlarged and almost encapsulating flagellomere 1.
- Facial ridge with lateroventrally directed setae.
- Postpronotal lobe swollen.
- Notopleuron not clearly demarcated.
- Hind coxa setose posteriorly.
- Wing base with costal margin convex.

All *Portschinskia* are bumblebee-like with very long, hair-like setae, which may be a further autapomorphy for *Portschinskia*. I am here omitting the median longitudinal impression of the facial plate listed by Pape (2001a), as additional species of *Portschinskia* studied after publication of the latter analysis revealed no (*Portschinskia magnifica* Pleske) or a very shallow (*Portschinskia himalayana* Grunin) impression, and the character state therefore cannot be upheld as a synapomorphy for *Portschinskia* spp. The last character state points to a characteristic convexity of the proximal part of wing vein costa, which herewith is proposed as an autapomorphy for *Portschinskia* spp. (e.g. Grunin, 1965, Figs. 91 and 96).

Oestroderma + *Oestromyia* + *Hypoderma* + *Pallasiomyia* + *Przhevalskiana* + *Pavlovskiata* + *Strobiloestrus*

- Egg with adhesive area on narrow stalk or petiole.
- Female terminalia long and telescopic.
- Female with epiproct apically rounded, arching on to ventral surface.

Oestroderma + Oestromyia
- Body with typical clothing setae (i.e. not hair-like).
- Wing fumose.
- Lower calypter with median margin diverging from scutellum.
- Epiphallus long.
- Wings at rest folded over the abdomen, wing tips not or only slightly separated.

Species of the *Oestroderma–Oestromyia* clade have clothing setae that are much more similar to the typical stiff (i.e. not hair-like) setae seen in most Calyptratae, and the setae are differentiated into clothing setae and bristles. While this is most probably autapomorphic for the *Oestroderma–Oestromyia* clade, it is interesting that the configuration fits the general calyptrate chaetotaxy with two primary notopleural bristles, acrostichals, dorsocentrals, intra-alars, supra-alars, scutellar marginals, etc., which here is taken as evidence for a reversal to the ancestral condition.

The last character state listed is taken from Grunin (1965, p. 4, see Fig. 189) and certainly apomorphic relative to the more widespread condition of widely separated wing tips.

Oestroderma Portschinsky, 1887

- Facial ridge broad.
- Triangular head profile.
- Wing with base, fore margin and lower calypter orange.
- Costa with black spot at base anterior to basicosta.

An *Oestroderma* sp. (China, possibly *Oestroderma sichuanense* Fan and Feng) examined for the present study was found to have no median impression of the facial plate, and this feature therefore cannot be upheld as part of the ground plan for the genus as done earlier (Pape, 2001a). Two additional autapomorphies for *Oestroderma* are added.

Oestromyia Brauer, 1860

- Larva III ecdysial scar of posterior spiracles without connection to peritreme.
- Proanepisternum with setae posteriorly.
- Phallic dorsolateral processes with apical swelling.
- Head yellowish or light brown, contrasting with black colour of body.

The head colour is here added as a further autapomorphy of *Oestromyia*. Note that the broad facial plate of *Oestromyia* is reminiscent of the facial plate seen in all the ungulate-parasitizing Hypodermatinae, which makes optimization ambiguous.

Hypoderma + Pallasiomyia + Pavlovskiata + Przhevalskiana + Strobiloestrus

- Larva III with sensory papillae of antennomaxillary lobe fused into one group.
- Larva I migrating subcutaneously.
- (Antenna blackish, contrasting with much lighter colour of anterior part of head.)

This clade consists of all the ungulate-parasitizing Hypodermatinae. Where the biology is known, the first-instar larva migrates to its final destination, which most probably is a derived character state for this clade. Note that Pape (2001a) by a lapsus had switched the apomorphic and plesiomorphic states of sensory papillae in his list of characters.

I have here added the contrasting antennal colour as an adult synapomorphy for members of this clade (given in parentheses pending the discovery of the adult *Pallasiomyia*). *Oestromyia* also possess this state, but this is considered a homoplasy. The shape of the facial plate, if considered different from that of *Oestromyia* spp., would be a further autapomorphy of this clade.

Hypoderma Latreille, 1818

• Antennal pedicel atomentose, shining.

Some authors consider *H. tarandi* (Linnaeus) in the monotypic genus *Oedemagena* Latreille because of the presence of vestigial palps (e.g. Grunin, 1965; Soós and Minár, 1986; Fan, 1992). Palpal rudiments occur in *Oestroderma* + *Oestromyia* but no trace of palps are seen in *Portschinskia*, indicating that this character is prone to homoplasy.

Present evidence is slightly in favour of a monophyletic *Hypoderma* including *H. tarandi*, and the furry bumblebee-like appearance may be further support for generic monophyly.

Pallasiomyia Rubtzov, 1939
Monotypic: *Pallasiomyia antilopum* (Pallas, 1771)

Przhevalskiana + *Pavlovskiata* + *Strobiloestrus*

• (First flagellomere deep black, contrasting with much lighter scape and pedicel.)
• (Pedicel short.)
• (Abdominal pattern checkerboard-like and changing with the incidence of light.)

The first two of the three possibly autapomorphic character states have not previously been proposed. The contrasting deep black colour of only the first flagellomere in species of this clade differs from the dark unicolorous antennae in the remaining Hypodermatinae. The illustration of a head of *Strobiloestrus* sp. reproduced by Grunin (1965, Fig. 2) shows the strikingly contrasting colour of the first flagellomere. The antennal pedicel is reduced to a mere strip in species of *Przhevalskiana* and *Pavlovskiata* and described as 'short' for *Strobiloestrus vanzyli* by Zumpt (1961).

The third character state was included in the analysis by Pape (2001a), and all states are given in parentheses as an exact interpretation will depend on data from the still unknown adult of *Pallasiomyia*.

Przhevalskiana Grunin, 1948
No evidence has so far been presented in support of the monophyly of *Przhevalskiana*, and the genus was maintained by Pape (2001a) for practical and historical reasons and anticipating further studies.

Pavlovskiata + *Strobiloestrus*

• (Ecdysial scar of posterior larval spiracles without connection to the peritreme.)
• (Mandibles reduced in second-instar larva but projecting in third-instar larva.)

Under successive approximations character weighting and weighting using implied weights, the genera *Pavlovskiata* and *Strobiloestrus* appeared as sister groups in the analysis of Pape (2001a). As already indicated by Pape (2001a), even the adults of *Pavlovskiata* and *Strobiloestrus* are morphologically similar in that species of both have shorter clothing hairs, contrasting to the more furry appearance of species of _Hypoderma_ and _Przhevalskiana._

Pavlovskiata Grunin, 1949
Monotypic: _Pavlovskiata subgutturosae_ Grunin, 1949

Strobiloestrus Brauer, 1892

- Larva I labrum absent.
- Larva I mandible tricuspidate.
- Larva II with paired fleshy projections on abdominal segments 1–2.
- Body sparsely beset with clothing setae.
- Notopleuron swollen.
- Wing fumose.
- Lower calypter almost encapsulating the haltere.
- Abdomen with low bumps or swellings.

Oestrinae

- Ventral surface of larva I flattened.
- Larva I with labrum absent.
- Larva I with several anterior ventrolateral spines, elongated and distinctly longer than the ventromedian spines.
- Larva I with discal ventrolateral spines elongated and slender, forming a distinct cluster.
- Larva I with spines on abdominal segment 8 hook-shaped and stouter than those of preceding segment.
- Larva I with (five pairs of) dorsal setae on abdominal segment 8.
- Larva I with posterior spiracles recessed in a dorsal cuticular invagination.
- Larva III with main position in nasopharyngeal cavities.
- Wing vein M with stump vein.
- Uterus with a bilobed incubatory pouch.
- Female ovo-larviparous.
- Phallic dorsolateral processes fused to wall of phallic tube in full length.

Pharyngomyia + _Cephenemyia_

- Larva III with a hood-like cuticular covering of the anterior spiracle.
- Puparium with lower valve detaching from puparium during eclosion.

Pharyngomyia Schiner, 1861

- Posterior spiracular plates of larva II–III diverging ventrally.

With one of the two known species of this genus, _Pharyngomyia dzerenae_ Grunin, known as third instars only, generic monophyly will be restricted to characters drawn from this stage.

Cephenemyia Latreille, 1818

- Larva I with elongate and hair-like spines posteriorly on segments.
- Lunule setose.

The long, hair-like setae covering the adult body may be a further synapomorphy for species of *Cephenemyia*.

Pharyngobolus + *Tracheomyia* + *Kirkioestrus* + *Cephalopina* + *Gedoelstia* + *Rhinoestrus* + *Oestrus*

- Antennal pedicel sparsely setulose or bare.
- Notopleuron swollen.
- Wing cell r_{4+5} closed.

Pharyngobolus Braucr, 1866
Monotypic: *Pharyngobolus africanus* Brauer, 1866

Tracheomyia + *Kirkioestrus* + *Cephalopina* + *Gedoelstia* + *Rhinoestrus* + *Oestrus*

- Larva III dorsally almost bare.

Tracheomyia Townsend, 1916
Monotypic: *T. macropi* (Frogatt, 1913)

Kirkioestrus + *Cephalopina* + *Gedoelstia* + *Rhinoestrus* + *Oestrus*

- Larva with its main position in nasal and frontal sinuses. Note the possible exception in *Cephalopina* as discussed under ovi- and larviposition habits above.

Kirkioestrus Rodhain and Bequaert, 1915

- Larva II with three groups of antennomaxillary sensory papillae.
- Larva III with large liplike lobe below posterior spiracles.

Cephalopina + *Gedoelstia* + *Rhinoestrus* + *Oestrus*

- Body with setiferous tubercles.
- Body sparsely setose.
- Wing vein M without stump vein.

Cephalopina Strand, 1928
Monotypic: *C. titillator* (Clark, 1816)

Gedoelstia + *Rhinoestrus* + *Oestrus*

- Ocelli on distinct tubercles.

Pape (2001a) listed as an additional autapomorphy fronto-orbital plates with setiferous tubercles, yet the sparse material and poor condition of *Gedoelstia* spp. examined does not allow a clear-cut decision.

Gedoelstia Rodhain and Bequaert, 1913

- Larva I with spines on abdominal segment 8 reduced.
- Larva III ecdysial scar of posterior spiracles vertically connected to peritreme.
- Prosternum bare.
- Metasternal area bare.
- Abdominal surface with abrupt, warty protuberances.
- Legs stout and fore tibia slightly curved.

I am here adding shape of leg as one (perhaps even two) additional autapomorphy. The apparently unique larviposition within the eyes of its host (Basson, 1962a,b) most probably will constitute a further autapomorphy of this genus. When I earlier (Pape, 2001a) noted the total absence of spines on the last abdominal segment of the first-instar *Gedoelstia* larva in Basson's (1962a) somewhat schematical original figures, I proceeded by coding this genus for presence of hook-shaped spines in the data matrix. Basson's illustrations indeed are schematical, yet I now consider it as doubtful that he would have missed well-developed hook-shaped spines, which, incidentally, are depicted on his illustration of the *O. ovis* larva (note that Basson's captions to larval figures were erroneously interchanged). The reduction could be directly related to the markedly different first-instar larval host entry and would be a further generic autapomorphy.

Rhinoestrus + Oestrus

• Larva I with only a few slightly elongated anterior ventrolateral spines.
• Facial carina with transverse impression.

Rhinoestrus Brauer, 1886

• Larva III with spines in regular transverse rows.

Oestrus Linnaeus, 1758

• Fronto-orbital plate with pit-like depressions.
• Ocellar tubercles fused.

4 Molecular Phylogeny and Identification

D. Otranto[1] and J.R. Stevens[2]

[1]University of Bari, Italy; [2]University of Exeter, UK

. . . multa fieri non posse priusquam sint facta iudicantur . . .
Caius Plinius Secundus, *Historia Naturalis*, 7:1,1

Introduction

The advent of molecular biological techniques has changed the face of parasitology research. This 'molecular revolution' has been made possible not only by new techniques (e.g. PCR and automated sequencing) but also by a better understanding of the many target genes and classes of gene (e.g. mitochondrial and ribosomal DNA) that can be used in biological and evolutionary investigations. Numerous applications of molecular biology have now been used in medical and veterinary entomology, leading to new insights into systematics (= taxonomy), phylogenetics (= evolution) and population genetics, as well as into more practical aspects, such as pathogenicity, diagnosis and control.

Despite the huge amount of sequence data and associated phylogenetic analyses of insects now widely available, few investigations based on molecular biology have focused on myiasis-causing larvae; for oestrids, in particular, there are very few molecular findings. Indeed, among the 100,000 insect sequences registered in GenBank (Caterino *et al.*, 2000), only 12 Oestridae genes have been studied (COI, COII, tRNA-Leu, tRNA-Ile and 12S (mtDNA), 16S rRNA, 28S rRNA and 18S rRNA (rDNA), glob-1 (*Gasterophilus intestinalis* haemoglobin), and hypodermins A, B and C (*Hypoderma* spp. serine proteases)), with only 48 nucleotide sequences registered since 1993 based on comparison with sequences in current gene databases. In contrast, a large number of studies on the sister groups Sarcophagidae and Calliphoridae, which are of interest in the field of forensic entomology (see molecular identification section), have been performed.

Despite this, evidence of a host immune response against oestrid larvae gathered over the past 30 years has spurred a range of studies focusing mainly on

©CAB International 2006. *The Oestrid Flies: Biology, Host–Parasite Relationships, Impact and Management* (eds D.D. Colwell, M.J.R. Hall and P.J. Scholl)

51

their control (recombinant vaccination strategies) and diagnosis (recombinant antigen techniques) (Pruett, 1999; Otranto, 2001). The small number of investigations involving the molecular identification of Oestridae may perhaps be accounted for by the fact that oestrids cause very specific/obligate myiases, which rarely cross-infect host animals. To date, comprehensive phylogenetic studies of oestrids are confined to the comparative morphological, ecological and behavioural studies recently presented by Pape (2001a, Chapter 3), using a cladistic approach.

The following section provides an overview of the commonly used PCR-based techniques, the genes used to date and those that are potentially useful for molecular phylogenetics and identification of Oestridae. Practical issues relating to the molecular study of these flies are also discussed.

In the second section, the results of phylogenetic studies of the mitochondrial gene encoding the subunit I of cytochrome oxidase (COI) of 18 Oestridae larvae are presented and their phylogenetic relationships discussed.

The third section deals with the molecular identification of some Oestridae with a special focus on the differentiation of larvae belonging to the genus *Hypoderma*, whose morphological identification at species level is sometimes difficult.

Molecular Techniques and the PCR Revolution: the Latest Fashion or a Reliable Tool?

Since the advent of PCR (Saiki *et al.*, 1985), few fields of the life sciences remain unaffected either theoretically or practically by the advent of molecular methodologies that have engendered a revolution in evolutionary and systematic thinking. Hence the question arises as to whether the molecular revolution is to be considered the latest fashion in science, without which no 'research can be claimed to be a research' (similar to the ELISA technique in the 1980s) or a reliable tool. The answer, of course, depends on one's own experiences; in our opinion, PCR-based technologies are an excellent, reliable tool if their use does not imply abandoning the scientific knowledge, techniques and methodologies required to tackle the issue under examination. Put differently: 'How could a scientist be a molecular entomologist without knowing anything of the structure of insects?'

Setting aside these philosophical considerations, two of the main advantages PCR offers in the field of entomology are the requirements for only very small amounts of template DNA for amplification and the ability to rapidly sequence novel genes and other informative DNA markers. By virtue of its high sensitivity and specificity, PCR has met with broad applicability in parasitological and entomological studies, e.g. taxonomic, phylogenetic and diagnostic. In particular, the advent of PCR has made it possible to study damaged and incomplete specimens that can be difficult or impossible to identify morphologically; such forms include degraded museum specimens, individual species of small size, species for which some stages are not yet morphologically described (e.g. *Hypoderma sinense* from yaks) and possibly ancient insect material.

One of the most commonly used PCR-based techniques is digestion of PCR products with restriction enzymes, a technique known as restriction fragment length polymorphism (RFLP) analysis. RFLP has previously been used to elucidate ambiguous taxonomy of a range of insects; however, since the advent of automated sequencing, it is mainly used for the identification and differentiation of sympatric parasite species or cryptic taxa (see 'Molecular Identification' section).

If the goal of a molecular analysis is to identify strains and/or species, then single strand conformation polymorphism (SSCP) analysis, which is a PCR-based mutation scanning method based on the different electrophoretic mobility of ssDNA in a non-denaturing gel, may be appropriate. SSCP can distinguish molecules differing by only a single nucleotide, which may be instrumental in identifying strains/species of otherwise indistinguishable organisms or in detecting population variations. In molecular entomology, SSCP analysis has been used to assess genetic distance and to identify haplotypes existing in insect populations belonging to the same species from different geographical areas (e.g. Gorrochotegui-Escalante *et al.*, 2000; Krafsur *et al.*, 2000; Marquez and Krafsur, 2002, 2003) or to identify morphologically indistinguishable species (e.g. Koekemoer *et al.*, 1999; Sharpe *et al.*, 1999).

For screening an entire genome quickly, without prior sequence information, random amplified polymorphic DNA (RAPD) offers a relatively inexpensive solution. RAPD has been used to study various myiasis-causing ectoparasites (e.g. *Lucilia sericata/Lucilia cuprina*, Stevens and Wall, 1997a,b; *Cochliomyia hominivorax*, Infante-Malachias *et al.*, 1999; Skoda *et al.*, 2002); however, the often poor reproducibility of RAPD fingerprints and the variation in fingerprint patterns between specimens and even between laboratories have proved to be significant limitations to the use of this technique.

Increasingly, the identification of strains and/or species needs to be supported by an accurate and detailed analysis of the genetic variations in DNA sequences not detected by PCR or RFLP analysis, i.e. primary sequence data. Automation of sequencing has reduced the costs of these procedures and the time required to perform them, making their use ever more widespread. Products from a variety of PCR-based techniques can now be sequenced in order to obtain information for taxonomic and diagnostic studies. Moreover, DNA sequences provide information fundamental to the design of internal primers, for identifying specific restriction endonuclease sites or comparing gene sequences of closely related taxa to study genetic diversity at both the inter- and the intraspecific levels. DNA sequencing may be performed directly or before a cloning step; direct DNA sequencing can detect the presence of polymorphic amplicons with different sequence types (alleles). However, if there is a high degree of heterogeneity within a PCR product, it may be necessary to clone the amplicon and then to sequence a number of clones representative of the population units (Vogler and DeSalle, 1994).

Having obtained suitable DNA sequences from the taxa under study, sequences can be compared along their length on a pairwise basis. In straightforward characterization studies where the presence or absence of a particular sequence motif, or even a single nucleotide, is sufficient to provide a definitive result (e.g. sequence-based species identification), a simple comparison of sequences, often against a database of existing data, may be all that is required.

However, where multiple sequence comparisons are required, e.g. for evolutionary studies, the often complex process of multiple sequence alignment must be undertaken, and the associated problem of identifying true homology between variable sites and portions of sequences must be addressed; this remains one of the most problematic areas of molecular phylogenetic analysis.

Alignment can be performed by one or a combination of three main approaches: (i) on the basis of secondary structural and functional domains, e.g. secondary structure in ribosomal sequences (Neefs *et al.*, 1990); (ii) using one of a range of specialist alignment programmes with various weighting options and gap penalties, e.g. ClustalX (Thompson *et al.*, 1997); or (iii) by eye, often in relation to previously aligned sequences. For meaningful evolutionary analysis, it is also necessary to include an appropriate outgroup. The definition of an outgroup and the associated placement of the tree's root sets the group of interest (the ingroup) in evolutionary context.

There are three main categories of phylogenetic analysis in widespread use with molecular data: distance methods (including some essentially phenetic methods), cladistics/parsimony and the ever more widely used maximum-likelihood analysis. The relative merits of a variety of methods within each category have now been explored by a range of simulation studies (e.g. Nei, 1991; Huelsenbeck, 1995; Wiens and Servedio, 1998).

The 'correctness' of a phylogenetic tree cannot be reliably interpreted without some form of statistical support for the evolutionary relationships presented. Bootstrap analysis is perhaps the most commonly employed method for providing such support (Felsenstein, 1985) and involves resampling the original dataset with replacement to produce a desired number of pseudo-replicate datasets; the number of times groupings defined within the original 'observed' tree are present with the replicate datasets is then taken as a measure of robustness of such groupings/relationships. Debate surrounding the non-linear nature of bootstrap support is considerable, although clarification of what such support means and how it can be interpreted continues to be improved (Hillis and Bull, 1993; Efron *et al.*, 1996).

Target Genes for Molecular Entomology

To date, the most common target regions in insect systematics and phylogenetics are mtDNA and the nuclear rDNA. Nuclear and mitochondrial genes and intergene regions accumulate mutations over the course of time at different rates, depending on function and mode of inheritance. Generally, introns and non-coding regions, such as the internal transcribed spacer (ITS) of rDNA, exhibit a high mutational rate compared, for example, with the (13) coding genes of mtDNA (with specific translational products) that evolve in accordance to their structure and function. Thus, if the aim of a research project is differentiation between species, then a target gene or region with a low intraspecific variability might be preferred to a gene with a high degree of variation. Conversely, if two larval populations of the same species, originating from two well-distinguished geographic areas, are to be compared, a target gene with a high level of intraspecific variation would be preferred.

Ribosomal DNA genes have also proven to be useful tools for identifying strains and/or species of parasites (Arnheim, 1983; Gasser, 1999); rDNA consists of arrays of tandemly repeated units containing spacers and associated rRNA genes (18S, 5.8S and 28S), with a sequence evolution rate varying across a repeating unit and even within a gene (Hillis and Dixon, 1991).

Numerous high-level phylogenetic studies have focused on the analysis of ITS, 18S and 28S rDNA genes, with ITS and 18S being most commonly used with Diptera and Hymenoptera; in the latter, 18S rDNA data yielded results that were both congruent with those of morphologically based phylogenetic analyses (Caterino *et al.*, 2000) and incongruent (Nirmala *et al.*, 2001). ITS-1 and ITS-2 of rDNA have proven useful for the identification of arthropods because of the low level of intraspecific sequence variation combined with higher levels of interspecific differences. Introns, non-coding regions within single-copy nuclear coding loci and tandemly repeated sequences or microsatellites (unit of repetition between one and five) have all proven suitably polymorphic for use in systematic studies and in interspecific and population investigations in various groups of insects (Vogler and DeSalle, 1994; Adamczyk *et al.*, 1996; Leebens-Mack *et al.*, 1998; Onyabe and Conn, 1999; De Barro *et al.*, 2000; Marcilla *et al.*, 2001).

In recent years, mtDNA has been widely used for taxonomic, population and evolutionary investigations in mammals as well as in arthropods because it is easy to isolate, has a high copy number and contains conserved sequences that make it possible to use universal primers (Kocher and Xiong, 1991). Mitochondrial DNA includes two rRNA genes, 13 protein-coding genes and 22 tRNA genes, which, with the exception of tRNAs, are highly conserved within vertebrates and insects (Wolstenholme, 1992). A useful review on the evolution, weighting and phylogenetic utility of mitochondrial gene sequences has been provided by Simon *et al.* (1994) together with a dataset of conserved primers; nevertheless, the existence of mitochondrial pseudogenes integrated into a nuclear genome may sometimes affect the reliability of PCR-based mitochondrial studies (Zhang and Hewitt, 1996a). These paralogous nuclear copies of mitochondrial genes, or nuclear mitochondrial pseudogenes (NUMTs), have been recently demonstrated to be potentially useful in evolutionary studies and for the study of spontaneous mutation in nuclear genomes (Bensasson *et al.*, 2001).

The mitochondrial gene encoding the subunit I of COI is the terminal catalyst in the respiratory mitochondrial chain and COI has proven to be particularly suitable as a molecular marker for the taxonomical differentiation and evolutionary studies of insects (Gaunt and Miles, 2002). It is used as a target gene for a number of molecular phylogenetic objectives because it is large in size and presents highly conserved and variable regions, with a different range of closely associated mutational rates (Lunt *et al.*, 1996). Studies on the insect COI gene revealed heterogeneity in sequences and nucleotide variability, making some regions useful for low- or high-level phylogenetic investigations (Zhang and Hewitt, 1996b).

The COI amino acid sequence is made up of 12 transmembrane helices (M1–M12), 6 external loops (E1–E6), 5 internal loops (I1–I5), 1 carboxyl (COOH) and 1 amino (NH_2) terminal. In particular the COI gene may have different rates of evolution in different lineages, which largely depend on functional constraints that may occur in specific regions. This variability has to be taken into account when selecting a region with a mutation rate suitable for the purpose of the investigation; highly variable, quickly evolving regions may not reveal the phylogeny of anciently divergent taxa, while well-conserved sequences are not helpful in establishing intraspecific phylogenetic relationships. The occurrence of different patterns of nucleotide and amino acid variability has been demonstrated in insects, with the COOH-terminal region being the most variable part, followed by E1, M3, E2, I2, I4, M9 and M12 regions (Lunt *et al.*, 1996). Similarly, Zhang and Hewitt (1996b) designed a set of ten conserved COI primers covering the most conserved regions across Insecta and assessed the usefulness of different amplicons for different phylogenetic questions. In particular, the evolutionary patterns and the utility of conserved COI primers have been investigated for the phylogenetic analysis of many insects such as Orthoptera (Harrison *et al.*, 1987; Zhang *et al.*, 1995; Lunt *et al.*, 1996), Diptera (Nigro *et al.*, 1991; Spicer, 1995; Lunt *et al.*, 1996), Hymenoptera (Lunt *et al.*, 1996), Coleoptera (Městrović *et al.*, 2000) and Lepidoptera (Brown *et al.*, 1994; Sperling and Hickey, 1994). Critically, however, the overall rate of molecular evolution in COI appears relatively uniform across Insecta (Gaunt and Miles, 2002).

Molecular Phylogenetics

In the past 20 years advances in molecular techniques have provided great insights into nucleotide and amino acid differences within and among species of insects in a phylogenetic context. In turn, this has led to a broader understanding of the structure and function of insect genes and their relation to distribution (phylogeography), physiology, development, evolution and ecology of insects, although the efforts of scientists in this field still need to be better concentrated (Caterino *et al.*, 2000).

To correctly analyse genes for phylogenetics, a number of factors need to be addressed: (i) the choice of a technique and gene for the specific purpose; where possible, the rationale of such choices requires validation by a pilot study; (ii) the data should be analysed with appropriate analytical and statistical (computer) models; and (iii) the results should be compared with different models and approaches (i.e. compared with the results of existing morphology-based phylogenies and with phylogenies based on other target genes).

Oestrids include 25 genera and 151 species of flies, ranked in four subfamilies: Cuterebrinae, Gasterophilinae, Hypodermatinae and Oestrinae (Wood, 1987; Hall and Wall, 1995; Pape, 2001a). These subfamilies have an enormous diversity in terms of lifestyle, spread, and parasite behaviour within the host. Up to now, relatively few studies have been carried out on the phylogenesis of Oestridae and of these, many have suffered from inadequate data and methodologies. Recently, however, an exhaustive cladistic analysis has been presented by Pape (2001a) who has performed phylogenetic analysis at the generic level, analysing 118 characters (morphological, ontological, physiological and behavioural), of which 15 were considered especially reliable because they were autapomorphic. Pape (2001a) provided new information on the monophyly of the family Oestridae within the Oestroidea and confirmed that this taxon is subordinate to the sister group Calliphoridae (Rognes, 1997).

Complementary to the morphology-based cladistic analysis of Pape (2001a), Oestridae has been analysed at the genus, subgenus and lower taxonomical levels using a molecular phylogenetic approach based on a variable region of the COI gene (Otranto *et al.*, 2000, 2003a,b, 2004). In the study of Otranto and Puccini (2000b), large fragments of the COI gene (1300 bp spanning from M3 to the terminal carboxylic region) of *Hypoderma bovis*, *Hypoderma lineatum*, *Oestrus ovis*, *Przhevalskiana silenus* and *Gasterophilus intestinalis* were sequenced, characterized and compared with analogous sequences of some Calliphoridae larvae available on the database. Distance matrices were calculated and a number of different tree-building methods were employed (e.g. neighbour-joining, Saitou and Nei, 1987); analysis was performed on a variety of subregions and on the whole 1300 bp sequence. The best trees were obtained with the whole fragment, probably because when analysing long sequences, the problem of variance in estimating distances is minimized. This study represented the first molecular characterization of mtDNA of Oestridae and some observed inconsistencies were probably due to the low number of larval species examined and possibly to the reportedly better resolution of COI for low-level (i.e. intraspecific) phylogenetic analyses.

Subsequently, characterization of COI from 18 species of Oestridae larvae was undertaken (Otranto *et al.*, 2003a). The aim of this project was to characterize the most variable part of the COI gene encoding the region spanning from E4 to the COOH-terminal sequence (688 bp), thus providing a dataset for their molecular identification and phylogenetic analysis. The COI region analysed is among the most variable in insects and is thus suitable for phylogenetic analysis of closely related species (Lunt *et al.*, 1996). See Table 4.1 for details (host, country of origin, accession number) of the 18 oestrid species examined. For the larvae in subfamily Hypodermatinae, at least ten individual larvae of each species were processed, while, due to the lower number of larvae, three specimens were examined for the larvae of all other subfamilies. Two PCRs were performed for each sample using a set of internal primers (Table 4.2) and sequences were confirmed twice in each direction.

Neither insertions nor deletions were detected in the sequences determined and no sequences exhibited any unusual mutations. As in other insect mtDNA, a typical base composition was observed with a high AT content (67.9%) and a strong bias in the third-codon position to adenine and thymine (82.5%). Out of 688 nucleotides, 303 variable sites were identified. Most of the third-codon positions were variable; the first and the second positions were more conserved with a high number of synonymous substitutions occurring in the third-codon position while most of the non-synonymous substitutions occurred in the first-codon position, the second position being highly conserved (Table 4.3). Such a pattern is typical of a sequence under strong functional constraints, where most of the substitutions accumulate at the third-codon position and usually do not affect amino acid composition (Howland and Hewitt, 1995).

Of the 11 amino acid regions inferred, four presented substantial variability and corresponded, on the basis of the topographical model of the COI protein within the mitochondrial membrane, to segments I4, I5 (internal loops on the matrix side of the mitochondrial membrane), M12 (membrane helices) and to the carboxyl-terminal region (Saraste, 1990). Figure 4.1 shows the mean variability

Table 4.1. Species of Oestridae larvae causing myiasis examined and ranked within subfamilies: number of specimens processed, animal parasitized, site of parasitism, country where they were collected and the collector, and GenBank accession number (the most representative sequence among those obtained for each species were registered in GenBank).

Species	Number of specimens processed	Animal	Site of parasitism	Collection locality	Collector	GenBank accession number
Hypodermatinae						
Hypoderma bovis	13	Cattle	Dermis and internal organs	Veneto, Italy	M. Pietrobelli	AF497761
Hypoderma lineatum	10	Cattle	Dermis and internal organs	Basilicata, Italy	D. Otranto	AF497762
Hypoderma diana	11	Roe deer	Dermis and internal organs	Sologne, France	C. Boulard	AF497763
Hypoderma tarandi	10	Reindeer	Dermis and internal organs	Island of Senja, Norway – Upsala, Sweden	A. Nilssen and J. Chirico	AF497764
Hypoderma actaeon	10	Red deer	Dermis and internal organs	Còrdoba, Spain	D. Reina	AF497765
Przhevalskiana silenus	12	Goat	Subcutaneous tissue	Calabria, Italy	D. Traversa	AF497766
Oestrinae						
Oestrus ovis	6	Sheep	Nasopharynx	Basilicata, Italy	D. Otranto	AF497767
Cephenemyia stimulator	3	Roe deer	Nasopharynx	Toulouse, France	P. Dorchies	AF497768
Cephenemyia trompe	3	Reindeer	Nasopharynx	Tromsø, Norway	A. Nilssen	AF497769
Cephenemyia ulrichii	3	Moose	Nasopharynx	Vindeln, Sweden	J. Chirico	AF497770

Species		Host	Location	Origin	Collector	Accession
Rhinoestrus usbekistanicus	3	Donkey	Nasopharynx	Nigeria	P. Dorchies	AF497771
Rhinoestrus phacochoeri	3	Warthog	Nasopharynx	Senegal	P. Dorchies	AF497772
Gasterophilinae						
Gasterophilus intestinalis	6	Horse	Digestive tract	Apulia, Italy	D. Traversa	AF497773
Gasterophilus haemorrhoidalis	6	Horse	Digestive tract	Apulia, Italy	D. Traversa	AF497774
Gasterophilus nasalis	3	Horse	Digestive tract	Apulia, Italy	V. Puccini	AF497775
Gasterophilus pecorum	3	Horse	Digestive tract	Apulia, Italy	V. Puccini	AF497776
Cuterebrinae						
Cuterebra baeri	4	Howler monkey	Dermis	Barro Colorado Island, Panama	D. Colwell	AF497777
Cuterebra jellisoni	4	Rabbit	Dermis	Owl Springs, USA	C. Baird	AF497778

Table 4.2. Sequences (5′ to 3′) of the conserved primers (UEA 7, UEA 8, UEA 9 and UEA 10) used to cover the region under examination. (From Lunt *et al.*, 1996.)

Common name	Standard name	Position	Sequence (5′ to 3′)
UEA 7	C1-j-2369	E4	TACAGTTGGAATA GACGTTGATAC
UEA 8	C1-N-2735	M11	AAAAATGTTGAGGG AAAAATGTTA
UEA 9	C1-N-2753	M11	GTAAACCTAACATTTTTT CCTCAACA
UEA 10	TL2-N3014	tRNAleu	TCCAATGCACTAATCTG CCATATTA

Table 4.3. Percentage of synonymous and non-synonymous substitutions and variable sites occurring in total sequences at different codon positions of the COI region examined.

Nucleotide codon position	Number of characters	Synonymous substitution (%)	Non-synonymous substitution (%)	Variable sites
Total sequence	688	—	—	303
First	230	9	28.4	75
Second	229	0	11.0	35
Third	229	86	10.2	193

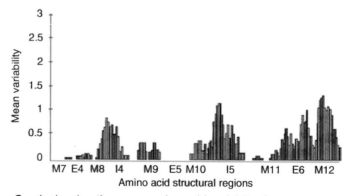

Fig. 4.1. Graph showing the mean amino acid variability (average numbers of amino acids per site observed in a given region) for the 11 structural regions of the Oestridae COI gene.

of amino acids within the COI of Oestridae, highlighting the sites with a greater variability, and Fig. 4.2 represents a bidimensional model of the terminal region examined. Phylogenetic analysis was performed with the minimum evolution methods, inferring the distances with the Kimura two-parameter model (Fig. 4.3).

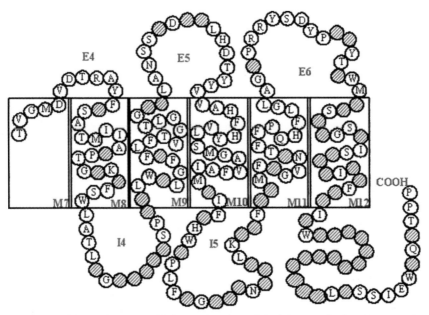

Fig. 4.2. COI of Oestridae: bidimensional model of the terminal region examined. Empty circles indicate amino acid residues while filled circles represent residues varying among oestrid species examined.

Overall, the phylogeny is consistent with classical morphology-based taxonomy. A strong divergence was observed among the four subfamilies, despite the low number of species examined in the Cuterebrinae subfamily. This clear splitting of the four subfamilies confirms their differentiation, as postulated by Wood (1987) and Pape (2001a, Chapter 3). Although the phylogenetic approach of Pape differed from our analysis in its aims (intergenus differentiation) and methodologies (morphological/cladistic), our results are in basic agreement with his. The phylogenetic signal of the COI gene confirms the fact that the Oestridae family is monophyletic, as had been previously established on the basis of morphological differences (Pape, 2001a).

Molecular Identification

Within the Oestroidea superfamily the molecular identification of larvae and adult flies has advanced for the Sarcophagidae and Calliphoridae largely for the purposes of forensic entomology (e.g. Sperling *et al.*, 1994; Stevens and Wall, 2001; Wells and Sperling, 2001; Wells *et al.*, 2001); Oestridae larvae, however, cause obligate myiasis and rarely cross-infect hosts, and consequently have received relatively little attention (Otranto and Stevens, 2002). The importance of molecular identification in forensic entomology is based mainly, but not only (Wells *et al.*, 2000), on the identification of carrion fly maggots at species level, in the course of investigations on murders, to elucidate some aspects such as the time, manner and place of death (Wallman and Donnellan, 2001).

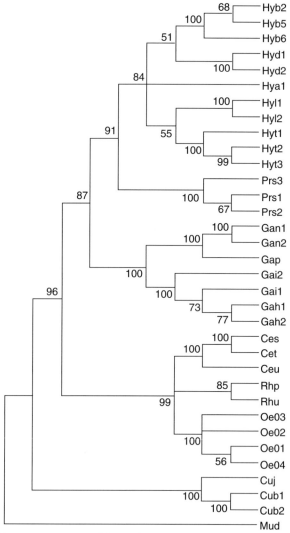

Fig. 4.3. Phylogeny constructed from all nucleotide sequences using the Kimura two-parameter model; branch lengths were calculated from pairwise distances according to the minimum evolution function. Bootstrap values (5000 pseudoreplicates) are indicated at the nodes. *Hypoderma bovis* (Hyb), *Hypoderma lineatum* (Hyl), *Hypoderma diana* (Hyd), *Hypoderma tarandi* (Hyt), *Hypoderma actaeon* (Hya), *Przhevalskiana silenus* (Prs), *Oestrus ovis* (Oeo), *Cephenemyia stimulator* (Ces), *Cephenemyia trompe* (Cet), *Cephenemyia ulrichii* (Ceu), *Rhinoestrus phacochoeri* (Rhp), *Rhinoestrus uzbekistanicus* (Rhu), *Gasterophilus pecorum* (Gap), *Gasterophilus haemorrhoidalis* (Gah), *Gasterophilus intestinalis* (Gai), *Gasterophilus nasalis* (Gan), *Cuterebra baeri* (Cub), *Cuterebra jellisoni* (Cuj), *Musca domestica* (outgroup) (Mud). The numbers associated with the species abbreviation indicate the sequences more represented within the single species.

Over the past 20 years, different mtDNA genes have been sequenced for the molecular identification of saprophagous larvae (Sperling *et al.*, 1994; Wallman and Donnellan, 2001) and recently regions within the 28S large subunit (lsu) rDNA have proven to be a reliable target gene for the differentiation of some Calliphoridae species (Stevens and Wall, 2001; Stevens, 2003). Unlike the immediate practical application to forensic science apparent with species of Sarcophagidae and Calliphoridae, the molecular identification of oestrid flies might erroneously appear to be a less important topic. However, while the morphology of oestrid species at subfamily and genus levels may be clear, this may not be the case if one examines the 151 fly species ranked within the family.

The major difficulties occurring in the morphological identification of oestrids are related to the small number of collections representing all developmental stages, bad specimen preservation (damage to many adult fly specimens), variations within specimens of larvae collected from different animals and countries, the absence of unitary morphological keys (beyond the great compendium of Zumpt, 1965), the broad span of knowledge the entomologist is required to have and the usual range of operator-dependent laboratory errors. Molecular identification may be achieved by two different approaches: restriction enzyme analysis and/or comparison of sequences with those available on the database. With the speed and efficacy of current sequencing techniques, while these two methodologies are comparable in terms of time consumption (< 48 h), restriction enzyme analysis requires greater operator skill. Furthermore, a limit to the restriction enzyme approach is that it utilizes only a small fraction of available apomorphic enzyme-target sites that may contain nucleotide variation, particularly when studying specimens of different populations coming from geographically different areas.

The first step towards the assessment of a molecular assay, when analysing similar and taxonomically close species, is the evaluation of intraspecific variation rate. Generally, a molecular identification assay will be considered reliable when the maximum level of intraspecific variation is below the lowest level of interspecific variation. Wells and Sperling (2001) reported an intraspecific level in the COI + II sequence of Chrysomyinae flies below 1% and greater than 3% between species.

For Oestridae we report a level of intraspecific variation ranging from 0.14% to 1.59% (for *Hypoderma diana* and *Hypoderma tarandi*, respectively) among the larvae examined, while the interspecific variation was 13.2%, 13.3%, 9.6% and 5.3% within the four subfamilies (Hypodermatinae, Oestrinae, Gasterophilinae and Cuterebrinae, respectively). The COI region used proved to be appropriate to address different questions concerning oestrid identification at the subfamily and interspecies level (Otranto *et al.*, 2003a); this analysis was also able to identify degraded larvae and to utilize specimens collected from well-separated geographical locations. A PCR–RFLP assay of the most common Italian species of Oestridae (i.e. *H. bovis, H. lineatum, G. intestinalis, P. silenus* and *O. ovis*) showed clear genetic differences among the genera examined (Otranto *et al.*, 2000), but only molecular characterization of the most variable region of COI was able to differentiate *H. bovis* and *H. lineatum* at species level (Otranto *et al.*, 2003b). Recently, sequences encoding for the E4-COOH region of COI from *H. bovis, H. lineatum, H. actaeon, H. diana* and *H. tarandi* were amplified by PCR. Subsequently, digestion of amplicons with enzyme *Bfa*I provided diagnostic profiles able to differentiate all *Hypoderma* spp. examined (Fig. 4.4) (Otranto *et al.*, 2003c). Results were also compared with SEM examination of surface ultrastructure of the same five species of *Hypoderma*.

Finally, the molecular characterization of the most variable region of the COI gene and of the ribosomal 28S gene of *Hypoderma* spp. third-stage larvae collected from cattle and yaks in China supported the designation of *H. sinense* as a proper species, thus providing new tools for unequivocal identification of this species and presenting key components for the evaluation of its endogenous cycle and pathogenicity in animals and humans (Otranto *et al.*, 2004).

Conclusion

This chapter presents numerous molecular applications relevant to the study of oestrids; of course, the applicability of many techniques to the phylogenetics, taxonomy and epidemiology of these flies remains to be explored. Phylogenetic approaches (cladistic and those employing a model of DNA evolution) have both

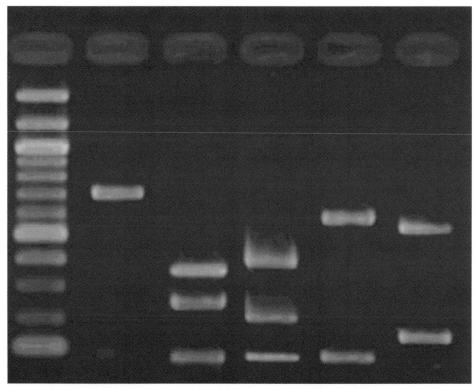

Fig. 4.4. PCR–RFLP analysis of the COI sequences encoding for the region spanning from E4 to COOH terminal of *Hypoderma bovis* (Lane 2), *Hypoderma lineatum* (Lane 3), *Hypoderma actaeon* (Lane 4), *Hypoderma diana* (Lane 5) and *Hypoderma tarandi* (Lane 6) by *Bfa*I restriction enzyme.

strengths and weaknesses that clearly need to be evaluated. Molecular phylogeny, based on one gene, reflects the evolution of that gene and it may be different from phylogenies based on other characters; of course, morphology-based phylogenies can also be misleading as morphological characters are also subject to significant convergent evolution, particularly at the larval stage due to simplification and/or loss of informative structures.

Our analysis of the COI sequences appears informative for the interspecific differentiation and phylogenetic resolution of Oestridae larvae, although it is dangerous to generalize the usefulness of a single gene for diverse groups of taxa. The high level of interspecific differences in larvae belonging to different subfamilies possibly reflects the fact that a long period of time has elapsed since their divergence from a common ancestor. This hypothesis is also supported by considerable morphological differences between species belonging to different subfamilies and by the high level of specialization in their life cycle and in the way they parasitize their host; this divergence among subfamilies also appears to be corroborated by the immune response of hosts. Hypodermins secreted by first-instar *H. bovis* and *H. lineatum* cross-react with antibodies against *H. diana* in roe deer, *H. tarandi* in reindeer and *P. silenus* in goats, all of which belong to subfamily Hypodermatinae, but do not cross-react with antibodies against *O. ovis* in sheep (subfamily Oestrinae) and antibodies against *G. intestinalis* in horses (subfamily Gasterophilinae) (Boulard *et al.*, 1996a).

Genetic identification of species coming from different geographical areas to colonize a new region and larvae retrieved in unusual hosts may suggest an adaptation of these species to a new environment and/or a change in host preference. In such cases, molecular phylogeny and identification may provide many insights into the molecular evolution of Oestridae in relation to their hosts, behaviour and ecology, whilst offering insights into their phylogeographic and demographic patterns. Such knowledge is essential to understanding the origin of Oestridae and their future (where these flies are going, led by their parasitism); such information is of particular importance when we consider that their biodiversity is largely 'endangered' by ongoing chemical and immunological pest control.

Acknowledgments

We thank C.R. Baird, C. Boulard, J. Chirico, D.D. Colwell, P. Dorchies, H. Yin, A. Nilssen, D. Reina, F. Borgsteede, K. Aasbakk and M. Pietrobelli for providing some of the specimens of Oestridae examined. We also wish to thank V. Puccini, V. Martella and D. Traversa for their useful advice and B. Guida and E. Tarsitano for their assistance during the laboratory work.

Appendix 4.1

Larval DNA preservation and extraction: practical advice and frequent constraints. According to several studies the most effective methods for the molecular preservation of insect DNA are ultracold freezing, ethanol (100%) preservation or freeze drying (i.e. silica gel, air-drying in desiccators, etc.) (Caterino *et al.*, 2000).

Bot fly larvae are sometimes not easy to retrieve in scientific collections, especially if they are stored in absolute or 70% ethanol or frozen (which is instrumental in molecular analysis).

Among the three larval development stages of Oestridae, the third instars (L3) are the most commonly detectable by practitioners and veterinarian inspectors at the slaughterhouse, and thus more frequently available for identification. We worked both on 70% and on absolute ethanol-preserved larvae without any difficulties in extracting the genome from larval internal organs.

In our experience, DNA extraction from L3 works fine mainly because internal organs are easy to separate from the external cuticle (rich in chitin) and a huge amount of tissue available for extraction and preservation of non-extracted tissue (which is instrumental in replicating the entire experiment). Developed L3 are also poorer in enzymes (that could impair the reliability of extraction) than first (L1) and second (L2) instars.

Nevertheless, a constraint that frequently occurs before DNA extraction from L3 is the difficulty in the morphological identification of larvae caused by the dark-brown chitinous colour (i.e. difficulties in identifying spine shape and distribution) and the presence of host tissue indurated by ethanol preservation within larval segments. Another constraint in processing ethanol-preserved L3 is that ethanol may not penetrate inside the larval body thus causing rotting of larval tissues and extraction failure. In order to avoid this, larvae have to be carefully washed in physiological saline (pH 7.3) as soon they are retrieved or delivered to the lab, and then identified using morphological keys (e.g. Zumpt, 1965) or more specific keys. Afterwards the larval cuticle is slightly cut and preserved in 70% or absolute ethanol. In our laboratory we obtained very good results by extracting DNA with commercial kits, rather than with the phenol-chloroform extraction procedure.

5 Life Cycle Strategies

D.D. COLWELL

Agriculture and Agri-Food Canada, Canada

Each oestrid subfamily is characterized by a distinctive life history (Fig. 5.1). General details of each of the life history patterns are provided in this chapter. In addition, egg development times, post-hatch longevity of first instars, larval migration routes, within-host larval growth and development and post-emergent behaviour of third instars are discussed. Differences between the subfamilies with regard to pupal biology are described in Chapter 9, those with regard to gonotrophic development in Chapters 9 and 10, those with regard to adult biology in Chapter 10 and those with regard to the nature of the host–parasite interactions in Chapter 11.

General Life Cycles

Hypodermatinae

Female warble flies oviposit directly on to the host (Fig. 5.1). Eggs are attached to individual hairs of the host's pelage using an attachment organ (see Chapters 7 and 10). Larvae hatch spontaneously following a short 3- to 7-day incubation period. They quickly move to and penetrate the skin where they remain at or near the site of penetration (e.g. *Hypoderma tarandi, Przhevalskiana* spp.) or they undergo an extensive and lengthy migration within host tissue before returning to subcutaneous sites (e.g. *Hypoderma bovis* and *Hypoderma lineatum*). These larvae have reduced sensory capacity (see Chapter 9), presumably related to the fact that they hatch very near their site of development (i.e. feeding site). At the subcutaneous sites the larvae are encapsulated by a granulomatous host reaction, but retain access to the surface, for respiration, through a hole maintained in the skin. Larvae feed on host cells and serous exudates provided by the host reaction. Growth of second and third instars is dramatic, particularly in the last instar where they require a great deal of the nutrient reserves to complete their adult life (see Otranto and Puccini, 2000).

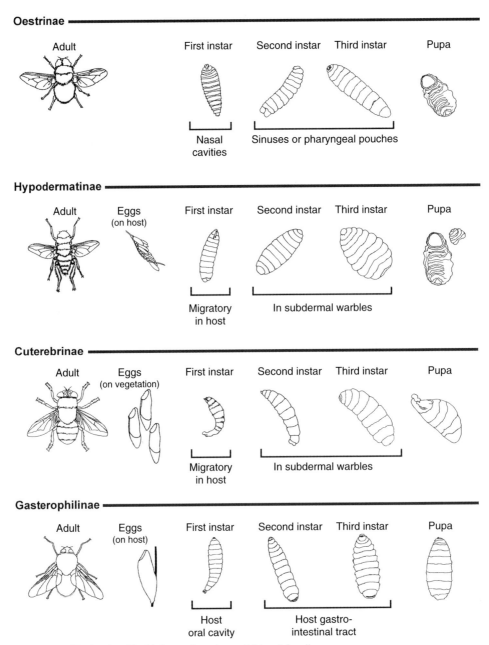

Fig. 5.1. Distinctive life history of each oestrid subfamily.

Newly hatched larvae are positively thermotactic, but show no evidence of phototaxis or geotaxis (Karter *et al.*, 1992). They move quickly to the host's skin, which is penetrated using regurgitated enzymes (proteinases and a collagenase) to dissolve the tissue. Migration within the host, when it occurs, is also accomplished with these enzymes (Chapter 10).

Mature third instars migrate from the warble and fall to the ground. The mature third instars are quite mobile and actively burrow into the grass and leaf litter. Within a short period of time after emerging from the host, usually less than 24 h, the larvae pupariate. Variation in environmental conditions, particularly temperature, influences the emergence of adult flies and thus affects the timing of the other parasitic life stages (see Chapter 9).

Oestrinae

Larvipositing females discharge groups of first instars towards the nose/mouth region or towards the eye of the host. These groups contain 2–20 larvae that are enclosed in a small amount of fluid that both provides moisture for the larvae and aids in their attachment to the host.

Newly deposited larvae migrate quickly into the nasal cavities of the host. At this site the development period can vary widely (see Chapter 11c), ranging from several days to several months. Individuals from the same fly may not all develop at the same rate. The factors regulating first instar development are not clearly understood, although there are indications that larval crowding, host immunity and climatic conditions may play a role in the onset of hypobiosis (see Chapter 11c). Second and third instar development usually takes place at a second site, e.g. the nasal sinuses or the pharyngeal pouch, and may be also quite variable in length.

Cuterebrinae

Female cuterebrids require several days after eclosion for egg development to be completed (Scholl et al., 1985). When this is complete they are receptive to mating and subsequently actively search for oviposition sites, which are usually highly specific places where host activity will bring them into close proximity with the eggs (e.g. rodent burrows by *Cuterebra polita*) (Catts, 1982). The females are guided in their choice of oviposition site by their ability to locate low-intensity host odours. The presence of large numbers of chemo-sensilla on the tarsi (e.g. ~5000/tarsus) presumably aids the females in locating low-intensity chemical cues indicative of the preferred host.

The eggs are usually firmly attached to the substrate (Colwell et al., 1999). Eggs require 5–7 days to fully embryonate after which the larvae emerge in response to the sudden increase in temperature and elevation in CO_2 associated with the presence of a potential host (Catts, 1982). Eggs in an individual cluster do not hatch simultaneously; possibly an adaptation that reduces larval mortality if the emergence is in response to an inappropriate host.

The small, active larvae are well endowed with cuticular sensilla that enable them to search for the host and find an appropriate site for entry into the host's body. Larvae of most species enter the host through a moist body opening (e.g. eyes, nares) or fresh skin lesion. The larvae then migrate internally for a brief period (4–6 days) before moving to the subdermal tissues where they open a small hole in the host's skin and complete their second and third instar development.

Larvae at the subdermal sites are quickly surrounded by a host granuloma, forming the characteristic warble (Cogley, 1991). The location of the subdermal sites where larvae develop are often species-specific (e.g. ~70% of the larvae of *Cuterebra* (= *Alouattamyia*) *baeri* are found on the throat/neck region of infested howler monkeys) (Milton, 1996).

Second and third instars, within the warble, feed on host sera and white blood cells (Colwell and Milton, 1998). This rich nutrient source allows dramatic and rapid growth. It has been estimated that overall increase in mass from the first instar to the mature third instar is approximately 100,000-fold (Catts, 1982).

Gasterophilinae

Eggs are deposited on the host and are either attached to hair shafts, using a unique attachment structure, on the skin surface (*Gyrostigma*) or attached at the base of the tusk (*Cobboldia*) (Patton, 1922a). One exception is *Gasterophilus pecorum* whose eggs are attached, in rows (groups of 10–15 eggs/batch), to grass or plant stems. Number of eggs per female varies dramatically, ranging from *G. pecorum* with 1300–2600 eggs/female to *Gasterophilus haemorrhoidalis* with 50–200 eggs/female.

First instars either hatch spontaneously or remain in the egg until an appropriate stimulus (e.g. increased moisture and friction associated with host feeding or grooming activity). Patton (1922a) remarked that eggs of *Cobboldia elephantis* hatched when immersed in 'warm' water. First-instar development takes place beneath the mucosa of the oral cavity or between the teeth. Each species of *Gasterophilus* has a characteristic and specific site of penetration and route of migration (see Table 5.1).

Moult to the second instar usually takes place at oral sites, but is followed by migration to the stomach or intestine where second- and third-instar development is completed. *Gasterophilus* spp. have high site specificity within the gut.

Egg Development (Embryogenesis) and Hatching

Development rates

Hypodermatinae

Development rates for eggs of the cattle grubs (*H. lineatum* and *H. bovis*) have been studied at both constant and fluctuating temperatures (Hadwen, 1919; Bishopp *et al.*, 1929; Pruett and Kunz, 1996; J. Weintraub, unpublished data). Embryonic development rates increase with temperature, but the developmental range is small. The lower threshold for development is 23°C and maximal development occurs at 35°C. Eggs are able to withstand periods of up to 3 days at temperatures below the developmental threshold. Prolonged exposure to temperatures below 20°C is lethal.

Temperatures alternating between levels near the lower developmental threshold and higher temperatures result in much higher percentage hatch (J. Weintraub,

Table 5.1. Life history features of *Gasterophilus* spp.

Species	Eggs/female	Oviposition site	Hatch	First-instar development	Second- and third-instar development
haemorrhoidalis	50–200	Singly on hair of lips	Stimulus	• Penetrate skin at hatching site, • migrate subcutaneously/submucosally, locate away from tongue and cheek	• Second instars move to stomach, attach in stomach or duodenum, • mature third instars migrate to rectum and re-attach
inermis	320–360	Hairs on cheeks	Spontaneous	• Penetrate skin at hatching site, • migrate subcutaneously/submucosally (produce 'summer dermatitis'), • reside subcutaneously at inner side of cheek	• Rectum (few details)
intestinalis	400–1000	Hairs on forelegs and chest	Stimulus	• Penetrate lingual mucosa, • migrate to stomach, • migrate to inter-dental space	• Moult to second instar at inter-dental sites, • attach at junction of glandular and non-glandular portion of stomach
lativentris	No data	No data	No data	No data	No data
meridionalis	No data	No data	No data	No data	• Attached to stomach mucosa,
nasalis	300–500	Hairs under chin	Spontaneous	• Migrate on surface to inter-dental spaces	• Moult to second instar at inter-dental sites, • migrate to duodenum, attach near pylorus

Continued

Table 5.1. *Continued.* Life history features of *Gasterophilus* spp.

Species	Eggs/female	Oviposition site	Hatch	First-instar development	Second- and third-instar development
nigricornis	300–350	Hairs of cheek or neck	Spontaneous	• Penetrate skin at hatching site, • migrate subcut-aneously/submucosally, • reside subcutaneously at inner side of cheek	• Moult to second instar in mouth, • migrate to duodenum, attach to mucosa and become encysted, • third instars leave cyst and re-attach to mucosal surface
pecorum	1300–2600	Grass or vegetation	Stimulus	• Hatch in mouth, penetrate mucosa, • migrate to base of tongue, the soft palate and occasionally the oesophagus	• Moult to second instar at oral site, • moult to third instar at oral site, • third instars migrate to stomach and attach to mucosa
ternicinctus	No data	No data	No data	No data	• Attached to stomach mucosa

unpublished data) than eggs maintained at constant temperatures. This type of temperature regime probably more closely resembles the environment on the host where cool night temperatures alternate with higher daytime temperatures that result from both increased ambient temperature and the heating associated with incident radiation. J. Weintraub (unpublished data) developed a model that allowed the prediction of hatch times based on a series of alternating temperatures. This model allowed control of larval hatch so that larvae could be harvested for use in a variety of studies including the establishment of artificial infestations via subcutaneous injection (see Colwell, 2002; Colwell and Jacobsen, 2002) and evaluation of *in vitro* culture approaches (Colwell, 1989a).

A thorough study of egg development of *H. tarandi* (Karter *et al.*, 1992) showed some differences from the cattle species. The developmental threshold was 16°C, approximately 6°C lower than that of the cattle grubs. The thermal optimum for development (i.e. fastest development rate) was 31°C, which is approximately 2°C lower than observed for the cattle grubs. Between this temperature and the thermal maximum (40°C) the rate of development did not have a linear relationship with temperature. However, rate studies were done at constant temperatures and the authors cautioned that this was unrealistic. High humidity was detrimental to egg development and the hatch of larvae.

Gasterophilinae

Despite a large volume of literature on control of the major species affecting domestic horses there are almost no reports on embryogenesis or the factors that regulate development and survival. Sukhapesna *et al.* (1975) studied the effect of temperature on the embryogenesis and survival of *Gasterophilus intestinalis* under laboratory conditions and on horses. Development took place at 10°C, but required 40 days and the eggs did not hatch when exposed to warm water (40°C). Development time decreased with temperature in a linear manner at temperatures between 15°C and 30°C. Fully embryonated eggs remained viable for up to 42 days when held at 0°C and for 57 days when held at 10°C. When held at higher temperatures the larvae retained viability for increasingly shorter times, indicating that while development times were shorter at higher temperatures the larval longevity was decreased. Survival of eggs laid on horses in October was tested between November and February. Percent hatch ranged from 56% in mid-November to 0% at the beginning of February, suggesting that viability declined with increasing age as well as with exposure to reduced ambient temperatures.

Cuterebrinae

Only a few species have been studied in detail. Development of *Cuterebra latifrons* eggs occurred at a typically temperature-dependent rate between 15°C and 35°C (Catts, 1967). The rate of development and the percent hatch declined at higher temperatures (36–40°C). No development took place at 5–10°C, but the eggs remained viable for up to 60 days. Incidental observations also suggest that eggs can overwinter and remain infective after experiencing temperatures as low as −27°C (Catts, 1967). Capelle (1970) reported that eggs of *C. polita* successfully embryonated when maintained at approximately 6°C.

Oestrinae

Development times have only been reported for *Oestrus ovis*. Scholl (reported in Cepeda-Palcios and Scholl, 1999) indicated that at alternating temperatures of 16°C and 28°C, fully developed first instars were present in the 'uteri' of females after 12 days. No indication was given of the duration of exposure to each temperature. Similarly, fully developed larvae were present in flies held at constant temperatures of 25°C after 16 days (D.D. Colwell, unpublished data).

Hatching

Hypodermatinae

Hatching occurs in the Hypodermatinae without a known stimulus. Laboratory studies with *H. lineatum* and *H. bovis* indicate that larvae hatch as soon as embryonic development is complete (J. Weintraub, unpublished data). Karter *et al.* (1992) noted that *H. tarandi* eggs did not hatch at temperatures below 20°C. This was also noted for eggs of *H. lineatum* (Pruett and Kunz, 1996).

Gasterophilinae

Spontaneous hatching, occurring as soon as embryonation is complete and temperatures are suitable, is known to occur in at least two species of *Gasterophilus* (see Table 5.1.). The remaining species for which information is available require a specific stimulus for the onset of hatch. The stimuli are usually associated with grooming or eating activity of the host. Sudden temperature increase, sudden humidity increase and mechanical stimuli form components of the host activity that result in hatching of eggs. This has been simulated by immersion of the fully embryonated eggs in warm water (40°C) (e.g. Sukhapesna *et al.*, 1975).

Cuterebrinae

Eggs of *Cuterebra* spp. hatch quickly in response to sudden slight increases in temperature (Catts, 1967; Colwell and Milton, 1998). Catts (1967) noted that a 2.5°C temperature increase for 30 s (at an ambient temperature of 20–25°C) stimulated hatching. Gradual increases in temperature and large incremental changes did not induce hatching. Hatching is also stimulated by 'breathing' on the eggs (Catts, 1964a; Capelle, 1970), which has led to suggestions that elevation in CO_2 levels is an important factor. Evidence from laboratory studies of Catts (1967) indicated that exposure to elevated CO_2 levels had no stimulatory effect and that temperature change was the only factor stimulating hatch.

Post-hatch First-instar Survival

Hypodermatinae

These delicate larvae survive for only a short period without entering a suitable host. Nelson and Weintraub (1972) indicated that newly hatched first instars of *H. lineatum* died if they were unable to penetrate host skin within 15 min.

Gasterophilinae

There has been no research on the survival of first instars of this group.

Cuterebrinae

These larvae are more robust than other oestrids and are able to survive for extended periods of time without entering a host. First instars of *C. latifrons* have been shown to survive in the laboratory for 5–7 days when held at 20–25°C and for 12–22 days when held on saline (Catts, 1967).

Oestrinae

There has been no research into the survival of first instars of this group.

Larval Migrations

Hypodermatinae

Larval migratory pattern in the two cattle-infesting species of *Hypoderma* appears unique. The first instars migrate within the deep tissues of the host following planes of connective tissue and nerves. This is in contrast to larvae of the reindeer warble (*H. tarandi*), which do not migrate to any appreciable extent from the site of initial penetration. Information on the migratory habits of the deer warbles (*Hypoderma actaeon* and *Hypoderma diana*) is limited although Zumpt cites work by Kettle and Utsi (1955) that suggests *H. diana* may have a migratory route similar to *H. bovis*. Larvae of other genera (e.g. *Strobiloestrus* and *Przhevalskiana*) develop in subcutaneous cysts without migration from the site of penetration.

The species infesting bovines have a more complex secretory enzyme complement than the reindeer warble (Boulard *et al.*, 1996a). Whether this has any relationship with the limited migration of the latter species is open to question.

Oestrinae

First instars quickly leave the external surfaces of the nares or eyes and migrate to the outer nasal cavities.

First instars of *Gedoelstia* spp. have a unique migration. They are deposited in the eyes from which they enter the cardiovascular system, often being recovered from the heart. After an unknown period at that location the larvae move, via the circulatory system, to the cephalic subdura. They then enter the nasal cavities by penetrating the foramen of the cribiform plates where they moult to second instars. The latter instars are completed in the nasal passages and sinus cavities.

Cuterebrinae

Larvae of the genus *Cuterebra* enter the host through moist body openings; e.g. nares, mouth, anus, and quickly enter the body cavity by penetrating through the mucosa (Gingrich, 1981). Internally, the larvae move over the surface of major organs, subsequently leaving the peritoneum and locating at subdermal sites where they complete development within the granuloma-like 'warble'.

Larvae of *Dermatobia* do not undergo migration from the original site of penetration.

Gasterophilinae

There is a great deal of variability in the migratory behaviour of *Gasterophilus* larvae (see Table 5.1). Eggs of *G. intestinalis* hatch as the host grooms the oviposition site and the newly emerged larvae quickly penetrate the lingual epidermis and migrate to the back of the tongue (Nelson, 1952).

Little is known about larval migration in the other genera although *Gyrostigma* and *Cobboldia* presumably have generally migrations similar to *Gasterophilus*. This may be surmised from similarities in oviposition and development of second and third instars in the stomach.

Larval Growth and Development

Oestrinae

Cepeda-Palacios *et al.* (1999) described the growth of *O. ovis* based on measurements of several hundred larvae recovered from an extensive survey of slaughtered goats. These authors developed a series of criteria to determine the physiological age of the larvae and then related this to changes in weight, length and width. A typical 'S-shaped' pattern of weight gain was observed with maximal rates of increase observed in the early third instar. Size of the mature third instar has been related to pupal survival (Cepeda-Palacios *et al.*, 1999, 2001). Data from artificial infestations using first instars recovered from laboratory-mated females (D.D. Colwell, unpublished data) showed similar growth patterns. Growth, measured as increase in length, was very rapid as second instars were observed as early as week 2 post infestation (PI) and third instars as early as week 3 PI. Maximum length of first instars was approximately 3 mm, recorded between weeks 4 and 6 PI. For second and third instars, maximum lengths were 13 and 23 mm, respectively. These data fit well with those provided by Cepeda-Palacios *et al.* (1999) for natural infestations.

Hypodermatinae

Growth of first-instar *H. lineatum* has been described for larvae reared *in vitro* for approximately 6 weeks (Chamberlain, 1970) and 90 days (Colwell, 1989a). In both studies larvae grew at the same rate during the first 6 weeks, increasing in length from 0.6 to 3.6 mm. During the subsequent 50 days in the latter study, larvae grew a further 1.4 mm. The growth, *in vitro*, seems somewhat greater than that reported from natural infestations where larvae recovered after 3–4 months were 3.4 mm long (Hadwen, 1919; Hadwen and Fulton, 1929). First instars recovered from subdermal sites on the back of cattle are up to 15 mm long, which probably is near the maximum as these larvae would be very near to the first moult. Growth of second and third instars has been documented for *Przhevalskiana sileunus* recovered from naturally infested goats slaughtered over a single season. Increase in length was almost twofold during the last two instars, with the majority occurring during the third instar (Otranto and Puccini, 2000).

Cuterebrinae

The growth rate, as reflected by weight gain, of second- and third-instar *Cuterebra emasculator* is biphasic (Bennett, 1955), with much more rapid growth in the later portion of the third instar than in the second instar and the early third instar. A similar trend was seen when larval volume was measured for *Cuterebra fontinella*, although the precision of this approach was not as fine (Cogley, 1991). Smith (1977) indicated that growth rate, as indicated by changes in weight, of *Cuterebra approximata* was also maximal during the third instar. Catts (1967) observed that growth, as indicated by change in body length, of *C. latifrons* was at a maximal rate during the late second instar, but slowed significantly during the later portions of the third instar. These observations seem at odds with the other work, but may be explained by the use of body length as the indicator of growth, particularly in the last instar where increase in length is relatively minor in comparison with increasing weight as a result of the accumulation of fat body and development of other tissues.

Gasterophilinae

Information on growth of larvae is completely lacking.

6 Morphology of Adult Oestridae

D.M. WOOD

Agriculture and Agri-Food Canada, Canada

Few small families of flies show such diversity of form among so few species as do oestrids, suggesting a long history of extinction as major mammalian groups have flourished, then disappeared. Few families have had such a controversial taxonomic history, perhaps as a result of their rarity in collections and consequent lack of understanding by most dipterists. Because of their universal need for a mammalian host, they are probably never very common in nature. They are also difficult to rear because their larvae, unlike those of most other calyptrates that can pupate successfully on a suboptimal amount of food, must be fully mature before they form viable puparia. If adults are reared at all, they are usually in poor condition, often deformed, teneral, shrivelled and greasy (a result of the large amount of stored fat), and are seldom properly prepared. Also, adults that frequent the vicinity of hosts, such as the elephant, rhinoceros or other large or fast-moving mammals, are not easily caught or even readily observed at close range. And not all oestrids frequent the vicinity of their hosts: all Cuterebrinae apparently oviposit on substrates with which the host is likely to come into contact, or in the case of *Dermatobia hominis* (Linnaeus Jr) on flies that themselves are attracted to mammals.

The total number of species of Oestridae in the world is only about 160. There are approximately 70 in the Old World (Zumpt, 1965), of which about 40 are in Africa, south of the Sahara (Pont, 1980a,b). In the Neotropics there are about 50 species, mostly in the genus *Cuterebra* (Guimarães, 1966) and there are some 40 species and subspecies of *Cuterebra* in North America (Sabrosky, 1986), along with five or six *Cephenemyia* species and a single native species of *Hypoderma*. Australia has only one, *Tracheomyia macropi* (Froggatt), whose larvae live in the tracheae of kangaroos. The total number is still in a state of flux, as some names have been proposed for larvae for which the adult remains unknown, while many others, especially Neotropical species of *Cuterebra*, are so poorly known that, of the more than 50 species of Cuterebrinae listed by Guimarães, the stated distributions of only ten include more than the type locality. This ratio suggests that more

©CAB International 2006. *The Oestrid Flies: Biology, Host–Parasite Relationships, Impact and Management* (eds D.D. Colwell, M.J.R. Hall and P.J. Scholl)

remain to be discovered, but also that the inventory may contain some synonyms yet to be recognized.

Oestrids have, at one time or another, been classified into three or four different families (Cuterebridae, Gasterophilidae, Oestridae and Hypodermatidae, the latter sometimes included in Oestridae), and have been aligned with various other families of calyptrate flies. Gasterophilidae were once placed next to the Muscidae, and even in the Acalyptratae, because of the lack of meral bristles, small calypters and the unbent vein M in a few species. Because of their great diversity of form, this taxonomic scattering has some morphological basis, but the demonstration of unifying features (shared derived character states) has shown that the group is probably monophyletic, and hence mammalian parasitism has presumably arisen once in the ancestor of all oestrids.

As the following description will show, however, each subfamily is quite heterogeneous, and shared derived character states for each of the subfamilies are not obvious, although members of each subfamily share similar life histories and larval habitats. Thus all Oestrinae pass their larval stages in the nasal and pharyngeal passages of their hosts, although these hosts encompass a wide range of mammals. The Hypodermatinae and Cuterebrinae all inhabit as larvae, boils or warbles the subdermal tissue, also of a variety of mammals. Larvae of the Gasterophilinae are found in the gut of elephants, rhinos and horses, with the exception of two parasites of the African elephant, *Ruttenia loxodontis* Rodhain and *Neocuterebra squamosa* Grünberg, which live in boils in the skin. Although these two species are usually placed in the Gasterophilinae (Zumpt, 1965; Pape, 2001a), they are not readily assignable to any subfamily on the basis of either morphology or life history. The first-instar larvae of many oestrids that have been well studied, including a few members of Hypodermatinae, Gasterophilinae and Cuterebrinae (except probably *D. hominis*), are capable of migrating through the tissues of their hosts to pass from their point of entry to the site of their final development – indeed such migrations seem to be an obligatory part of their life cycle (for *Cuterebra* see Hunter and Webster, 1973b). Only first instars of Oestrinae apparently do not make such migrations, but may not need to as they are deposited directly into or near the nostrils, eyes or mouth of the victim.

Adult oestrids vary considerably in size, from 8 to 35 mm in length. *T. macropi* and *Cephalopina titillator* (Clark) of the Oestrinae and *Przhevalskiana* spp. of the Hypodermatinae are among the smallest, while the species of *Gyrostigma* and some *Cuterebra* are the largest. All oestrids have small and compact, or atrophied, mouthparts. Most also have small antennae, and all lack bristles, so characteristic of other calyptrates, although most are extensively pilose as well as pruinose (microtomentose). A few species also have some stouter hairs that resemble small bristles.

Mimicry of aculeate Hymenoptera seems to have been an important aspect of their evolution, perhaps conferring some protection to species that probably always have existed in low numbers relative to other insects. In this respect, their abundant pilosity, which provides them with much of their coloration, may have been an inherent advantage in developing mimicry. For example, members of the genus *Gasterophilus*, all Old World in origin, resemble honeybees, which also originated in the Old World until distributed worldwide by man. Most of the species of *Cuterebra*, all New World in origin, resemble bumblebees or carpenter bees, or perhaps in the case of the males of *Cuterebra semiater* (Wiedemann), mutillid wasps. *Cephenemyia* and *Hypoderma*, found throughout the northern hemisphere, as well as *Portschinskia* of the Old World, also resemble bumblebees. Some African species may resemble wasps (*Cobboldia loxodontis* Brauer is bluish black with a rather striking orange head, *Cobboldia chrysidiformis* Rodhain and Bequaert is bright metallic green, while *Gyrostigma* spp. are yellow or orange and brown or black); all appear more wasp- than bee-like, but my ignorance of African wasps, and African mimicry patterns in general, makes this conclusion entirely speculative. In stark contrast, members of the oestrine genera *Oestrus*, *Gedoelstia*, *Kirkioestrus* and *Rhinoestrus*, with their mottled patterns of black, brown, gold and white, look superficially similar to one another (although they differ considerably in microscopic details) and seem to resemble bird droppings. Although rather different in appearance, *Strobiloestrus vanzyli* Zumpt is also somewhat similarly mottled and, with its enormous white lower calypters, may also resemble a bird dropping. The inflated yellow head of *Oestromyia leporina* (Pallas) (suggesting the piophilid genus *Thyreophora* which may also have been a mimic), and the orange wing bases and calyptrae of *Oestroderma potanini* Portschinski may also play some role in mimicry. In other respects *Oestromyia* and *Oestroderma* are much alike, and, except for their quite different heads, appear to be sister taxa (Pape, 2001a).

Head

Among the most prominent features of the majority of oestrids are the relatively small eyes and small antennae, and the enormously widened ventral part of the head (see Pape, 2001a). The latter condition is a result either of an enlarged genal region and correspondingly narrowed face (most species) or conversely, in Hypodermatinae (except *Portschinskia* and *Oestroderma*), of a greatly widened face and correspondingly reduced genal region.

The head of most oestrids is as wide as the thorax, even wider in some species of *Cuterebra*. In *Andinocuterebra*, the head, although about as wide as the thorax, appears narrower because of the presence of a large tuft of erect yellow hair on the anepisternum. In *Portschinskia*, however, the head is distinctly narrower than the thorax. The eye is apparently bare in all oestrids.

The frons at the vertex of almost all oestrids is quite broad, especially in *D. hominis*, in which the frontal width in both sexes appears to be about the same, whereas in most other oestrids it is narrower in the male than in the female. In a few oestrids (*Pharyngobolus africanus* Brauer (Zumpt, 1965, p. 14), *Portschinskia*, *Oestroderma*) the male frons is strongly narrowed, as it is in many other muscoid flies that wait for females and chase away competing males.

The frontal vitta, conspicuous in all calyptrates because of the presence of a strip of longitudinally wrinkled cuticle that is usually darker in colour and lacking pruinosity, is readily recognizable in all oestrids, although in Cuterebrinae and Gasterophilinae, it is smooth and velvety, bearing soft erect hairs, and is not as clearly delineated from the parafrontals as in other oestrids.

The lunule in calyptrates is usually shiny and bare, and this is true of most oestrids as well. However, some Hypodermatinae (*Hypoderma*, *Oestroderma*,

Pavlovskiata and *Przhevalskiana*) as well as *Cephenemyia* (Oestrinae) have small hairs arising on the lunule (Pape, 2001a).

Three ocelli are present in all oestrids except *Andinocuterebra*, *Gyrostigma* and *Tracheomyia*. Ocelli are relatively large in *Dermatobia*, and apparently also in *Ruttenia* (Zumpt, 1965, p. 139, Fig. 177). In *Oestrus*, *Kirkioestrus*, *Gedoelstia* and *Rhinoestrus*, and evidently also in *Cephalopina* (Zumpt, 1965, p. 188, Fig. 269), each ocellus is raised on its own black spherical protuberance; the ocelli thus are directed forward or sideways, not upwards, and are clearly visible from the front or the side. The three protuberances may be clearly separate from one another (*Gedoelstia*) or their bases may be partly or entirely in contact, and the degree of contact seems to vary, even intraspecifically, particularly in *Oestrus ovis* Linnaeus. The three ocelli may also be mounted on a single rounded raised tubercle, as in *Hypoderma* and perhaps some other members of the Hypodermatinae, but not in *Oestroderma*. In all other oestrids, however, ocelli are directed upwards or at a slight angle to the horizontal, and are not situated on a tubercle or tubercles.

The ocellar triangle is present in all oestrids, even in those lacking ocelli; it may be scarcely any larger than the tubercle bearing the ocelli (*Hypoderma*) or may be much more extensive, with a long narrow pointed angle extending forward medially into the frontal vitta, and distinguishable from the latter by being shiny (*Dermatobia*, *Cuterebra* and *Gasterophilus*).

One of the most striking aspects of most oestrids is the relatively minute antennae, especially in comparison with the majority of other muscoid flies. The scape and pedicel are usually short and inconspicuous and are usually haired (bare, or with minuscule setae in *Cobboldia* and *Strobiloestrus*, as well as in Oestrinae, except *Cephenemyia*), and the pedicel is bare and shiny (devoid of pruinosity) in *Hypoderma* and *Strobiloestrus*. In *Hypoderma* the globular first flagellomere is partially hidden within the cup-shaped apex of the pedicel; in *Portschinskia* it is completely hidden (Zumpt, 1965). Each antenna is sunken into a deep pit, especially in *Stobiloestrus*, so that only the arista is visible in profile. Even in frontal view little but the first flagellomere and arista are visible. In *Gyrostigma* (Gasterophilinae), the first flagellomere is almost completely hidden in the anterior view, with only the arista protruding, and the pedicel is adorned apically with setose projections.

In most oestrids the first flagellomere is scarcely twice as long as it is wide, while in the Cuterebrinae, it is longer, usually two to three times as long as wide in species of *Cuterebra*, and about five times as long in *Dermatobia*. The first flagellomere is also similarly elongate in *Cobboldia elephantis* (Steel) (Zumpt, 1965, p. 6, Fig. 1d). The longest antenna is found in *Andinocuterebra*, where the first flagellomere is seven or eight times as long as it is broad, nearly as long as the face. This fly, looking very like a species of the syrphid genus *Microdon*, resembles a bee even more than other oestrids, and because of this may have been overlooked by most fly collectors.

The arista is typically composed of three aristomeres in all calyptrate flies, but in oestrids the first (basal) aristomere appears to be completely hidden within the first flagellomere, and only the second aristomere can be observed in most specimens. Even if it is extremely small and scarcely discernible; the arista thus appears to consist of at most only two aristomeres, and often only one. The shape of the terminal aristomere differs considerably between Cuterebrinae and all other oestrids (except *Ruttenia*, in which it is enlarged racquet-like towards the apex); in the Cuterebrinae it tapers evenly to the apex, is somewhat flattened side to side and is pectinate, either dorsally or dorsally and ventrally, whereas in other oestrids it is bulbous at the base and tapers abruptly to a hairlike apex. In Cuterebrinae, the pectinations of the arista may arise only on the dorsal surface (*Dermatobia* and Neotropical *Cuterebra* (subgenus *Metacuterebra*)); or on the dorsal surface, with a few strands on the ventral side as well, towards the apex, as in *Andinocuterebra* and *Cuterebra* (subgenus *Cuterebra*), or from both dorsal and ventral surfaces (*C. semiater*); in the latter species the aristal hairs are long and branched, imparting a bushy aspect to the arista.

In most oestrids the face is concave behind the antennae, forming a single median 'antennal cavity', or a pair of cavities separated by a broad, flattened, facial carina, each cavity partially to nearly completely enclosing the antenna so that little is visible in profile (as in *Cephalopina*, *Gedoelstia*, *Oestromyia*, *Pavlovskiata* and *Strobiloestrus*). In *Oestrus* and *Rhinoestrus* the carina is present between lunule and antennal bases, but is interrupted below the level of the scape to form a median cavity. In most oestrids, the face below the antennal cavity is abruptly narrowed, extending to the lower cranial margin as a narrow median strip between the facial ridges, or (as in most Hypodermatinae, viz. *Oestromyia*, *Przhevalskiana*, *Pavlovskiata*, *Strobiloestrus* and *Hypoderma* itself) is abruptly widened below the antennae into a U-shaped or pyriform shield-like plate that occupies much of the lower half of the front of the head. In *Andinocuterebra* (Cuterebrinae) the face, instead of being abruptly narrowed or widened below the antennae, is slightly widened, then evenly tapered ventrally to the lower cranial margin, and is impressed with a median longitudinal groove. Although the face is abruptly narrowed ventrally in *Oestroderma* and *Portschinskia* (Hypodermatinae), the median part of the face is raised medially and impressed with a similar median groove.

In almost all calyptrates, the facial ridge, or lateral margin of the face, is raised. Often, it also bears a row of stout bristles, of which the most ventral, the vibrissa, are particularly large. In oestrids, however, the facial ridge is flattened and usually bears appressed hairs that are directed ventrally, and anything resembling a vibrissa is absent. In *Oestroderma* and *Portschinskia himalayana* Grunin (as shown by Zumpt, 1965, p. 193, Fig. 279), and in *Andinocuterebra* as well, these hairs are directed medially along the median half of the facial ridge, but laterally along the lateral half. This condition is not typical of all species of *Portschinskia*, however, for the hairs are all directed laterally in *Portschinskia magnifica* Pleske (as shown by Zumpt, 1965, p. 193, Fig. 279; Pape, 2001a, p. 145, Fig. 2D).

Although extremely variable, the parafacial of the majority of oestrids is haired, sometimes densely so (*Hypoderma*, *Cephenemyia*) and is usually quite narrow. It also may be broad and sparsely haired (*Gasterophilus*, *Cuterebra*), moderately broad and with a few scattered hairs (*Dermatobia*, *Gedoelstia*), or narrow and bare (*Oestroderma*, *Oestromyia*, *Portschinskia* and *Strobiloestrus*). In *Oestrus*, the parafacial is sparsely haired, each hair arising from a small pit, similar to but smaller than those on the parafrontal; in *Rhinoestrus*, the parafacial is similarly clothed, but the few sparse hairs each arise from a small pustule, homologous with the larger pustules of the parafrontal. The parafacial of *C. loxodontis* is polished orange and entirely bare. *D. hominis* and species of *Cuterebra* have a relatively bare shiny black swollen area on the upper part of the parafacial, extending almost over the entire parafacial in *C. semiater* and a few other Neotropical species. Although present, this shiny bulge is least developed in *Andinocuterebra*, although it appears to be present in all Cuterebrinae that have been examined.

The genal groove (the membranous area at the ventral end of the ptilinal suture between parafacial, face and gena) is quite extensive in most oestrids, especially in *Gasterophilus*, where it occupies at least the lower third of the front of the head. Being less sclerotized than the rest of the head, and flexible at eclosion, it is subject to shrinkage and distortion, and perhaps may be thus distorted in most of the known specimens of *Pharyngobolus* and *Tracheomyia*. It is usually less pruinose than the surrounding areas, and thus readily distinguishable, but in some species of *Cuterebra* it is as extensively pruinose as the adjacent sclerites and appears as a continuum with the rest of the head. In *Andinocuterebra*, *Oestroderma* and *Portschinskia*, it is black and strongly wrinkled.

In most muscoids, the anterodorsal margin of the gena is convex, bowed upward to form the genal dilation. This dilation, however, is absent in *Gasterophilus* and virtually absent in *Hypoderma*, and in other oestrids is never particularly well developed. In *Gedoelstia* the genal dilation is not clearly delineated from the genal groove. The gena is always haired, the hairs being directed anteriorly and/or medially towards the base of the proboscis.

Mouthparts

Although always undersized relative to the rest of the head, the mouthparts of Oestridae are usually present, even though in reduced or rudimentary form. The proboscis is best developed in *Cobboldia* (Gasterophilinae) and in Cuterebrinae, all of whose members have a small stout prementum or prostheca. In most specimens of *Cuterebra* and *Andinocuterebra*, however, the proboscis is concealed amongst the hairs that line the edges of the subcranial cavity, but it is clearly visible in *Dermatobia* and *C. semiater*. Palpi are well developed in *Cobboldia* (longer than the prementum in *C. loxodontis*). Palpi are present in most other taxa as well, but they are always reduced to small rounded knobs (Zumpt, 1965, p. 6, Fig. 1). They are absent in Cuterebrinae. In the Hypodermatinae the proboscis is best developed in *Oestroderma* and *Oestromyia*, in which a short pointed proboscis is flanked by a pair of small rounded palpi generously covered with black hair; even the clypeus is distinct, as a triangular black sclerite. None of the species of *Hypoderma* has a proboscis; this is reduced to a small black featureless knob, although tiny rounded palpi are present in *Hypoderma tarandi* (Linnaeus). In Oestrinae, functional mouthparts are present in *Cephenemyia*: adults of this genus have been observed drinking water (Catts, 1982) and they may be present also in *Pharyngomyia*, although in other members of the subfamily the proboscis is reduced to a minute black knob with no external opening. The labella, when present, are always reduced, their diameter being no wider than the diameter of the prementum, suggesting that the proboscis can imbibe liquids but cannot be used effectively (or at least is not habitually used) to soften and sponge up dried sugars as in other muscoid flies. Evidently, proboscis and palpus reduction has proceeded independently in each of the several oestrid lineages.

Thorax

The muscoid flies, in general, have a pruinose thorax (the submicroscopic vestiture that provides coloration and water repellency), with an overlay of sparse, erect
or recumbent hairs, and with much larger erect bristles arranged in a characteristic chaetotaxy (McAlpine, 1981). Oestrids, in contrast, have a pruinose thorax,
that is also extensively pilose, of which the pile is usually denser than in other muscoid flies and resembles that found in some members of the Syrphidae. The pile
may be dense and erect (*Gasterophilus*, *Hypoderma* and *Cephenemyia*) or sparser and
recumbent. The barest oestrids are probably *Strobiloestrus* and *Gedoelstia*, but even
in these genera most of the thorax is clothed in a short recumbent pile. The quality of the vestiture of pile is probably governed in many oestrids by selection for
mimicry, as different species apparently related to one another may differ considerably in their degree of pilosity (e.g. the species of *Cephenemyia* are pilose and
resemble bumblebees, while *Pharyngomyia picta* (Meigen) is grey pruinose and thus
looks superficially entirely different). Bristles, as found in other muscoids, are
almost entirely lacking, although in *Dermatobia*, *Oestroderma* and *Oestromyia*, the hairs
along the hind margin of the anepisternum, on the meron and along the lateral
edge of the scutum and scutellum, especially on the notopleuron, are stouter and
distinctly more bristle-like, although still small by calyptrate standards.

The prosternum is pilose (bare in *Gedoelstia*), not just along its lateral margins,
but often over its entire surface, and the membrane lateral to it as well (which is
sparsely haired in *Gedoelstia*). The proanepisternum is usually also covered with
pile, although it is rather variable, and Pape (2001a, p. 147) has divided it into
four character states. *Cuterebra*, *Dermatobia*, *Cobboldia*, *Tracheomyia*, and some species
of *Hypoderma*, have setae scattered over the central part of the sclerite, while
Gasterophilus and *Gyrostigma* have setae on the lower part. The proanepisternum is
bare in *Oestroderma*, except for a well-developed proanepisternal hair (or small bris-
tle) at its ventral margin. Although difficult to observe because it is obscured by
the edge of the head, the proanepisternum is also bare in all Oestrinae, except
Tracheomyia, and is also bare in *Pavlovskiata*, *Przhevalskiana* and *Strobiloestrus*
(Hypodermatinae) (Pape, 2001a). The postpronotum in *C. loxodontis* is rather
swollen, and this is also true of *Portschinskia* (Pape, 2001a), but it is not enlarged in
other oestrids. The postpronotum, scutum and scutellum are almost entirely covered with pile, usually recumbent in Oestrinae (except *Cephenemyia*), and in
Dermatobia and *Cobboldia*, and erect in *Gasterophilus*, *Hypoderma*, *Cuterebra* and
Cephenemyia. In *Hypoderma*, as well as in *Gedoelstia cristata* Rodhain and Bequaert,
Oestrus bassoni Zumpt and some species of *Rhinoestrus* (as shown by Zumpt, 1965),
the scutum bears two shiny black longitudinal acrostical stripes just behind the
head, flanked by a pair of shiny spots in the dorsocentral region, and a transverse
row of four shiny spots behind the transverse suture. In *Hypoderma* and *G. cristata*
the dorsocentral postsutural spots are elongate. A similar arrangement of stripes
and spots is also found in *Oestromyia* and *Oestroderma*, except that the spots are matt

black on a grey background. These stripes are also apparently present in *Tracheomyia* (Zumpt, 1965, p. 141). The rather similar pattern of spots on the scutum of several members of both Oestrinae and Hypodermatinae may constitute evidence of a sister-group relationship between the two subfamilies. In *Oestrus* and *Rhinoestrus* each hair arises from a small black tubercle; in *O. ovis* the tubercles become larger towards the scutellum, with the largest shiny black areas at the apex of the scutellum.

The notopleuron is barely distinguishable from the scutum in Cuterebrinae and Gasterophilinae, but has become greatly enlarged in *Strobiloestrus* and *Gedoelstia* (and also in *Cephalopina*, *Pharyngobolus* and *Tracheomyia* according to Pape, 2001a) to form a gnarled conical projection, capped with a tuft of pile. It is also somewhat enlarged in *Oestrus* and *Rhinoestrus*, but is not gnarled or conical. The notopleuron is not differentiated in *Portschinskia* (Pape, 2001a). The postalar wall is pilose along its dorsal margin in *Gasterophilus* and *Cobboldia*, or its hind margin (*Dermatobia*); in Oestrinae (except *Cephenemyia*) it bears some sparse pile (*Kirkioestrus*), a small tuft of short pile in the middle (*Rhinoestrus*) or is fairly extensively pilose (*Oestrus*), while in *Hypoderma* and *Cuterebra* it is extensively haired. The suprasquamal ridge is also pilose in the Cuterebrinae but is bare in other oestrids (Pape, 2001a).

The scutellum of *Gasterophilus*, *Andinocuterebra*, *H. tarandi* and also of *Gyrostigma rhinocerontis* (Hope), according to Pape (2001a), is covered with erect pile, similar to that of the scutum, which complements their bee-like appearance. Other species of *Hypoderma* also have erect pile on the dorsum of the scutellum, except that the apex is bare and slightly indented medially. In other oestrids, this pile is more appressed or recumbent, except along the apical margin where the hairs are longer and extend backwards, especially conspicuous in *Kirkioestrus*. In *Rhinoestrus*, and to a lesser extent in *Oestrus*, the margin is set with tiny medially directed, bristle-like setae; these apical hairs are presumably the homologues of the scutellar bristles of other calyptrates. The scutellum of *Cuterebra* is exceptionally long, extending back over the mediotergite and base of the abdomen like a shelf, and is more than half as long as wide. It is usually pilose on its underside as well as above and the hairs extend back from both directions at the apex of the scutellum (except in the male of *C. semiater*, in which the long bright orange pile, concolorous with that of the abdomen, is confined to the upper side of the scutellum).

The subscutellum, which is always bare, varies from, at most, a narrow transverse strip in the Cuterebrinae, *Oestroderma* and *Gasterophilus intestinalis* (De Geer), to a distinctly wider and more convex strip in *Gasterophilus nasalis* (Linnaeus) and *Hypoderma*, to a relatively enormous bulge in both *Gedoelstia* and *Strobiloestrus*, where it is as large as in even the most well endowed tachinid. It is also well developed in *Ruttenia* (Zumpt, 1965, p. 139). The subscutellum is pruinose in *Gedoelstia*, contrasting with the shiny black postnotum, but shiny black like the postnotum in *Strobiloestrus*.

The anterior spiracle is a small oval slit in *Gasterophilus*, with only a slight suggestion of a fringe along the anterior edge. In *Cobboldia*, and in the Cuterebrinae, the opening is larger and is entirely obscured by a long fringe arising from either side of the spiracle. In Oestrinae and Hypodermatinae (except *Oestroderma*, in which no fringe is apparent) the fringe only partly obscures the opening.

The anepisternum (mesopleuron) and katepisternum (sternopleuron) are typically almost entirely covered with pile. In bumblebee-like species the pile is erect, and is usually black and yellow, while it is appressed in those species not resembling bees. The anepimeron (pteropleuron) is less extensively pilose, while the katepimeron (barette) may be pilose (*Cobboldia, Cephenemyia, Oestrus, Rhinoestrus, Kirkioestrus, Strobiloestrus, Hypoderma* and all of the Cuterebrinae), or bare.

The meron typically bears a patch of erect hair only on its posterior half or less, often with longer hairs along its posterior border, suggestive of meral bristles. In *G. intestinalis*, however, the meron is bare, and only a few hairs are present in *Cobboldia*. The mediotergite and laterotergites (each side subdivided into an anterior anatergite and posterior katatergite) of the postnotum are always bare. *Gyrostigma* is unique in having a cone-like anatergite (Pape, 2001a).

The posterior spiracle, located between the meron and the halter, is usually more circular than the anterior spiracle, although it is oval in *Neocuterebra* (a vertically oriented oval) and in *Gyrostigma* (horizontally oriented). Pape (2001a) has coded its shape, and the nature of the encircling fringes, as three different character states. It is fringed similarly to the anterior spiracle, as described above. Thus, in *Gasterophilus* there is only a small to moderate fringe along the anterior border (Pape, 2001a, Fig. 5C); in Oestrinae and Hypodermatinae, and in *Neocuterebra* and *Ruttenia* both anterior and posterior edges are fringed, but the fringes do not close the opening (Pape, 2001a, Fig. 5A, B and F), while in *Cobboldia, Pharyngomyia* and the Cuterebrinae the posterior fringe is organized into a cover, or 'operculum', and both are long enough to obscure the opening entirely (as in Pape, 2001a, Fig. 5E). In *Gyrostigma* the posterior spiracle is enormous, with a row of fimbriate processes along both dorsal and ventral margins (Pape, 2001a, Fig. 5D).

Wing

In all oestrids, veins Sc, R1, R2+3 and R4+5 are located just behind the leading edge of the wing in typical calyptrate fashion. Vein M, however, is much more variable; it may be straight (lacking a bend) or slightly curved posteriorly to end in the wing membrane (*Gyrostigma*) or to end in the wing margin well behind the wing tip (*Gasterophilus*) or curving evenly forward to end in the costa in front of the wing tip and just behind the apex of vein R4+5, as in all Cuterebrinae and Hypodermatinae, and in *Cephenemyia* and *Pharyngomyia* (Oestrinae), as well as in *Neocuterebra, C. loxodontis* and *C. elephantis* (Gasterophilidae). In all remaining Oestrinae, and in *C. chrysidiformis* and *C. loxodontis*, vein M curves obliquely forward to end in R4+5. In *Cephenemyia, Pharyngomyia, Pharyngobolus, Kirkioestrus* and *Tracheomyia*, vein M appears to extend a short distance into the membrane as a 'stump' vein, at the position of the bend of M; the apex of vein M beyond this stump vein thus looks like a cross-vein. The apex of vein M may be straight (*Tracheomyia*), slightly sinuous (*Oestrus*) or strongly sinuous (*Cephalopina, Gedoelstia* and *Kirkioestrus*). The most unusual condition, unique in the Diptera, is found in *Ruttenia*, in which vein M curves evenly round to meet the apex of R4+5 within the wing membrane, without touching the wing margin. In all oestrids the cross-vein dm-cu is in the usual position, closer to the wing margin than to cross-vein r-m, except in *Gasterophilus* and *Gyrostigma*, in which it is retracted to or near the level of r-m; it is absent in some species of *Gasterophilus* (Zumpt, 1965, p. 112, as 'lower marginal cross-vein (tp)'). The wings of most oestrids are hyaline, with a vague pattern of darker colour in some species of *Gasterophilus*. The wings of *C. loxodontis* and *C. elephantis*, of all *Gyrostigma*, and of many, perhaps all, species of *Cuterebra* are dark brown. Although paler, wings of *Oestromyia* and *Oestroderma* are also infuscated.

The alula is not particularly large in most oestrids but it is well developed in Cuterebrinae. In *Strobiloestrus*, the alula has a distinctive black spot at its distal corner. The calypters, both upper and lower, are present in all oestrids and in most are similar in structure to their counterparts in other calyptrates. However, they are small relative to the body in *Gasterophilus*, and are even further reduced in *G. intestinalis*. They are enlarged in *Gedoelstia*, and enormously enlarged in *Strobiloestrus*, in which they are white, wrinkled and, in lateral view, extend down to the hind coxa with an inflated lobe along their posterior margin and another inflated lobe projecting medially so that the subscutellum is partly hidden in dorsal view. In *Oestroderma*, on the other hand, they are long and narrow, extending back to the middle of syntergite I+2 but widely separated from each other medially so that the entire postnotum is visible from above; in addition they are bright orange, matching the colour of veins C, R1 and R2+3 and the costal cell and cells r1 and r2+3 of the wing, in striking contrast to the dark grey of the body. Among Cuterebrinae, the calypters are pale in *Dermatobia*, but are dark in most *Cuterebra*. They are sometimes conspicuously waxy white, and are thus in striking contrast to the darker wings and/or body (*C. elephantis* and *Gedoelstia*, as well as *Strobiloestrus*). In *Gasterophilus* and *Gyrostigma* the hind edge of each calypter bears a fringe of long pale hairs; in other oestrids these hairs are microscopic.

Legs

The legs of most oestrids are of moderate size and with few exceptions are covered with appressed hair. In *Cuterebra*, however, the legs are rather short and quite stout, and the tarsomeres are broad and dorsoventrally flattened; each tarsomere is expanded laterally and medially, with long hairs along the edges of the flanges. The tibiae of *Cobboldia loxodontis* are bare and shiny black on their dorsal surfaces. The leg hairs tend to be short and bristly, especially on the lower surfaces of the tibiae, and the sides of the tarsi.

Abdomen

Some species of Oestridae have rather stout hairy abdomens, especially the species that seem to mimic bumblebees or other bees, and the pattern is provided by the hairs. In contrast, all species of *Oestrus*, *Kirkioestrus*, *Gedoelstia* and *Rhinoestrus* have characteristic mottled patterns of white, gold and brown or black pruinosity. In addition, tergites 3–5 of both species of *Gedoelstia* are each adorned with a median raised area bearing a pair of tubercles that each bear thorn-like projections on their dorsal surface. Some of these thorn-like projections each bear a small seta. Most species of *Rhinoestrus* are also adorned with seta-bearing tubercles (Zumpt, 1965) on the dorsum of the abdomen. The abdomens of *Cobboldia* are shiny metallic dark blue or green, or black, and rather rugose.

As is typical of all calyptrates, the first abdominal tergite is greatly reduced and fused to tergite 2 to form syntergite 1+2. The fusion is not well marked in *Gyrostigma*, so that both tergites can be easily distinguished (Pape, 2001a). This syntergite is usually no larger than the remaining tergites, but in the female of *Oestrus variolosus* (Loew) (but not in the male) it extends back nearly to the middle of the abdomen, displacing the remaining tergites posteriorly. Tergites 3–5 are usually about the same length, each narrowing progressively to the apex. Tergite 6 is rather prominent in some oestrids, especially in *Gedoelstia* in which the apex of the abdomen, which is formed of tergite 6, is broadly truncate. Each of the first six tergites bears a spiracle on each side near its anteroventral corner. The spiracles in most oestrids are rather small and inconspicuous; in Cuterebrinae, however, they are larger and more prominent, and in some species of *Cuterebra* are each ringed with a circle of dark erect hairs.

The sternites are entirely exposed, as in Sarcophagidae, Calliphoridae and Rhinophoridae, not covered laterally by the tergites, as in Tachinidae. In *Oestroderma*, the sternites are relatively large, and extend laterally nearly to the edges of their respective tergites, but in most other oestrids they are widely separated from the tergites by an expanse of leathery, sometimes hairy, membrane. Sternites 1 and 2 are separate from each other, not fused to one another as are their respective tergites; sternite 1 is usually larger than the others and is U-shaped, while sternites 2, 3 and 4 are rather small and rounded, or somewhat square or rectangular. In *Gasterophilus*, only sternite 1 is distinctly sclerotized, the remainder appear to be undifferentiated.

Male Genitalia

In relation to the size of the fly, the hypopygium of oestrids appears to be disproportionally small, and carried rather far forward below a relatively large tergite 6 and sternite 8, and dissection suggests that there is little mobility of the parts in relation to the surrounding sclerites. The phallus is also short and stout (except in *Dermatobia*). The surstyli tend to be rather stout and are widely separated from one another. In the details, however, especially in the arrangement of the pregonites (lobes of the gonocoxites) and postgonites (gonostyli), and bacilliform sclerites, there is surprising diversity. Sternite 5 appears to be undifferentiated as a distinct sclerite in *Gasterophilus*. In *Oestroderma* it is rather deeply notched apically but in other oestrids it is weakly notched to almost truncate apically, in contrast to the complex median notch found in Tachinidae.

Tergite 6 occupies the usual dorsal position, between tergite 5 and the remainder of the genitalia; it is symmetrical, and is usually sparsely haired. It is relatively large, especially in *Gedoelstia*, in which it forms the truncated apex of the abdomen. The remaining sclerotized parts of segments 7 and 8, usually referred to as syntergosternite 7+8, form a more or less symmetrical ring behind tergite 6. In oestrids the dorsal part of the ring, usually interpreted as sternite 8, is tightly associated with tergite 6, forming a smaller concentric ring directly behind it. The ventral part of the ring is at most a narrow band, sometimes attached to the dorsal part of syntergosternite 7+8 (as in *Cephenemyia*) but usually free from it, and forms an insignificant transverse arch dorsal to sternite 5. Spiracles 6 and 7 are closely associated with each other, at least in *Cuterebra*; in which spiracle 6 is located at the lateral extremity of tergite 6, but, as a result of circumversion, spiracle 7 is wedged between spiracle 6 and the lateral extremity of sternite 8. These spiracles can best be observed in *Cuterebra* because they are relatively large and enhanced by a ring of black hairs; they are hidden beneath tergite 5, by folds of integument, or by hairs, in most other oestrids. The epandrium is encircled by, and rather firmly attached to, syntergosternite 7+8.

The cerci in *Gasterophilus* and *Gyrostigma* are rather widely separated, weakly sclerotized, setose lobes, rounded apically, and are independent from each other, unlike most calyptrates. The cerci are in contact with each other medially in *Cephenemyia*, *Oestrus*, *Dermatobia* and *Cuterebra*, although their apices are separate, and the median line of fusion between them is clearly visible. In *C. loxodontis*, and in *Hypoderma*, however, the cerci are completely fused to form a syncercus, and in *Hypoderma* no line of median contact is visible.

Surstyli are always well developed in oestrids, and are particularly massive in *Gasterophilus* and *Gyrostigma* (an analogous situation occurs in the tachinid genus *Cylindromyia*, in which the cerci are fleshy and pad-like while the surstyli are massive). Normally in calyptrates, the surstyli are free from the hypandrium, while in *Hypoderma* the surstyli are fixed immovably to the epandrium. A unique arrangement of cerci, surstyli and hypandrium occurs in *C. loxodontis*. The surstyli are massive basally and are connected directly to the arms of the hypandrium, without the intervention of a bacilliform sclerite, and have a curved, hooked apex. At the base of each surstylus, next to the syncercus, there is a small basal lobe. Furthermore the anteromedian part of the syncercus extends forward between the bases of the syrstyli.

The hypandrium is generally U-shaped, sometimes rather square anteriorly (*Gasterophilus*). The arms of the hypandrium are closely approximated medially in *Cobboldia* and *Hypoderma*, moderately so in *Gasterophilus*, *Oestrus*, *Cephenemyia* and *Cuterebra*, but widely separated medially in *Dermatobia*,

Pregonites (lobes of the gonocoxites) in *Gasterophilus* and *Cobboldia* (Gasterophilinae), as well as in *Cephenemyia*, and *Oestrus* (Oestrinae), arise laterally from the arms of the hypandrium. In *Cobboldia* the pregonites are widely separated from each other, with a long slender connecting link between them passing as a bridge in front of the phallus, supporting the latter. They also arise rather far back, near the point of junction of hypandrium and epandrium, leaving a wide membranous space within the anterior U-shaped ring of the hypandrium. In *Hypoderma* and *Cuterebra*, in contrast, pregonites arise close to each other medially, and farther forward, leaving no membranous area within the U of the hypandrium. In *Dermatobia* they also arise anteriorly, but each lacks a ventrally projecting appendage. Pregonites of *Gasterophilus* support small hairy pads. In *Hypoderma* the pregonites are fused medially to each other to form a trough supporting the phallus, and their ventrally projecting appendages are very large, like a pair of heavily sclerotized tusks that are longer than the postgonites. *Cuterebra* also has well-developed conical projections on the pregonites, although they are smaller than those of *Hypoderma*.

In *Dermatobia* and *Cuterebra* the postgonites are long, pointed and sickle-shaped. They are notched apically in *Hypoderma*, but in other genera are simple apically. Bacilliform sclerites, which are a pair of bars or plates connecting the arms of the hypandrium with the base of each surstylus, are also quite variable. As mentioned above, they are absent in *C. loxodontis*, because the surstyli are connected directly to the hypandrium, but present in other genera. In *Cuterebra* each forms a rather wide, distinctly sclerotized bar. They are distinct but broader in *Dermatobia* and *Oestrus*, and broadest in *Gasterophilus* and *Hypoderma*, where they are fused together medially, forming a plate extending across the entire space behind the phallus.

The phallus (usually called aedeagus, but as shown by Sinclair *et al.* (1994) to actually consist in all Brachycera of the fused parameres enclosing the true aedeagus) is usually short, stout and heavily sclerotized in oestrids. In *Gasterophilus* it is exceptionally short, and does not project beyond the postgonites. In *Dermatobia*, it is exceptionally long and strongly curved. The epiphallus is short, blunt and massive in *Cuterebra*, huge and leaf-like, and flattened side to side, in *C. loxodontis*, and extremely small in *Hypoderma* (presumably it is vestigial). It is also well developed in *Oestroderma* and *Oestromyia* (Pape, 2001a). It appears to be lacking in *Gasterophilus*, *Oestrus*, *Cephenemyia* and *Dermatobia* and is also apparently greatly reduced or absent in *Gyrostigma* and *Cephalopina* (Pape, 2001a).

The aedeagal and ejaculatory apodemes, in most oestrids, are not particularly different from their counterparts in other calyptrates. In *Gasterophilus*, however, the aedeagal apodeme is not free anteriorly but fused to the anterior rim of the hypandrium.

Female Genitalia

The external female genitalia consist of the tergites and sternites of segments 6–8, plus the epiproct and hypoproct (the dorsal and ventral sclerites, respectively, of the remnant of the proctiger, which is usually regarded as segment 9 but I believe is postsegmental in origin) and the cerci (the paired appendages of the proctiger). In some oestrids (Hypodermatinae) these segments are elongate and telescoped to form a retractable 'ovipositor', but in their simplest form these three segments become successively smaller distally. The least modified genitalia appear to be found in *Dermatobia* and *Cuterebra*, of the Cuterebrinae, a group that does not oviposit on the host but on substrates contacted by the host. These three segments, 6–8, are telescoped within the much larger and longer segment 5 in the form of concentric rings, and only their outer edges are visible unless extracted. Only segment 6 bears spiracles in the female of *Dermatobia* and *Cuterebra*, each located in the ventral corner of the tergite (the male of *Cuterebra* has a seventh pair of spiracles). In *Cuterebra*, sternites 6 and 7 articulate with the tergites laterally but are rather weakly sclerotized medially. Sternite 8 is subdivided medially, and the two lateral halves are papillate and resemble a second, smaller pair of cerci. All three sternites are weakly sclerotized in *Dermatobia*; although sternite 8 is also medially split the two halves are flat and unmodified. Aside from these small details, there appear to be no fundamental differences between the female genitalia of *Dermatobia* and those of *Cuterebra*, indicating that no special structures have evolved to facilitate deposition of eggs on blood-sucking flies. If anything, the genitalia of *Dermatobia*, in which the sternites are scarcely differentiated, are even simpler than those of *Cuterebra*.

In *Hypoderma* (and evidently in other Hypodermatinae as well), the postabdomen, consisting of segments 6–8, is tubular and telescopic and is at least partially withdrawn into the abdomen. Tergites and sternites 6 and 7 are long and narrow but otherwise not especially modified, both tergite and sternite 8 appear to be subdivided medially into two halves, but again are not particularly modified. The proctiger has undergone the most modification: the epiproct is quite large, relative to its homologue in the Cuterebrinae, somewhat hexagonal and with a clavate extension from its dorsal surface that curves ventrally and extends beyond the base of the epiproct. Although this extension was labelled 'cercus' by Wood (1987, p. 1155, Fig. 14), it appears more likely that the cerci are the structures ventrolateral to it that appear to be extensions of the hypoproct. The hypoproct itself is subdivided medially as is sternite 8 ahead of it. These modifications are presumably to aid in the attachment of the egg holdfast to a hair of the host.

In *G. intestinalis*, the postabdomen is not telescopic or withdrawn out of sight within the abdomen, but is flexed under the abdomen at rest. The tergite and sternite of segment 6 are distinctly separate from each other, but those of segment 7 are fused to form a large conspicuous tube, swollen in the middle, with a mid-dorsal weakening of membrane near the apex. Sternite 8 forms a horse-shoe-shaped keel, divided medially except at the rounded apex. Large invaginated sacs (accessory glands?) lying within segment 7 open on either side of the base of tergite 8. Tergite 8 is split medially into two halves; each is drawn out to form a slender, pilose point. The cerci are long and attenuate. All of these sclerites, including the hypoproct between the cerci, are pilose, sternite 8 being the least setose.

In *C. loxodontis*, tergite and sternite 6 are separate from each other, and tergite 6 bears a spiracle on each side. Tergite and sternite 7 are fused to each other, as in *Gasterophilus*, with no evidence of a line of separation between them. Tergite 7 is as long as tergite 6, although sternite 7 is strongly narrowed midventrally, and is V-shaped, with the opening of the V directed posteriorly. Sternite 8 is not split medially and is closely associated with tergite 8, although a line of separation between them is clearly visible. The proctiger (including cerci) is unusually long, as long as segment 8. The epiproct is a small semicircular sclerite, setose at its apex, while the hypoproct is strongly pointed apically, flanked by the long slender cerci.

Conclusions

The proboscis is perhaps best developed in the Cuterebrinae, although palpi are lacking in this group. The proboscis is also developed, if very small, in *Cobboldia*, and palpi are relatively large. In *Gasterophilus* these structures are represented by a nodule. Tiny palpi are also present in *Ruttenia*, *Neocuterebra* and *Gyrostigma* of the

Gasterophilinae (Pape, 2001a: Figs 1 and 2), as well as in *O. potanini* and *H. tarandi* (Hypodermatinae).

The arista is flattened and setose (along its dorsal and/or ventral margin) only in Cuterebrinae. In other oestrids it is cylindrical and hairlike, thickened only at its base. The first flagellomere is longest in Cuterebrinae and in *Cobboldia*, but in other oestrids it is extremely small. The suprasquamal ridge is pilose only in Cuterebrinae. In female Gasterophilidae (including *Ruttenia* and *Neocuterebra*, but not *C. elephantis* (Pape, 2001a)), tergite and sternite 7 are fused to form a ring, but are separate in other oestrids.

7 Egg Morphology

D.D. Colwell

Agriculture and Agri-Food Canada, Canada

Eggs of three oestrid subfamilies are well adapted to maximize the survival and aid entry into the host by emerging larvae. Within each subfamily characteristic features define the relationship between oviposition and interaction with the host.

Hypodermatinae

Eggs of the Hypodermatinae are robust and pale yellow, with a smooth chorion. A distinct operculum is absent, but larvae emerge from the eggs through a slit developing in the anterior end of the egg. Hypodermatid eggs are firmly attached to host hairs (Cogley *et al.*, 1981), although there are several genera for which no information is available (e.g. *Pallasomyia*, *Przhevalskiana*). Attachment is achieved through a unique structure located on the ventral terminus of the egg. Detailed description of the structure of this feature has been made for *Hypoderma lineatum* and *Hypoderma bovis* from cattle as well as *Hypoderma tarandi* from reindeer and *Strobiloestrus vanzyli* from the Lechwe (Cogley *et al.*, 1981) (Fig. 7.1). While the structures were remarkably similar, minor differences were noted in the length of the clasper that attaches to the hair shaft and the arrangement of the attachment organ relative to the petiole that connects the attachment organ to the egg proper.

Variation in structure of the attachment organ is seen in *Oestromyia* and *Oestroderma* where the terminal portions are highly elaborate star-shaped structures (Reitschel, 1980). Although detailed examination of these unusual structures has not been undertaken the mode of action of the organ relative to attachment of eggs to the hair shafts is consistent with other members of the subfamily.

During oviposition the attachment organ is deformed by the ovipositor such that the clasper can surround the hair shaft in a 'weiner and bun' manner. Once the ovipositor releases the egg the attachment organ springs back to its original shape and the hair shaft is enclosed. The clasper is lined with a semi-fluid adhe-

©CAB International 2006. *The Oestrid Flies: Biology, Host–Parasite Relationships, Impact and Management* (eds D.D. Colwell, M.J.R. Hall and P.J. Scholl)

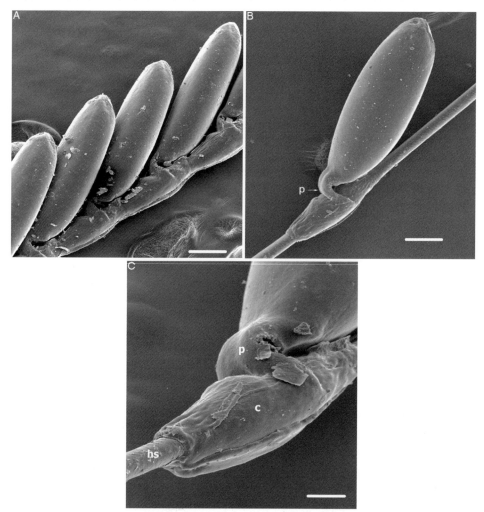

Fig. 7.1. Scanning electron micrographs of *Hypoderma lineatum* eggs: (A) String of eggs attached to a single hair shaft (bar = 0.15 mm); (B) Individual egg: clasper and petiole (p) constitute the attachment organ (bar = 0.19 mm); (C) High-magnification view of attachment organ composed of the clasper (c) and the petiole (p) attached to the hair shaft (hs) (bar = 60 μm).

sive layer, produced by the follicular cells, which flows into surface features of the host hair aiding the attachment.

Cuterebrinae

Eggs of the Cuterebrinae are dark brown to black, with a highly sculptured chorion and a well-developed subterminal operculum on the dorsal surface. The ovoid operculum is usually completely lost as the larvae hatch. Cuterebrid eggs are generally ovid in cross-section, with both ends slightly tapered. The eggs have a terminal micropyle on the same end as the operculum. The ventral surface of most cuterebrid eggs is flattened, presumably to assist adherence of the eggs to the substrate (Colwell *et al.*, 1999). Eggs of *Dermatobia hominis* appear somewhat distinct with a flattened ovoid cross-section and with little flattening of the ventral surface evident (Leite, 1988). Eggs are deposited and attached to the surface, whether an abiotic substrate or a porter insect in the case of *D. hominis*, with the aid of an uncharacterized adhesive.

Sculpturing of the chorion is elaborate and varies with the environment to which the eggs are exposed (Colwell *et al.*, 1999). Tropical cuterebrids have deep chorionic sculpturing while those from temperate climates tend to have sculpturing that is less deeply indented, tending to almost smooth in some desert species. Perhaps as suggested by Colwell *et al.* (1999) the deep sculpturing will have a plastron-like effect, trapping air close to the egg and preventing drowning when these tropical eggs are inundated by precipitation or condensation.

Outlines of the cells that deposit the chorion are evident on the egg surface. The shape of these cells has been used, in conjunction with the depth of the sculpturing, to identify species of *Cuterebra* (Fig. 7.2; Colwell *et al.*, 1999). Examination of a larger number of species is required to test the general applicability of this technique.

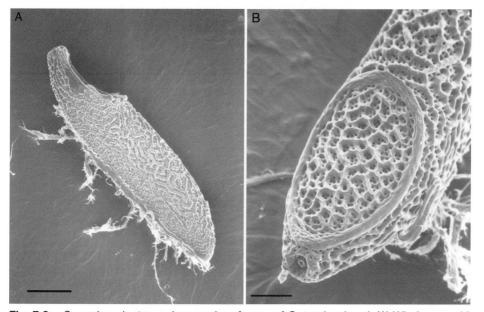

Fig. 7.2. Scanning electron micrographs of eggs of *Cuterebra baeri*: (A) Whole egg with operculum removed. Note the highly sculptured chorion and the flattened ventral surface with adherent material (bar = 0.22 mm); (B) Close-up of an intact egg with an undisturbed operculum (bar = 0.75 mm).

Gasterophilinae

Eggs of the Gasterophilinae are either creamy white or brown-black with a distinct, subterminal operculum on the dorsal surface. The egg surface is lightly, transversely striated (Fig. 7.3) except for *Gasterophilus pecorum* where the surface is marked by light polygonal-shaped 'cells' (Cogley, 1991). These eggs are also attached firmly to the host. *G. pecorum* is the exception as females attach their eggs to the surface of grasses. All species have a ventral attachment organ (Cogley and Anderson, 1983; Cogley, 1990), but substantial variation in structure is noted. Those species in which the egg is attached to the host hair have an attachment organ that encloses the hair shaft in a manner similar to that of the Hypodermatinae. Unlike the eggs of Hypodermatinae there is no thin petiole separating the attachment organ from the egg. Here the attachment organ is more intimately associated with the egg (Fig. 7.3b). Length of the attachment organ and its position relative to the egg proper varies among the species of *Gasterophilus* (Cogley and Anderson, 1983; Cogley, 1990). Similar to the Hypodermatinae, the attachment organ is deformed by the ovipositor so that it can encompass the hair shaft and clamp firmly around the hair after deposition is complete. Also present is an undefined adhesive layer that completes fixation of the egg in place.

Distinct attachment organs are present on eggs of *G. pecorum*, whose eggs are attached to leaves of grass, and the rhinoceros bot flies *Gyrostigma conjugens* and

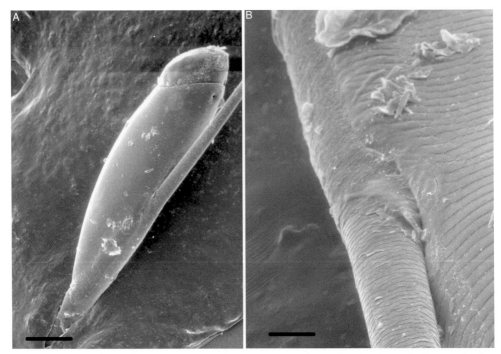

Fig. 7.3. Scanning electron micrographs of the egg of *Gasterophilus intestinalis* attached to a host hair shaft: (A) Whole egg (bar = 0.25 mm); (B) Close-up of attachment of egg to hair shaft (bar = 125 μm).

Gyrostigma rhinocerontis (= *pavesii*), whose exact oviposition sites are not known, but are thought to be on the host skin surface. These attachment organs are terminally located and are composed of highly developed endochorionic filaments coated with an adhesive (Cogley and Anderson, 1983). The filaments are pliable and conform to the surface features such as plant leaf trichromes. This characteristic, in combination with the adhesive, results in a firm attachment of the egg to the substrate.

8 Larval Morphology

D.D. COLWELL

Agriculture and Agri-Food Canada, Canada

Oestrid larvae generally resemble other muscoid larvae (Chu-Wang and Axtell, 1971; Teskey, 1981; Colwell, 1986; Colwell and Scholl, 1995). However, the long evolutionary history of the group (see Chapter 3), coupled with the obligatory parasitic lifestyle, has resulted in adaptational changes in external structures as well as in some internal organs and tissues.

First-instar nasal bots (Oestrinae) live for various lengths of time in a specialized oviduct before deposition on the host. Subsequently, the first instars are exposed, albeit briefly, to the rigours of a free-living existence as they search for openings through which to enter into the host. This has driven sensory adaptations as well as adaptations in cuticular structure. In addition, structural adaptations of the Malpighian tubules have been observed in larvae resident in the uterine structure. Structure of the midgut has been altered in some larvae that have developed unique migratory habits within the host. Unique adaptations in respiratory structures are also noted.

First-instar Hypodermatinae hatch spontaneously and, responding to thermal stimuli, move quickly to the host skin. Larvae penetrate directly into the skin with the aid of enzymes. Sensory complement is reduced in comparison with other oestrid larvae, presumably as a consequence of the reduced need for host location.

First-instar Cuterebrinae hatch only in response to rapid changes in environmental conditions that may indicate the presence of a suitable host. Their cuticular sensory complement is well developed for locating the nearby potential host and for finding a suitable site of entry.

First-instar Gasterophilinae hatch either spontaneously or in response to a host stimulus, which might be moisture from licking, or a combination of heat and moisture as the host consumes the plant to which the eggs have been attached (*Gasterophilus pecorum*). Subsequently they penetrate the epithelium and migrate to sites in the oral cavity, although in one species there is a superficial migration followed by mucosal penetration (see Table 5.1). Cuticular sensillae are well developed, presumably to aid migration in the oral cavity, although only a few species have been examined in detail.

Body Form and Cuticular Spines

Larval oestrids are generally fusiform, although the posterior end is usually rounded or blunt. This may vary among instars and with adaptations of the respiratory structures. Segmentation is clearly visible and is usually accompanied by one or more rows of spines on either or both the leading and trailing edge of each segment. Size varies considerably although the first instars are generally small, 1–2 mm in length, while, because of tremendous growth the third instars may reach 45–50 mm in length.

Oestrinae

First instars are small, generally white in colour and are distinctly dorsoventrally flattened (Fig. 8.1a). Well-developed clusters of spines are evident on the trailing edge of each thoracic and abdominal segment (Colwell and Scholl, 1995; Guitton et al., 1996). These spines are more abundant on the ventral and lateral surfaces (Fig. 8.2a). The thoracic and abdominal spines presumably aid attachment to and migration on host mucosal surfaces following larviposition. Gedoelstia, with its unusual migration within the host, lacks the degree of spination (Fig. 8.3a). A distinctive cluster of spines is present on the terminal abdominal segment of Oestrus ovis (Fig. 8.3b), but is absent from several other genera in the subfamily (Zumpt, 1965; D.D. Colwell and I.G. Horak, unpublished data) (Fig. 8.3c).

Second instars are white to creamy white in colour and exhibit a substantial variation in size. They are dorsoventrally flattened, although the dorsal surface is rounded, giving an overall ovoid cross-sectional appearance (see Guitton et al., 1996). Second instars have spine distribution similar to that of the first instars, although there is a reduction in the lateral groups. The cluster of spines on the terminal abdominal segment is variable, present in some genera, but absent in others.

Early third instars are creamy white in colour, but as they mature increasing amounts of melanization occur, generally in bands on each segment (Fig. 8.4). The body form is similar to that of the second instars. The cuticle is generally smooth or slightly reticulated between the rows of spines (Fig. 8.5a–d). Spine size and distribution varies among the genera and perhaps reflects the habitat occupied at this stage; e.g. O. ovis and Gedoelstia sp. found in nasal cavities, have relatively few short spines, with large areas of the thoracic and abdominal segments devoid of spination (Fig. 8.5a and b); in contrast Cephenemyia and Pharyngomyia third instars, found in pharyngeal pouches, have high densities of relatively large spines (Fig. 8.5c and d). In these genera there is presumably greater need for assistance in maintaining position in the host.

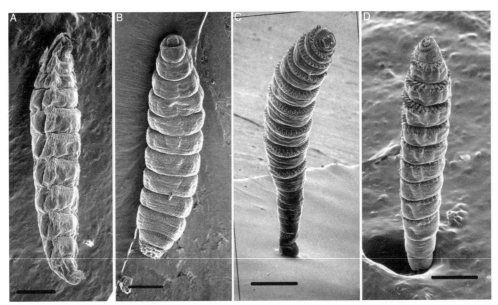

Fig. 8.1. Scanning electron micrographs of whole first instars representative of each of the subfamilies. All specimens were obtained from the uterus or from eggs laid by laboratory-reared and mated flies. Larvae were prepared for electron microscopy by freeze substitution (Colwell and Kokko, 1986). (A) *Oestrus ovis*: note that the larvae are dorsoventrally flattened, segmentally arranged spines are present only on lateral and ventral surfaces, the presence of a terminal cluster of spines (bar = 120 µm). (B) *Hypoderma lineatum*: note the uniform distribution of spines (bar = 120 µm). (C) *Cuterebra fontinella*: note the terminal adhesive sac, the segmental arrangement of the spines, with the last abdominal segments nearly bare (bar = 120 µm). (D) *Gasterophilus intestinalis*: note cylindrical body form and segmentally arranged spines, spine numbers decline on the last few abdominal segments (bar = 120 µm).

Hypodermatinae

First instars are small, 1–2 mm, generally white to creamy white in colour. The body is nearly cylindrical with only slight tapering at anterior and posterior ends (Fig. 8.1b) (Colwell, 1986). Spines are small, simple in form and are sparsely but uniformly distributed (Fig. 8.2b). As the first instars grow they become relatively smaller and more sparse.

Second instars are white in colour with a body form similar to that of the first instars. This stage is moderate in size ranging from 15 to 20 mm in length. *Strobiloestrus* are distinctive, having a series of large, fleshy lobes on all thoracic and abdominal segments (Fig. 8.6).

Third instars are large, reaching 20–30 mm in length. Early third instars are creamy in colour, but gradually become melanized and when mature are completely black. The cuticle is reticulated and usually has a granular appearance (Fig. 8.7a and b). This stage has relatively few large spines that are distributed in bands at the leading and trailing edges of thoracic and abdominal segments (Fig. 8.7a and b). These spines tend to have simple tapered-cylindrical or flat blade

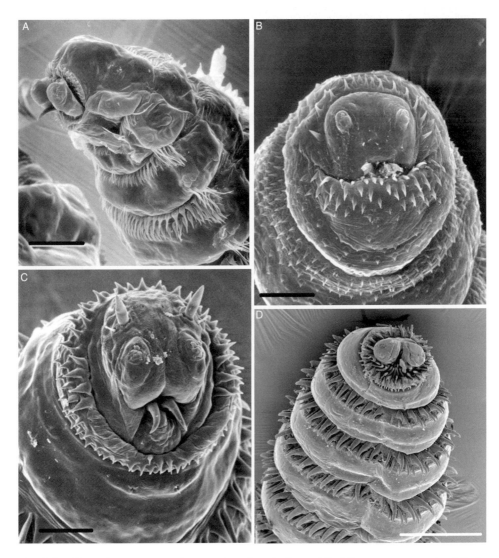

Fig. 8.2. Scanning electron micrographs of cephalic and first thoracic segments of first instars representative of each of the subfamilies. All specimens were obtained from the uterus or from eggs laid by laboratory-reared and mated flies. Larvae were prepared for electron microscopy by freeze substitution (Colwell and Kokko, 1986). (A) *Oestrus ovis*: note the prominent mouth hooks, well-defined antennal lobes and the arrangement of spines on the thoracic segments (bar = 75 µm). (B) *Hypoderma lineatum*: note the poorly developed antennal lobes and the small mouth hooks (bar = 60 µm). (C) *Cuterebra fontinella*: note the well-developed mouth hooks, and the prominent cephalic sensillae (bar = 75 µm). (D) *Gasterophilus intestinalis*: note the mouth hooks are not evident, the well-developed segmentally arranged spines (bar = 100 µm).

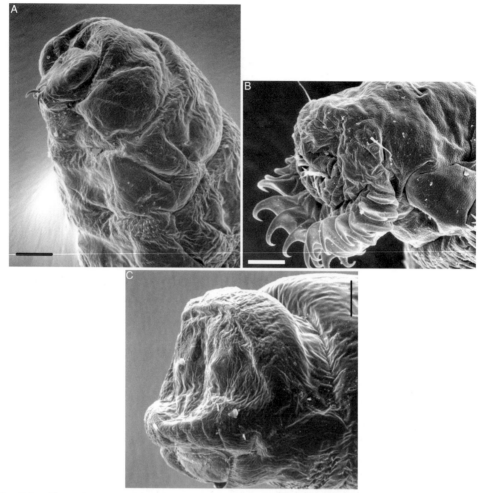

Fig. 8.3. Scanning electron micrographs of the whole body and terminal abdominal segments of first-instar *Gedoelstia* spp. and *Oestrus ovis*. *Gedoelstia* spp. specimens were obtained from the eye of a bontebok collected in South Africa (courtesy of I.G. Horak). (A) *Gedoelstia* spp. whole body. (B) *Oestrus ovis* terminal segments: note the well-developed cluster of ventrally directed spines (bar = 25 μm). (C) *Gedoelstia* spp. terminal segments: note the absence of the terminal spines (bar = 125 μm).

shape, often with a single point, but occasionally with two or more (Otranto *et al.*, 2003). The sharpness of the points may vary among segments. On most segments the spines are oriented caudally and presumably aid the larvae in maintaining their position within the cyst. There are usually a group of cephalad-oriented spines on the cephalic and terminal segments (see Colwell *et al.*, 1998; Otranto *et al.*, 2003).

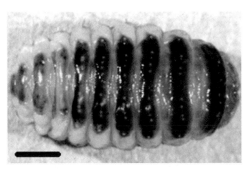

Fig. 8.4. Light micrograph of third-instar *Rhinoestrus purpureus* (bar = 5 mm).

Hypodermatid larvae also have small groups of simple spines located on the cephalic segments, the presence and arrangement of which have been used to distinguish between various species of *Hypoderma* (Colwell *et al.*, 1998; Colwell, 2001; Otranto *et al.*, 2003, 2004). The cephalic spines may aid the feeding of third instars as they appear to be used to abrade host cells from the proximal surfaces of the cyst. In addition, the structure of spines on the thoracic segments has been used to distinguish between three species of *Hypoderma* from bovid hosts and also between the six species from bovid and cervid hosts (Otranto *et al.*, 2003, 2004).

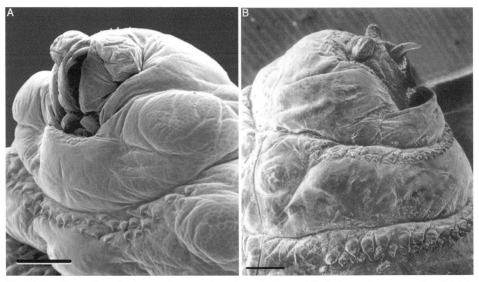

Fig. 8.5. Scanning electron micrographs of cephalic and first thoracic segments of third instars of several genera of the subfamily Oestrinae showing variation in degree of spination: (A) *Oestrus ovis*, (B) *Gedoelstia* spp., (C) *Pharyngomyia auribarbis*, (D) *Cephenemyia trompe*, (E) *Cephalopina titillator* (bar = 0.5 cm).

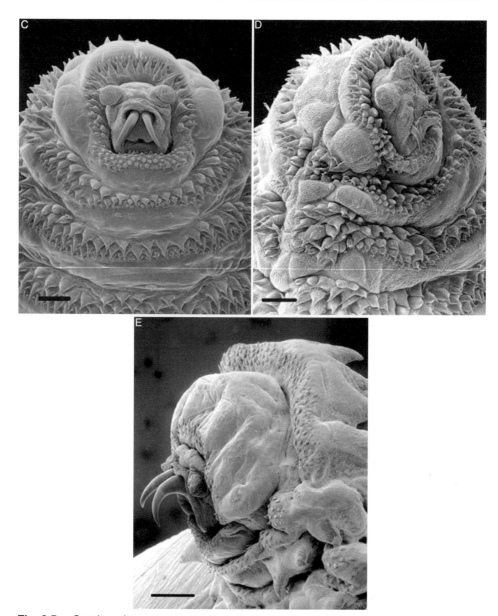

Fig. 8.5. *Continued.*

Cuterebrinae

First instars are small (1–1.5 mm in length), white and fusiform (Fig. 8.1c). The cephalic segment is well defined with prominent sensillae and mouth hooks (Fig. 8.2c); these larvae are unique in possessing a terminal sac that aids the larvae during their migration to the host and subsequently in their search for an appropriate point of entry (Colwell, 1986; de Filippis and Leite, 1997). First instars

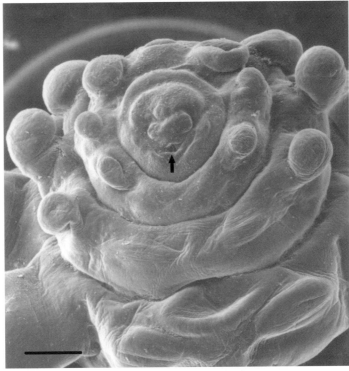

Fig. 8.6. Scanning electron micrograph of second-instar *Strobiloestrus vanzyli*. Note the distinctive fleshy lobes. Arrow indicates mouth (bar = 0.5 mm).

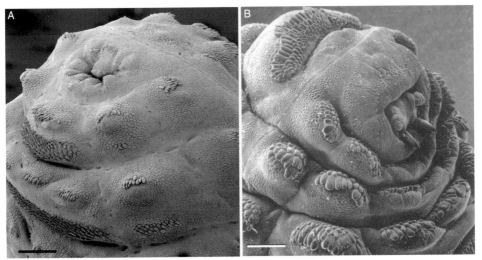

Fig. 8.7. Scanning electron micrographs of cephalic and thoracic segments of third-instar (A) *Hypoderma lineatum* (bar = 0.68 mm) and (B) *Strobiloestrus vanzyli* (bar = 0.88 mm), illustrating differences in spines as well as in the size and prominence of the mouth hooks.

have large, complex spines usually distributed in bands on the cephalad and caudal edge of each thoracic and abdominal segment (Colwell, 1986; de Filippis and Leite, 1997; Colwell and Milton, 2002). Both dorsal and ventral surfaces have similar distributions although some minor differences in spine shape may be evident when comparing the two surfaces. Differences in structure along the length of the body are also noted, with spines on the thoracic segments and the upper abdominal segments being more complex in shape. In contrast first-instar *Dermatobia hominis* have spines with rather simple structure, concentrated primarily on the thoracic and anterior abdominal segments (Leite, 1988). On the thoracic segments the spines are uniform in distribution, but on the later segments they become clustered in three to four rows on the leading edge of each segment. The spines are uniform in size except on the first three abdominal segments where large spines are interspersed among the smaller ones (Leite, 1988).

Second instars are creamy white in colour and are much larger than the previous stage (10–15 mm). The cephalic segment has grown comparatively less than the others and the sensory features have become less prominent (Fig. 8.8). Spine structure has become much less complex (Fig. 8.8) (see Colwell and Milton, 2002) and the distribution has become more uniform, actually suggesting an increase in the total number. The reduction in spine complexity is more evident in the second instars of *D. hominis* (de Filippis and Leite, 1998). In addition, the number of spines is relatively unchanged, giving a sparse appearance in comparison with first instars.

Third instars are large, cylindrical and only slightly tapering at the anterior end. They are also cream-coloured, but become completely black as they reach maturity. These larvae have a dense covering of broad, flat spines that are uniformly distributed

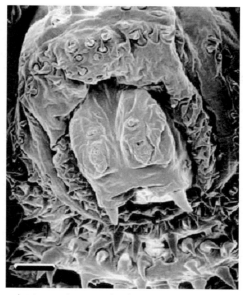

Fig. 8.8. Scanning electron micrograph of cephalic and the first thoracic segments of second-instar *Cuterebra baeri*. Note the relative reduction in size of the cephalic segment compared with the first instar (see Fig. 8.1) (bar = 0.50 mm).

(Fig. 8.9a), with little evidence of concentration at leading or trailing edges of individual segments (Colwell and Milton, 2002; D.D. Colwell and C.R. Baird, unpublished data). The spines clearly function to hold the larvae within the tissue cyst; this may be critical because the rodent and lagomorph hosts can be adept at reaching for and attempting to remove the larvae with their teeth. The spines vary in shape among the segments, with variation in sharpness and the degree of trailing edge irregularity. There appear to be specific characteristics to the spines that will allow specific identification of third instars (D.D. Colwell and C.R. Baird, unpublished data), although comparison of a larger number of species from widely disparate species groups is required to determine the generality of that idea. On third-instar *D. hominis* the number of spines are generally unchanged from the previous stage (Fig. 8.9b; there is usually a single row of large spines on each thoracic and abdominal segment, although these may be absent from the last three abdominal segments (de Filippis and Leite, 1998)). On the latter segments there is a relatively dense, uniform covering of small, cephalad-oriented spines, often with multiple points (de Filippis and Leite, 1998).

Gasterophilinae

First instars are small (1–2 mm in length), white in colour and generally fusiform (Fig. 8.1d). Cephalic segments are well developed with prominent sensillae and mouth hooks (Fig. 8.2d). The spines tend to be simple in structure and, while distributed along the body, are concentrated in bands at the leading edge of thoracic

Fig. 8.9. Scanning electron micrographs of cephalic and thoracic segments of (A) third-instar *Cuterebra baeri* (bar = 0.5 mm) and (B) *Dermatobia hominis* (bar = 0.5 cm). Note the differences in spine number and shape.

and abdominal segments. Some minor differences in shape between dorsal and ventral spines have been noted (Colwell and Scholl, 1995; T. Pape, D.D. Colwell and C. Dewhirst, unpublished data). While most segments have spines the number and density decrease towards the terminal end; the last segment may be completely devoid of spines (Colwell and Scholl, 1995; T. Pape, D.D. Colwell and C. Dewhirst, unpublished data).

Second instars are less fusiform, having the appearance of a long cone. The diameter of the larvae increases slightly with each posterior segment (Leite and Scott, 1999). The terminal abdominal segment is blunt. Spines are present in small numbers, arranged in two to three rows at the leading edge of each segment, except the last three abdominal segments, which are devoid of spines.

Third instars have a body form similar to the earlier stage, but in several species of *Gasterophilus* the body is almost cylindrical with a very small cephalic segment perched atop the thoracic segments. Multiple rows of small, simple spines are arrayed on the cephalic segment around the mouth hooks and the antenomaxillary lobes. Spines on the thoracic segments are arranged in one to two rows at the leading edge of the segment. Each row of spines generally has two lateral gaps. The spines are generally simple in structure with broad bases and pointed tips. There are distinct differences among the five species from horses that have been examined (D.D. Colwell, I.G. Horak and D. Otranto, unpublished data). Spines of the abdominal segments are arranged in single or double rows at the leading edge of each segment, with a tendency for the number of spines to become smaller towards the terminal segments.

Mouth Hooks (Mandibles and Maxillae)

The mouth hooks (mandibles) vary in size, form and function among the four subfamilies and among the instars. They function to abrade mucosal and epidermal surfaces for acquisition of host cellular debris as nutrient. They also function as holdfast structures to help larvae maintain position in the host.

Oestrinae

Mouth hooks are robust and prominent in all instars (Fig. 8.10). They are used during migration in the first instars. In later instars they are used to abrade the host mucosa as an aid to feeding activities. They function as holdfast structures, particularly in genera such as *Pharyngomyia* and *Cephenemyia* where maintenance of position is critical.

Hypodermatinae

Mouth hooks are small and relatively inconspicuous in first instars (Fig. 8.2b). In later instars they become more robust, in keeping with their role in abrading the surface of the 'warble' loosening host cells for ingestion (Fig. 8.7b).

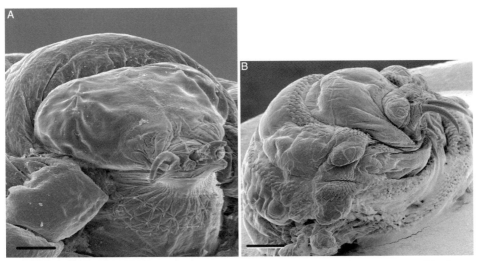

Fig. 8.10. Scanning electron micrographs of cephalic and thoracic segments of instars of *Oestrus variolosus* and *Tracheomycia macropi* showing the changes in mouth hook orientation and structure (A) second-instar *O. variolosus* (bar = 50 μm) (B) third-instar *T. macropi* (bar = 0.6 mm).

Cuterebrinae

Mouth hooks of first instars are robust and conspicuous, aiding the migration of the larvae on to the host and into a body opening. These structures remain prominent in the later instars (Fig. 8.9), fulfilling a function similar to that described for the Hypodermatinae.

Gasterophilinae

First instars have prominent, robust mouth hooks (maxillae). Second instars have robust prominent mouth hooks and medially located mandibles (here a dispute of terminology arises as Wood (1987) follows Teskey (1981) in referring to the mouth hooks as mandibles whereas Leite and Scott (1999) refer to the mouth hooks as maxillae and the associated structures located medially as the mandibles). Third instars of *Gasterophilus* and *Gyrostigma* have robust laterally directed mouth hooks used for attachment to the host mucosa (Fig. 8.11b). The inner surfaces are textured, primarily with a series of lateral striae resembling scales, to aid in maintaining attachment (Fig. 8.11c) (Erzinclioglu, 1990; Otranto *et al.*, 2004). Sensory structures have been described in grooves or pits on the outer surface of *Gasterophilus* spp. (see below). These features are absent from third-instar *Gyrostigma* (D.D. Colwell, T. Pape and C. Dewhirst, unpublished data). The mouth hooks of third-instar *Cobboldia* spp. are less robust than the other genera and they are oriented posteriorly (D.D. Colwell and M.J.R. Hall, unpublished data) (Fig. 8.11a).

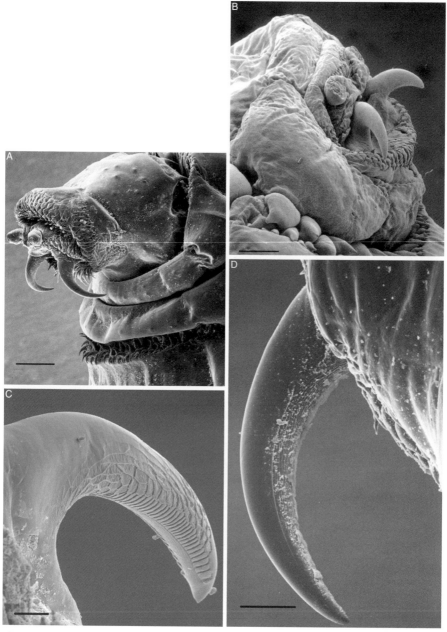

Fig. 8.11. Scanning electron micrographs of cephalic and thoracic segments of third-instar *Cobboldia loxodontis* and *Gasterophilus intestinalis*. (A) *Cobboldia loxodontis*: note the mouth hooks oriented in the same anterior–posterior plane as the body (bar = 0.5 cm). (B) *Gasterophilus intestinalis*: note the mouth hooks are oriented at an angle to the anterior–posterior plane of the body and are used to anchor the larvae to the gut wall (bar = 0.5 cm). (C) *Gasterophilus intestinalis* mouth hook: note the reticulated surface and the ridges on the inner surface that will aid in gripping the host gut wall (bar = 125 μm). (D) *Cobboldia loxodontis* mouth hook: note the smooth outer surface and the longitudinal striations of the inner surface (bar = 125 μm).

The inner surface lacks lateral striae, but has a series of longitudinal striations, which is in keeping with the unattached habit of these larvae (Fig. 8.11d).

Larval oestrids lack the oral ridges associated with the mouth in most free-living (e.g. *Haematobia irritans*, Baker, 1986) and some other parasitic Diptera such as *Lucilia cuprina* (Sandeman *et al.*, 1987). These structures are used to direct liquid or semi-liquid food towards the mouth.

Cuticle

The cuticle provides protection against both abiotic and biotic components of the larval environment. First instars have a thin, unsclerotized and unmelanized cuticle that is relatively incapable of resisting desiccation (Colwell, 1989a; Innocenti *et al.*, 1997). As a consequence, first instars have a relatively short time in which to gain access to the host environment. Although very little work has been done to characterize the cuticle of early first instars and to examine potential differences between oestrids with distinctly different life history strategies there has been limited investigation of two groups of veterinary importance. Cattle grubs (*Hypoderma* spp.) and sheep nose bots (*O. ovis*) have distinctly different life histories and interactions with the host (Chapters 5 and 11). Studies of *Hypoderma* spp. and *O. ovis* have demonstrated differences in cuticular ultrastructure, the most curious of which is the presence of large vacuoles in the procuticle of *O. ovis* (Innocenti *et al.*, 1997; Colwell and Leggett, 1998). These structures are present in larvae removed from the uterus of females (Fig. 8.12) as well as in larvae recovered from hosts. Larvae recovered from hosts are usually three to four times larger than those recovered from the fly and thus the vacuoles do not seem to be present in order to allow expansion of the larval size without addition of cuticle. Other functions of these structures have not been postulated.

Within the host, the cuticle becomes a first line of defence against inflammatory responses as well as against immune responses in sensitized hosts. Innocenti *et al.* (1997) have shown that components of the first-instar cuticle are strong immunogens, recognized by the host, against which antibodies are produced. Unfortunately, similar work with other species is lacking.

One mechanism to reduce host recognition and limit the inflammatory response has been noted for first-instar *Hypoderma* spp. Colwell (1989a) described the presence of a ruthenium red positive layer, probably an acid mucopolysaccharide, on the surface of newly hatched larvae. These positively charged substances on the surface are thought both to reduce activation of inflammatory mediators and to limit immune system recognition. This may be particularly important in modulating early host responses to the larvae as they migrate through the upper, highly reactive layers of the host dermis. This type of evasion would be of particular benefit to those larvae migrating deep within the host tissues and thus exposed to the full host immune response. Hypodermatids not undertaking extensive deep tissue migrations, remaining at or near the skin surface (e.g. *Przhevalskiana silenus*, *Oestromyia* sp.), may not require elaborate means to avoid host responses as

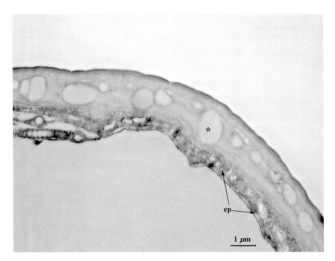

Fig. 8.12. Transmission electron microscope image of a cross-section through the cuticle of a first-instar *Oestrus ovis* recovered from the uterus of a laboratory-reared fly. Note the thin epithelial layer (ep) and the large 'vacuoles' in the endocuticle (asterisk) (bar = 1 μm).

it would be to their advantage to stimulate rapid production of the host granuloma that provides protection and nutrient for the larvae.

Examination of the structure of the cuticle in later instars is extremely limited. Kennaugh (1972) described light microscope observations of two species of *Hypoderma*. He disputed earlier work by Enigk and Pfaff (1954) regarding the organization of the cuticular layers and concluded that there were substantial similarities with larvae of other higher Diptera.

Cuticular Sensillae

Oestrid larvae have an array of cuticular sensillae distributed over the body. These are similar in structure to those known from other muscoid-type larvae (e.g. *Musca domestica*); however, in comparison with their free-living counterparts the parasitic maggots generally have a much-reduced complement of sensillae, both in number and in structural diversity. There are also differences among the subfamilies in number and type of sensillae that are present. This is particularly evident in the cephalic and thoracic segments.

Cephalic

The cephalic sensillae of muscomorph larvae are distinctive and two of the most prominent features, the dorsal organ and the terminal organ (see Chu-Wang and Axtell, 1971, 1972a,b), have been assigned functional roles as olfactory and gustatory sensors, respectively (Cobb, 1999; Oppliger *et al.*, 2000). Ultrastructural studies have shown large numbers of neurons associated with these features, which is in keeping with the functional role.

The dorsal organ is absent from first-instar Hypodermatinae (Colwell, 1986) (Fig. 8.13a) and Oestrinae (Colwell and Scholl, 1995). In addition, ultrastructural studies of first-instar *Hypoderma* indicate that innervation of the terminal organ is simple with approximately 15 pairs of neurons terminating in this structure (Fig. 8.13b). An additional two to four neurons are associated with pit-like sensillae dorsal to the dorsal organ (Fig. 8.13c). These bear a resemblance to known hygro-receptors in other insects (Colwell, 1989a).

The dorsal organ is present on first instars of Gasterophilinae (Colwell and Scholl, 1995) (Fig. 8.14) and Cuterebrinae (Colwell, 1986; Colwell and Milton, 2002). In the Cuterebrinae, including representatives of the genera *Cuterebra* (Fig. 8.2c) and *Dermatobia*, the dorsal organ is large and unique in shape (Baker, 1986; Colwell, 1986; Leite, 1988; Leite and Williams, 1989; de Filippis and Leite, 1997; Colwell and Milton, 2002), while in the Gasterophilinae the structure is much less developed (Colwell and Scholl, 1995; T. Pape, D.D. Colwell and C. Dewhirst, unpublished data). Ultrastructural studies have not been done to compare the degree and complexity of innervation with that seen in the free-living species.

The significance of olfaction in migratory behaviour of the first instars is of considerable interest. Larvae of one species each of *Gasterophilus* and *Gyrostigma* have been examined (Fig. 8.14); *Gasterophilus intestinalis* larvae leave the egg in response to moisture from the host's grooming activity and migrate to the tongue and gum area of the host's mouth. This behaviour clearly requires a large amount of sensory input. Unfortunately, no information is available on the hatching stimulus required by *Gyrostigma rhinocerontis* or on its post-hatch migratory behaviour. It appears that the presence of the cephalic olfactory sensillae is associated with larvae that must undergo a more or less extensive migration to find a suitable site for development within the host. This requirement is found also in the cuterebrid larvae that have been studied. Examination of larvae from species that have different hatching stimuli and migratory requirements might give additional clues as to the importance of the olfactory sensillae in these behaviours.

Olfactory sensillae are absent from first-instar Hypodermatinae (e.g. *Hypoderma* spp.) and Oestrinae (e.g. *O. ovis, Gedoelstia* sp. (D.D. Colwell, M.J.R. Hall and I.G. Horak, unpublished data) and *Rhinoestrus usbekistanicus* (Guitton *et al.*, 1996)), which suggests a reduction in sensory requirements is associated with the less demanding migration of these larvae. While reduced sensory requirement is quite evident for larvae of the cattle grubs, which seem to require only a thermal gradient to direct their movement to the skin surface that they then penetrate, it is less clear for the Oestrinae. Larvae deposited on the nares presumably must locate the nasal passage and migrate inward towards the sinuses, a route that appears to be much more demanding than that of the cattle grubs.

Sensory requirements for the first instars as they migrate within host tissues or organs are not well defined or understood. As with many tissue-dwelling parasites, the oestrid larvae have well-defined site specificities, but the nature of the stimuli they use to locate and accept the sites remain undefined.

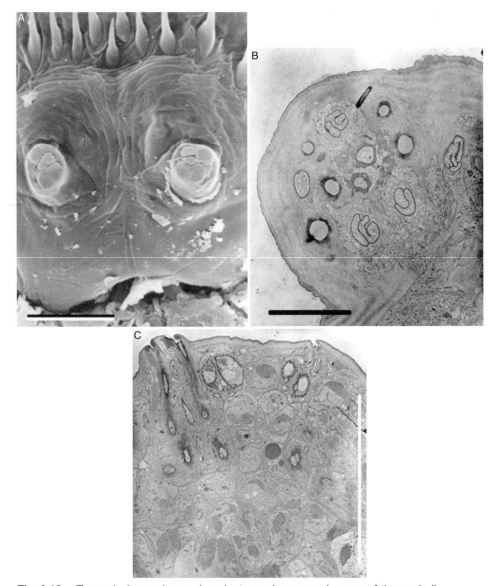

Fig. 8.13. Transmission and scanning electron microscope images of the cephalic sensillae of first-instar *Hypoderma lineatum*: (A) scanning electron micrographs showing the ventral cluster (olfactory sensilla) and the medially located pit sensillae (bar = 5 µm), (B) transmission electron microscope image (TEM) of an oblique section through the olfactory lobe showing the presence of 12 neurons (bar = 2.5 µm), (C) TEM of longitudinal section through the cephalic segment showing the olfactory lobe and the neurons associated with the pit sensilla (bar = 2.0 µm).

Later instars of the subcutaneous cyst-dwelling oestrid larvae (Hypodermatinae and Cuterebrinae) are not migratory and presumably have little requirement for olfactory sensillae. Similarly, gustatory requirements would seem to be limited as the host delivers a constant, invariable, supply of nutrient directly to the larva vis-à-vis

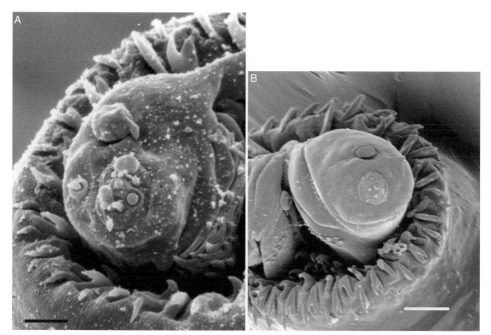

Fig. 8.14. Scanning electron micrographs of the cephalic segments of first-instar (A) *Gasterophilus intestinalis* (bar = 50 µm) and (B) *Gyrostigma rhinocerontis* (bar = 15 µm). Note the presence in both specimens of a large domed sensillum (olfactory sensillum) and the complex sensillum below (gustatory sensillum).

the serum and cells in the granulomatous cyst. Scanning electron microscope (SEM) studies of *Hypoderma* spp. (Colwell *et al.*, 1998; Otranto *et al.*, 2003) and *Cuterebra* (Colwell and Milton, 2002) and *Strobiloestrus vanzyli* (Fig. 8.15a) third instars show an absence of the dorsal organ (olfactory sensory structures), which supports this contention.

Second- and third-instar Gasterophilinae are generally more mobile than those of the previously discussed subfamilies. They generally undergo migrations from the oral cavities to the stomach or intestine and despite being firmly attached to the mucosal surface can be somewhat mobile within those sites. Hence the well-developed cephalic sensory complement is maintained (Leite and Scott, 1999; Otranto *et al.*, 2004; D.D. Colwell, T. Pape and C. Dewhirst, unpublished data). *Cobboldia* spp. larvae, living unattached in the stomach lumen, also have well-developed cephalic sensilla (Fig. 8.15b).

SEM examination of the mouth hooks of *G. intestinalis* and *Gasterophilus nasalis* has shown putative sensillae on the anterior surface (i.e. the surface, deeply embedded in the host mucosa) (Cogley, 1999). The purported sensillae are located in long narrow or oval-shaped pits on each hook. Leite and Scott (1999) and Leite *et al.* (1999) have shown the same structures, but did not ascribe a sensory function. The presence of sensillae on the mouth hooks may be logical and the well-established connection between external form and function in insect sensilla (Zacharuk and

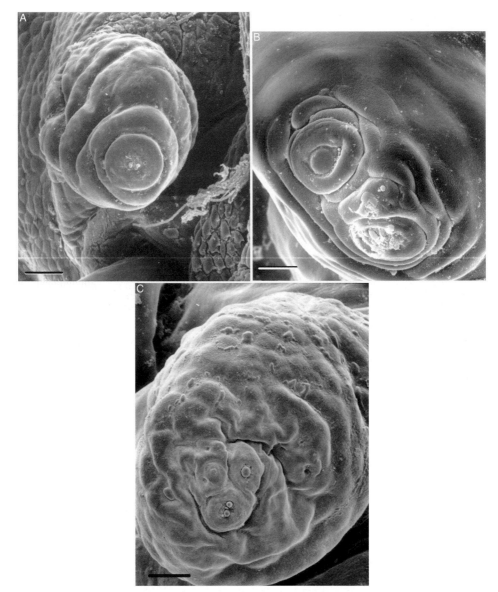

Fig. 8.15. Scanning electron micrographs of the antennal lobes of third-instar
(A) *Strobiloestrus vanzyli*: only the gustatory group of sensillae are present (this genus is
unique in having distinct antennal lobes) (bar = 75 µm), (B) *Cobboldia loxodontis*: both
olfactory (dorsal organ) and gustatory sensory clusters are evident (bar = 19 µm), and
(C) *Pharyngomyia picta*: both olfactory and gustatory clusters of sensillae are evident
(bar = 25 µm).

Shields, 1991) makes the case strong. However, the structural details shown by
Cogley (1999) are not convincing and unless transmission electron microscope
(TEM) studies can confirm the presence of neurons the function of these features
is in doubt. In addition, these features are absent from the mouth hooks of third

instars of *G. rhinocerontis* (Fig. 8.16a) as well as from *Gasterophilus inermis*, *G. pecorum* (Otranto *et al.*, 2004), *Gasterophilus meridionalis* (D.D. Colwell and I.G. Horak, unpublished data) and *Cobboldia* spp. (Fig. 8.16b), giving rise to doubt regarding the function ascribed to the structures by Cogley (1999).

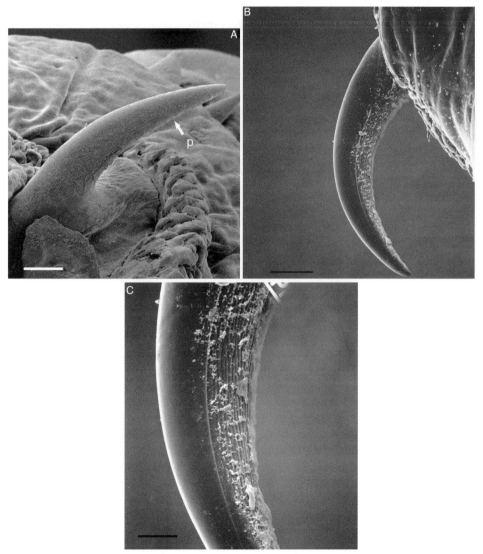

Fig. 8.16. Scanning electron micrographs of the maxillae (mouth hooks) of third-instar *Gyrostigma rhinocerontis* and *Cobboldia loxodontis*. (A) Maxilla of *Gyrostigma rhinocerontis*. Note the anterior and lateral surfaces have no ridges or striations whereas the posterior surface (p) has scale-like striations (bar = 0.15 mm), (B) details of *Gyrostigma rhinocerontis* maxilla lateral and posterior surface showing scale-like striations (bar = 30 μm), and (C) details of the maxillae of *Cobboldia loxodontis*. Note the longitudinal grooves on the inner surface (bar = 40 μm).

Second- and third-instar Oestrinae living in the nasal cavities do not undergo significant migrations and persist in having a poorly developed cephalic sensory array. They lack the dorsal organ, as in the first instar (Fig. 8.15c).

Thoracic and Abdominal

Sensillae on the thoracic and abdominal segments of oestrid larvae are structurally similar to those described from their free-living muscomorph counterparts (see Colwell, 1986; Colwell and Scholl, 1995). The morphotypes include trichoid, campaniform and coeloconic sensilla arrayed on thoracic and abdominal segments, often arranged in groups. The parasitic lifestyle is, however, associated with a distinct reduction in the total number of sensillae present (Colwell, 1986) relative to free-living species. This trend is evident throughout the family, regardless of the differences in life histories.

Variation in the numbers and distribution of sensilla is seen among the subfamilies. A particular example is the presence of tri-setae trichoid sensillae on the ventral surface of the thoracic segments of all species that have been studied except those in the subfamily Hypodermatinae. The absence in the latter subfamily perhaps reflects the limited requirement for migration in the first instars as well as in the later stages.

Sensillae numbers also vary within family groups depending on habitat differences. This is shown in the Gasterophilinae where larvae of *Cobboldia* have a greater diversity and number of sensillae than either *Gasterophilus* or *Gyrostigma*.

Midgut and the Peritrophic Matrix

Ultrastructural study of the first-instar midgut has been limited to species of veterinary importance. Colwell (1989a) described a simple midgut structure for first instars of *Hypoderma bovis* and *Hypoderma lineatum*, with no evidence of regionalized differences in cell morphology. The distinct absence of a peritrophic matrix, and indeed the absence of cardia associated with secretion of the matrix, in that study confirmed the earlier observations by Boulard (1969). This feature appears to be unique among the larval oestrids and is associated with a closed midgut (see below). In contrast, first-instar *O. ovis*, recovered from the 'uterus' of gravid females, have a thin peritrophic matrix (Colwell and Leggett, 1998) and newly hatched first instars of *D. hominis* have a distinct cardia region at the anterior of the midgut (L.C. Evangelista and A.C.R. Leite, unpublished data). Although a peritrophic matrix is absent in these larvae, presumably secretion starts shortly after the larvae begin penetration. This suggests three quite distinct approaches to the parasitic lifestyle by first instars in these three subfamilies. More detailed information is required to determine whether these trends appear generally throughout the subfamilies. The situation in *Gasterophilus* spp. also needs to be elucidated.

Like other dipteran larvae, first-instar *H. bovis* and *H. lineatum* produce digestive enzymes in midgut cells as membrane-bound entities that are later secreted into the gut lumen (Terra *et al.*, 1996; D.D. Colwell and F.L. Leggett, unpublished data). Production of these enzymes ceases when the larvae moult to the second instar (Moiré *et al.*, 1994). The enzymes are regurgitated into the host tissue where they break down the connective tissue producing a 'soup' subsequently ingested by the larvae. This nutrient 'soup' includes not only digested host tissue but also immunoglobulins that are ingested and absorbed into the haemolymph as recognizable fragments (Colwell and Leggett, 2003). Absorption of large proteins has been suggested as a mechanism by which the larvae reduce energetic costs of protein synthesis (Ono, 1933).

The third-instar midgut has no regional specialization as has been noted for other fly larvae (see Terra *et al.*, 1988, 1996), although some suggestion of the presence of gastric caecae has been made for *H. bovis* (Boulard, 1969) and *Gasterophilus* sp. (Keilin, 1944). Evangelista and Leite (2003) showed no variation in cell types along the midgut of third-instar *D. hominis*, although they did indicate the presence of regenerative cells. Preliminary studies of *H. lineatum* suggest that there are two distinct cell types present: highly active exocrine secretory cells and absorptive cells (Fig. 8.17). It has been suggested that the lack of regional and cellular differentiation is a result of the high nutrient requirements associated with rapid growth of these larvae (Evangelista and Leite, 2003). These authors also suggest the continuous acquisition of nutrients during the last instar requires continuous digestion and absorption, which requires an undifferentiated gut.

A peritrophic membrane is present in the later instars of *Hypoderma* spp. (Boulard, 1975b) and is present in all instars of other species that have been examined, including *O. ovis*, *D. hominis* (Evangelista and Leite, 2003) and the parasite of howler monkeys *Cuterebra* (= *Alouattamyia*) *baeri* (Colwell and Milton, 1998). In all species studied the peritrophic matrix is of type II, being secreted continuously from the specialized cells of the cardiaca at the proximal end of the midgut.

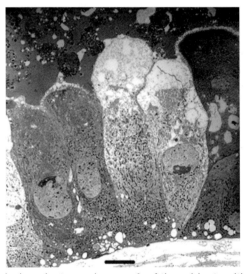

Fig. 8.17. Transmission electron micrograph of the midgut epithelium of third-instar *Hypoderma lineatum*.

Malpighian Tubules

Oestrid larvae have four Malpighian tubules that join the gut at the midgut–hindgut junction. There is little regional differentiation along the length of the tubules although only three species have been studied: *H. bovis*, *H. lineatum* (Colwell, 1989a) and *O. ovis* (Colwell and Legget, 1998).

Ultrastructure of the Malpighian tubules in first-instar *Hypoderma* spp. is characteristically Dipteran (Colwell, 1989a). The Malpighian tubules of first-instar *O. ovis*, removed from the uterus of the female, are filled with storage granules (Colwell and Legget, 1998). This is presumably an adaptation to the larviparous life history of this subfamily.

Respiratory Systems

External respiratory structures are well described in all oestrid subfamilies. The spiracular arrangement is amphineustic or metapneustic although the presence of anterior (prothoracic) spiracles is variable. Features of the posterior spiracles and their position relative to the last abdominal segment are diagnostic for distinguishing amongst the families (see James, 1947; Zumpt, 1965) and are used within several genera to distinguish amongst the species (e.g. *Hypoderma*) (Otranto *et al.*, 2004).

Prothoracic Spiracles

Prothoracic spiracles have been described in the Gasterophilinae and Cuterebrinae. In these subfamilies the spiracles are clearly extruded at the end of the third instar, during formation of the puparium. The presence of prothoracic spiracles in the Oestrinae appears inconsistent, but they are evident in SEM images of third-instar *Cephenemyia auribarbis* and *Pharyngomyia picta* (Fig. 8.18). There is no evidence of these spiracles in the Hypodermatinae.

Posterior Spiracles

Posterior spiracles are present either as perforated plates (Hypdermatinae, Oestrinae) or as sinuous sclerotized ridges (Gasterophilinae, Cuterebrinae). The size and configuration of the posterior spiracles varies among instars (see Fig. 8.19).

In first instars there may be only two small openings, each with a cuticular rima that may or may not bear spines (Colwell, 1986, 1989b). The posterior spiracles of first-instar cuterebrids are subterminal, being located dorsal and anterior to the adhesive sac (Colwell, 1986). The posterior spiracles of *Gasterophilus* and *Gyrostigma* are postterminal, being on a pair of extensible protrusions (Colwell and Scholl, 1995; D.D. Colwell, T. Pape and C. Dewhirst, unpublished data). Each protrusion has two spiracular openings.

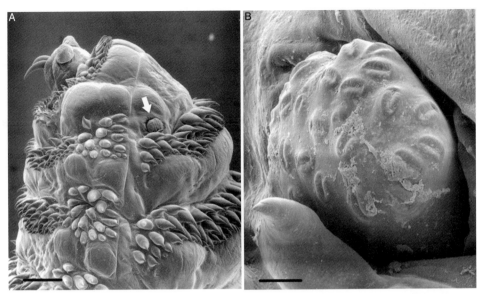

Fig. 8.18. Scanning electron micrographs of the lateral surface of the cephalic and thoracic segments of a third-instar *Pharyngomyia picta* showing (A) an anterior spiracle (arrow) (bar = 0.70 mm) and (B) details of the anterior spiracle showing the numerous individual spiracular openings (bar = 50 μm).

Second-instar posterior spiracles have multiple openings. In cuterebrids there are two pairs of slit-like openings (de Filippis and Leite, 1998), each leading to a felt chamber that contains a sponge-like mass of chitinous strands. Each opening is surrounded by a heavy chitinous rima. The posterior spiracles of the gasterophilids are very similar (Leite and Scott, 1999). The paired second-instar posterior spiracles of Oestrinae are plate-like with multiple openings in addition to the medially located ecdysal scar (Fig. 8.19). Each opening is surrounded by a cuticular rima and leads to a small chamber. All the small chambers of a single spiracular plate connect with the larger felt chamber, which, in turn, is connected with the major tracheal trunk (Colwell, 1989b). Hypodermatinae second instars have paired clusters of 10–20 spiracular openings arrayed around the two ecdysal scars (Fig. 8.20) (see also Colwell, 1989b). Each opening connects with a small antechamber, the group of which connects with the larger felt chamber.

Third-instar posterior spiracles of Cuterebrinae are composed of two pairs of three slit-like openings that are structurally similar to the previous instar. Gasterophilid third-instar posterior spiracles are composed of paired plates, each with three sinuous openings. Each opening is surrounded by a rigid cuticular rima. Third instars of both Oestrinae and Hypodermatinae have paired spiracular plates, each with large numbers of openings (Colwell, 1989b). The rima surrounding each opening may be adorned with small spines (Colwell, 1989b, Otranto *et al.*, 2003).

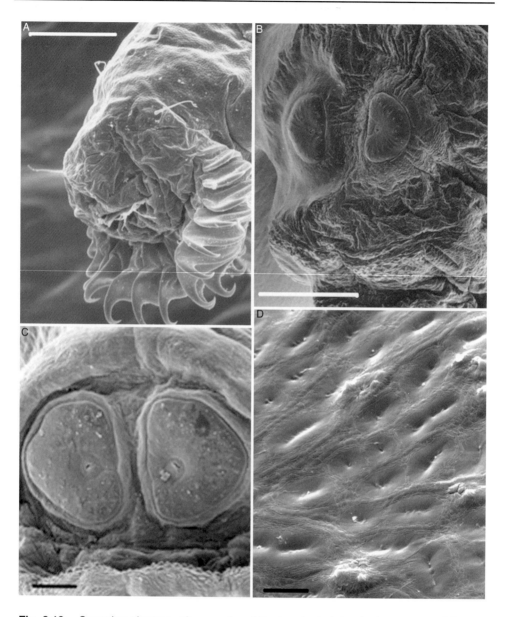

Fig. 8.19. Scanning electron micrographs of the terminal abdominal segments of first-, second- and third-instar *Oestrus ovis* showing changes in the structure of the posterior spiracles: (A) first instar recovered *in utero*: note the spiracular openings are not evident (bar = 0.5 mm), (B) second instar recovered from an artificially infested sheep: note the spiracular plates are well developed with the ecdysal scar near the inner edge of each plate (bar = 0.5 mm), (C) third instar recovered from an artificially infested sheep: note the increase in size of the spiracular plates and the ecdysal scar that is more medially located (bar = 0.5 mm), (D) third instar showing detail of the spiracular openings in a spiracular plate (bar = 0.5 mm).

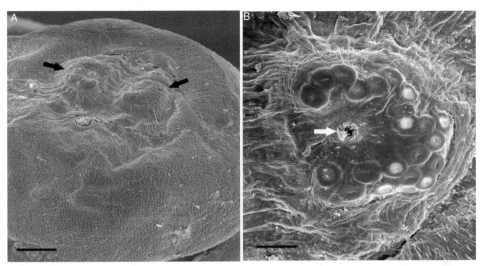

Fig. 8.20. Scanning electron micrographs of the posterior spiracles of second-instar *Strobiloestrus vanzyli* (A) terminal abdominal segment showing the two clusters of spiracular openings (arrows) (bar = 0.19 mm), (B) individual cluster of spiracular openings surrounding an ecdysal scar (arrow) (bar = 50 µm).

Posterior spiracles of the second- and third-instar Gasterophilinae are concealed within the spiracular atrium that allows exclusion from the external environment (see Chapter 14). In the other subfamilies the posterior spiracles are completely exposed. Ecdysal scars are evident on the posterior spiracular plates of all members of the Oestrinae and Hypodermatinae, but are absent from the remaining two subfamilies.

Internally, the respiratory system conforms to the general structural arrangement common to most Diptera (Teskey, 1981). Details of the tracheal system in third-instar *G. intestinalis* have been illustrated by Tatchell (1960).

Perhaps the most unique adaptation to the respiratory system of the oestrid larvae is the presence of haemoglobin associated with the oenocytes in the haemolymph of the gut-dwelling Gasterophilinae. Other adaptations are structural related to the spiracles, both prothoracic and posterior. Details of the variations in external structure have been described by Principato (1987) and Principato and Tosti (1988). The slit-like openings of the post-abdominal spiracles are flexible and provide primary protection against contamination of the tracheal system. Similar to other members of the family, and other Diptera, there is a dense network of cuticular material beneath the surface of the spiracular openings that functions as a filter to further prevent flooding or contamination of the tracheal system (Principato and Tosti, 1988).

9

Pupal Biology and Metamorphosis Behaviour

A.C. Nilssen

University of Tromsø, Norway

According to Fraenkel and Bhaskaran (1973) and the review by Denlinger and Zdárek (1994), metamorphosis in higher flies, the cyclorrhaphous Diptera, which includes Oestridae, is complicated by the addition of an extra metamorphic event, *pupariation*. A rigid structure, the *puparium*, is formed from the outer cuticle of the third instar and provides an encasement for the developing individual. Functionally, this puparium is similar to the cocoon found in other insect taxa in that both protect the pupa, but the cocoon has a completely different origin. The fly pupates within the tight confines of the puparium, and the developing adult is then encapsulated and concealed within the cuticles of both the pupa and the third-instar larva. At eclosion, the emerging adult must break through both layers.

The Postfeeding Larva and Pupariation

It is important to distinguish between pupariation and true pupation because they are two different processes and separated in time (Fraenkel and Bhaskaran, 1973; Cepeda-Palacios and Scholl, 2000b). The first change signalling the start of pupariation is that the larva gradually becomes immobile, but there are also structural and morphological changes. The larva shortens and the outer part of the cuticle becomes hard and dark (many oestrids have blackish puparia). It becomes more or less barrel-shaped. Even if the puparia in oestrids are smoother and more rounded than the mobile larvae, Grunin (1965) stated that the puparia keep so many of the morphological characteristics of the larvae that it is easy to decide to what species a puparium belongs.

Denlinger and Zdárek (1994) have given a timetable of major events from the beginning of the third instar to adult eclosion in higher flies, and a slightly modified timetable is shown in Fig. 9.1. Detailed studies on puparial/pupal development have been performed on the following oestrids: *Cephenemyia phobifera* (Bennett, 1962), *Cuterebra tenebrosa* (Baird, 1972b, 1975), *Dermatobia hominis* (Lello *et al.*, 1985),

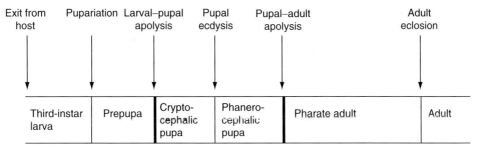

Fig. 9.1. Timetable of major events from exit from host to adult eclosion in oestrids (modified after Denlinger and Zdárek, 1994). Arrows indicate landmarks for the start and end of each developmental stage. In the period from pupariation to adult eclosion the developing individual is enclosed in the puparium, which is formed from the outer cuticle of the third instar.

Hypoderma lineatum and *Hypoderma bovis* (Scholl and Weintraub, 1988), *Cuterebra fontinella* (Scholl, 1991), *Cuterebra austeni* (Baird, 1997) and *Oestrus ovis* (Cepeda-Palacios and Scholl, 1999, 2000b). These studies show that oestrids follow the general picture of metamorphosis behaviour in higher flies.

After cessation of feeding the third instars (the postfeeding larvae) enter a wandering phase. In oestrids, this starts with departure from the host. Some studies show that the larvae leave the host in the early morning (*H. lineatum*, Bishopp *et al.*, 1926; *H. bovis*, Minár and Breev, 1982; *D. hominis*, Catts, 1982), whereas Solopov (1989b) found that departure rate of *H. tarandi* from reindeer was highest from 0900 to 1500 h, i.e. at the warmest and lightest part of the day.

The seasonal departure rate of dropped larvae has been studied in oestrids from semi-domestic reindeer by Solopov (1989b) (*H. tarandi*) and Nilssen and Haugerud (1994) (*H. tarandi* and *Cephenemyia trompe*); they showed that the dropping period lasted about 2 months. The uneven dropping rate observed by Nilssen and Haugerud (1994) appeared to be due to the occurrence of successive periods of infection caused by separate periods of weather that were favourable for mass attack by the flies, i.e. early and late infection (the previous year) gives early and late dropping, respectively. Gomoyunova (1973) found that *H. bovis* larvae from calves dropped out between the end of April and the end of July; those from cows dropped between early March and the end of August.

Two techniques to collect dropped larvae (the 'collection cape technique' and the 'grating technique') have been described by Bishopp *et al.* (1926), Gregson (1958), Simco and Lancaster (1964), Minár (1974), Pfadt *et al.* (1975), Barrett (1981), Minár and Breev (1982) and Nilssen and Haugerud (1994).

The dropped larvae may crawl several metres (*H. lineatum* and *H. bovis* up to 3 and 3.6 m, respectively (Bishopp *et al.*, 1926)) and the crawling may last for hours or days. The duration of the crawling period depends on how soon they find a suitable substratum to hide in and on temperature. Biggs *et al.* (1998) found that the larvae of *O. ovis* were poor diggers, and very slow and clumsy. The larvae were negatively phototrophic and therefore moved to the shade and burrowed into the substrate. Cepeda-Palacios and Scholl (2000b) described further details of the behaviour of newly emerged larvae of *O. ovis*:

When they were placed in contact with sand, they wandered from < 10 min to several hours, after which they began to burrow into the sand with the aid of their mouth hook. When the larvae arrived at the desired depth (1–10 cm), they turned the anterior portion upward, retracted the cephalopharyngeal apparatus, invaginated the head segment into the second segment and collapsed the posterior spiracular disk against the anal tubercle. Muscular longitudinal contraction caused the larval body to adopt an arched position and shrink to 70% of the original length.

Cepeda-Palacios and Scholl (2006).

Many oestrid researchers have noted the pupariation time (time from when larvae leave the host to hardening of the puparia), and an overview is given in Table 9.1. Apparently, *Cephenemyia* spp. and Cuterebrinae pupariate faster than do *Hypoderma* spp. The process before completion of pupariation, however, depends on temperature and may be prolonged at low temperatures. Nilssen (1997b) found

Table 9.1. Pupariation time (time from when larvae leave the host to hardening of the puparia) in some oestrids. In most cases, the temperature is not given, but the observations were either done in the field or in the laboratory, and a temperature of 20–25°C is most likely.

Species	Time (h)	Temperature (°C)	Reference
Cephenemyia phobifera	24–30	25	Bennett (1962)
Cephenemyia trompe	5–6	–	Hadwen (1926)
Cephenemyia trompe	<24	–	Bergman (1917)
Cephenemyia trompe	4	–	Natvig (1916)
Cephenemyia trompe	5–10	8	Nilssen (1997b)
Cephenemyia trompe	3–5	22	Nilssen (1997b)
Cuterebra austeni	<24	–	Baird (1997)
Cuterebra horripilum	<1–72	–	Haas and Dicke (1958)
Cuterebra latifrons	48	–	Catts (1967)
Cuterebra tenebrosa	<12	–	Baird (1975)
Hypoderma bovis	24–48	15–20	Minár and Breev (1982)
Hypoderma bovis	72	–	Gregson (1958)
Hypoderma lineatum	24–168	–	Chamberlain and Scholl (1991)
Hypoderma lineatum	6–168	–	Pfadt (1947)
Hypoderma tarandi	24–168	–	Breyev and Karazeeva (1957)
Hypoderma tarandi	24	–	Natvig (1916)
Hypoderma tarandi	264–432	8	Nilssen (1997b)
Hypoderma tarandi	72–144	12	Nilssen (1997b)
Hypoderma tarandi	<24	20	Nilssen (1997b)
Oestrus ovis	1–120		Cobbet and Mitchell (1941)
Oestrus ovis	2–46	16–32	Cepeda-Palacios and Scholl (2000a)
Rhinoestrus purpureus	0.5	–	Zayed (1992)

in laboratory experiments that *H. tarandi* at 4°C did not pupariate at all and some larvae were found crawling up to 47 days after exit from the host. In nature, pupariation will be postponed if the larvae of temperate species are dropped in early spring when the temperature is low (some species may be dropped on snow or experience frost), but development will resume when the temperature increases. Physiologically, the larvae that leave the warm body of the host (temperature close to 40°C) are subjected to quite a shock when they land on frozen ground. So far, there seem to have been no investigations on how the oestrid larvae manage such an extreme change in temperature.

If the larvae find a suitable substrate, they burrow superficially or to a depth of a few centimetres, but this is obviously not a prerequisite for pupariation. Larvae are easy to rear in empty vials (Nilssen, 1997b). Cepeda-Palacios and Scholl (2000b) found that if *O. ovis* larvae were disturbed in an early phase of pupariation, they were able to reverse the process and resume wandering or burrowing.

It has been observed that if larvae are taken from the host before they are 'mature', they often fail to pupariate (*H. bovis*, Minár, 1974; Minár and Breev, 1982; *H. tarandi*, Nachlupin and Pawlowsky, 1932; *Cuterebra emasculator*, Bennett, 1972; *H. tarandi* and *C. trompe*, A.C. Nilssen, unpublished data). Obviously, there are some physiological commitments required before the larvae can enter the wandering stage and start metamorphosis. Likely, hormones are involved, but it may also be crucial to have obtained a critical weight.

When pupariation is completed, the insect is developmentally still a third instar, although a larva that now is immobile, contracted and with a heavily sclerotized and tanned cuticle. It is still not a proper pupa, but a *prepupa* inside a puparium (Fraenkel and Bhaskaran, 1973; Denlinger and Zdárek, 1994).

The Pupal and Pharate Adult Stages

The next phase is the larval–pupal *apolysis* (a term used for separation of the cuticle from the underlying epidermis), in which the new pupal cuticle starts to separate from the old larval cuticle. Then, the fly enters the *cryptocephalic* ('hidden head') pupal stage followed by a *phanerocephalic* ('visible head') pupal stage. The transformation from the cryptocephalic to the phanerocephalic pupa is accomplished through a short violent process consisting of head evagination, expansion of thoracic appendages and withdrawal of the larval tracheal lining (Denlinger and Zdárek, 1994). Catts (1967) found that at the phanerocephalic pupal stage, mortality did not result when opercula were removed. It has therefore been possible to do detailed studies on morphological events on live individuals from this stage until eclosion (Catts, 1967; Baird, 1972b, 1975, 1997). Denlinger and Zdárek (1994) reported that if there is a pupal diapause in higher flies, development stops at the phanerocephalic pupal stage.

The next developmental stage, which starts with pupal–adult apolysis, is *pharate adult* development. This stage, which involves development of adult structures, occupies the largest portion of the interval between pupariation and adult eclosion. This development in oestrids has been described in detail by Baird (1972b, 1975, 1997), Scholl and Weintraub (1988) and Cepeda-Palacios and Scholl (2000b).

Adult Emergence (Eclosion)

The final stage is adult *eclosion*, which starts with the *extrication* behaviour, by which the fly opens the puparial valve, also called the operculum. It is found at the anterior end of the puparium and consists of two parts (valves). The lower valve often detaches only partly (*Gasterophilus, Cobboldia* and *Gyrostigma*) or not at all (Cuterebrinae and Hypodermatinae, *Kirkioestrus, Cephalopina, Oestrus* and *Rhinoestrus*), whereas *Pharyngomyia* and *Cephenemyia* have a puparium with both upper and lower valves that detach. In *Hypoderma* and *Przhevalskiana*, the dorsal valve is completely flattened and fits as a lid on to the enlarged ventral valve (Wood, 1987; Pape, 2001a).

Extrication involves two stereotypic movement patterns: forward movement and obstacle removal. The former behaviour consists of peristaltic contractions similar to larval crawling, whereas the latter has strong repetitive simultaneous contractions of abdominal and thoracic muscles (Denlinger and Zdárek, 1994). As a result, the ptilinum inflates repeatedly and results in forceful directed movements at the front of the puparium and the shedding of the operculum. Extrication goes on until the fly is completely released from the puparium.

After extrication the adult expansion begins. According to Denlinger and Zdárek (1994) this is a process during which the body inflates and the imago attains its final form. It begins with reorientation of the legs to walking position and irreversible retraction of the *ptilinum,* an eversible sac at the front of the head. The ptilinum can be expanded in the newly emerged fly by blood forced into the head from the thorax and abdomen. It assists in their escape from the puparium, but is also used by the fly to dig its way to the surface. Once the fly's cuticle has hardened, the ptilinum is no longer eversible and its muscles degenerate, and the only remnant of it in an adult fly is the ptilinal suture.

Once the fly is free, it starts activities that include walking, grooming, air intake and muscular contractions that increase the haemocoelic pressure. This facilitates extension of the wings, abdomen and other body parts. When the whole body has expanded, the swallowed air is released and the haemocoelic pressure declines. The meconium fluid (the waste products of pupal metabolism accumulated in the rectum) is also liberated; in *H. bovis* this occurred within 1–2 h (Gregson, 1958; Minár and Breev, 1982). Cepeda-Palacios and Scholl (2000b) described the eclosion of *O. ovis* in this way: 'Aided by the ptilinal sac, the emerging fly broke the opercular window and exited the puparium in ≈3 min. Extension of the wings took 5–15 min, the hardening of wings and initiation of flight took ≈30 min.' Mating may take place a short time after emergence, and in *Gasterophilus intestinalis* males have been observed to attempt to copulate with females that had not fully emerged from the puparium (Cogley and Cogley, 2000).

Some studies show that eclosion occurs in the morning (e.g. Bishopp *et al.* (1926) for *H. lineatum* and *H. bovis*; Catts (1982) and Baird (1997) for *Cuterebra* spp.), possibly as a response to increasing temperatures. In laboratory rearing at constant temperatures, such a diel eclosion pattern cannot be expected and was not observed in experiments with the reindeer oestrids *H. tarandi* and *C. trompe* (A.C. Nilssen, unpublished data).

Abnormal Pupariation Sites

Some nasopharyngeal oestrids are observed to pupariate within the host (Cameron, 1932; McMahon and Bunch, 1989; Ruíz-Martínes and Palomares, 1993). Ruíz-Martínes and Palomares (1993) found that 15.5% of *Cephenemyia auribarbis* and 1.4% of *Pharyngomyia picta* found in red deer in Spain were pupae; these pupae were located both in the oesophagotracheal and in the nasal-olfactory area. It was hypothesized that puparia formed in the oesophagotracheal area have increased chances of survival. Cameron (1932) suggested that such pupae may exit the host with faeces through anus. Bennett (1962) observed one larvae of *C. phobifera* under the tongue of a dead deer host. Nilssen and Haugerud (1995) found mature, dead third-instar *C. trompe* 'trapped' in the sinuses of reindeer, but never pupae. If, however, pupae are expelled in faeces, specific methods are required to find them. Consequently, it is an open question how common intrahost pupariation is. Species of Gasterophilinae are normally expelled via faeces, but so far nobody seems to have found that they are expelled as pupae.

The timetables for the different stages from emergence from the host to adult eclosion have been published for a few species: *C. tenebrosa* (Baird, 1972b), *C. austeni* (Baird, 1975, 1997), *H. bovis* and *H. lineatum* (Scholl and Weintraub, 1988), and *O. ovis* (Cepeda-Palacios and Scholl, 2000b). There is some variation between the species, but the phenology of the intrapuparial development is essentially similar to each other and to other cyclorrhaphous flies. Figure 9.2 illustrates the phenology in *H. bovis* and *H. lineatum* (from Scholl and Weintraub, 1988).

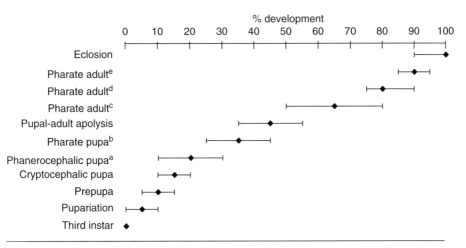

[a] Eyes white-yellow

[b] Eyes yellow-orange

[c] Eyes brown

[d] Frons almost black, and hairs appeared on thorax and abdomen

[e] Head and thorax had a silvery appearance as pupal cuticle separates from adult in preparation for emergence

Fig. 9.2. Percent development (mean and range) of *Hypoderma lineatum* and *H. bovis* from third instar to eclosion (after a table in Scholl and Weintraub, 1988). Note the long duration of the pharate adult stage.

Gonotrophic development, which starts in the late third instar or in the pupal stage, has been studied in detail in *D. hominis* (Lello *et al.*, 1985), *H. lineatum* and *H. bovis* (Scholl and Weintraub, 1988), *C. fontinella* (Scholl, 1991) and *O. ovis* (Cepeda-Palacios and Scholl, 1999).

Pupal Duration

Temperature is the most important factor influencing the developmental rate of insects even if other factors like humidity may be important, but mostly as a mortality factor. Apart from those species that have a diapause (see below), developmental rate in insects is more or less proportional to temperature. Many models have been published to describe this relationship, from the simple linear (day-degree) model to more complicated non-linear models (see Briere *et al.*, 1999). In nature, the temperature fluctuates, making it difficult to find the necessary parameters in a model. Therefore, use of a variety of constant temperatures in the laboratory is the best way to create a reliable model. This has also been done in the following oestrids: *Cuterebra latifrons* (Catts, 1967), *C. emasculator* (Bennett, 1972), *Rhinoestrus purpureus* (Zayed, 1992), *C. trompe* (Nilssen, 1997b), *O. ovis* (Breev *et al.*, 1980), *H. lineatum* and *H. bovis* (Pfadt *et al.*, 1975) and *H. tarandi* (Nilssen, 1997b). Data from these published papers were used to construct Fig. 9.3, which compares pupal duration as a function of temperature. The new rate model of temperature-dependent development published by Briere *et al.* (1999) was used:

$$R(T) = aT(T - T_0) \sqrt{T_L - T}$$

where R is the rate of development (1/days) as a function of temperature, T is temperature in centigrade, T_0 is the low temperature developmental threshold, T_L the lethal upper temperature and a an empirical constant. This non-linear regression equation was fitted by the Nonlin module in SYSTAT (1992). To draw the curves in Fig. 9.3, the rate values obtained were back-transformed from 1/days to days to give the pupal duration as a function of temperature.

The figure is self-explanatory, but a few comments can be given. We see that *C. trompe* is the species with the shortest pupal period. This can be regarded as an adaptation to a subarctic climate because the species parasitizes the northernmost ungulates, reindeer and caribou. When looking at the curves of the three *Hypoderma* species, however, we see that *H. bovis*, the northern ox warble fly, develops more slowly than the more southerly distributed *H. lineatum*, i.e. the opposite of what is expected. *H. tarandi* is a more northern (subarctic) species than both *H. lineatum* and *H. bovis* (Wood, 1987), and we might expect a more rapid pupal development for *H. tarandi*. The comparison shows, however, that the curves are very similar for *H. tarandi* and *H. lineatum*, but both develop faster than *H. bovis*. The similarity in pupal development between *H. tarandi* and *H. lineatum* was also observed by Breyev and Karazeeva (1957).

In Fig. 9.3, the pupal duration of males and females is pooled, but it has invariably been shown that on average males eclose before females, and the difference decreases with increasing rearing temperature (Nilssen, 1997b). The reason

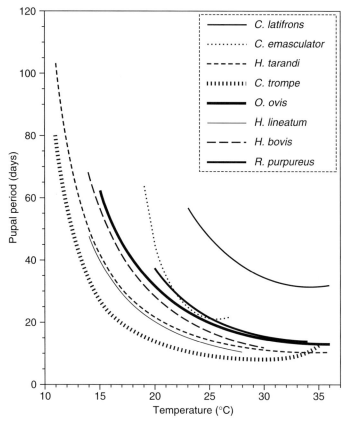

Fig. 9.3. Pupal duration as a function of temperature in some oestrids: *Cuterebra latifrons* (Catts, 1967), *Cuterebra emasculator* (Bennett, 1972), *Hypoderma tarandi* and *Cephenemyia trompe* (Nilssen, 1997b), *Oestrus ovis* (Breev *et al.*, 1980), *H. lineatum* and *H. bovis* (Pfadt *et al.*, 1975) and *R. purpureus* (Zayed, 1992). For the *Cuterebra* species, pupal duration is given for non-diapausing individuals.

for this phenomenon, called *protandry* (Nylin *et al.*, 1993), may be physiological, e.g. that the gonotrophic development is more complicated and needs more time in females (see Breev *et al.*, 1980), but there also may have been a stronger selection for early eclosion in males than in females due to the mating strategy. In most oestrid papers reviewed, the sex ratio of reared individuals was approximately 1:1. Smith (1977), however, found a sex ratio of 2:1 in favour of males for *Cuterebra approximata*.

Pupal duration may have an important effect on the population dynamics of an oestrid species. This was shown by Nilssen (1997b) for the two reindeer oestrids *H. tarandi* and *C. trompe*. Eclosion dates were very variable between years and between localities, and in cold summers and cold localities eclosion occurred so late that the mating and host-searching activities were limiting factors, possibly explaining the differences in larval infestation levels observed among years.

Diapause in the Pupal Stage

Diapause is a delay in development in response to regularly recurring periods of adverse environmental conditions (Chapman, 1998). Diapause facilitates winter survival or surviving regularly occurring dry seasons, and contributes to the synchronization of adult emergence. Photoperiod is the most important stimulus initiating diapause. In oestrids, diapause has so far been most convincingly observed in North American *Cuterebra* species and in *Oestromyia leporina*. Admittedly, Rogers and Knapp (1973) reported that *O. ovis* pupae kept at 16°C did not develop into adults, and, in their own words 'remained in a dormant state' or were 'apparently in diapause', but later studies on *O. ovis* (e.g. Breev *et al.*, 1980; Biggs *et al.*, 1998; Cepeda-Palacios and Scholl, 2000b) do not seem to have corroborated the occurrence of a pupal diapause in this species. On the other hand, a facultative pupal diapause as an overwintering strategy is well documented in several *Cuterebra* species. This feature increases the difficulty of laboratory investigation of the genus.

Catts (1967) found that in autumn and winter months 18% of the pupae of *C. latifrons* entered diapause and Bennett (1972) showed that in *C. emasculator* there was a greater success of metamorphosis at lower temperatures without retardation of the developmental period, suggesting the possibility of a diapause (Fig. 9.4). Presumably 120 days at 6°C was optimal. He hypothesized that diapause contributes

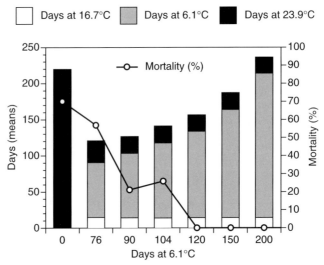

Fig. 9.4. Effect of varying periods at 6.1°C on pupal mortality and postrefrigeration developmental time in *Cuterebra emasculator* (after a table in Bennett, 1972). All refrigerated puparia received a prerefrigeration period of 14 days at 16.7°C. Note that if the puparia were exposed only to 23.9°C (the bar at the left), they needed a mean period of 219 days to eclose, probably because they were in diapause. The experiment shows that a period of cold temperature contributes greatly to both a lowered pupal mortality and a remarkable uniformity in the postrefrigeration developmental time. Apparently, in this species speed of diapause development is enhanced by cold temperatures.

to a concentration of the fly population in time. Baird (1972b) showed that *C. tene-brosa* pupae entered a pupal diapause that could last from 6 to 12 months. Diapause occurred after the larval–pupal moult, which takes place 7–10 days after the puparium is formed, i.e. slightly before the cryptocephalic stage (see Scholl, 1991). Denlinger and Zdárek (1994) reported that in pupal diapause in higher flies, development typically stops at the phanerocephalic pupal stage, and later Baird (1997) found the same stage for diapausing *C. austeni* pupae. Pupal diapause could be terminated by injections of 0.1–0.6 μg of the hormone ecdysterone (Baird, 1972b). Baird (1975) showed that photoperiod in the larval period was the initiator of diapause in the pupal stage. He demonstrated sophisticatedly that *C. tene-brosa* larvae from woodrat hosts kept on a long day length (>15 h light) gave rise to a high percentage of non-diapausing pupae whereas larvae reared on short day lengths produced diapausing pupae (Fig. 9.5). Non-diapausing pupae emerged in 52 days, whereas diapausing pupae required up to 2 years to emerge. Attempts to terminate pupal diapause by temperature (cooling) or photoperiod manipulations during the *pupal* stage were unsuccessful in terminating the diapause.

Baird (1997) showed for *C. austeni* why this kind of pupal diapause has evolved and why it is linked to photoperiod and the fact that there are two generations per year. Early developing larvae (which experience long day length) tended to develop without diapause, whereas late developing larvae (with shorter day length) had a higher prevalence of diapausing pupae. Non-diapausing pupae eclosed late that summer or autumn as a second-generation adult, whereas the diapausing pupae overwinter and become the spring generation of adults. If individuals of the latter generation do not enter diapause, they might eclose at a time with temperatures unfavourable for mating and host searching. By postponing the eclosion till next spring, the individuals are secured warm weather and synchronized occurrence. However, pupal diapause also has been found in a species (*C. approximata*)

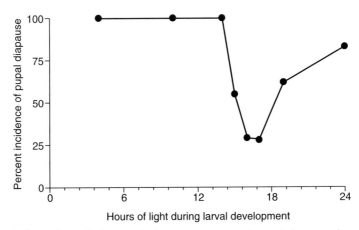

Fig. 9.5. Effect of larval photoperiod on occurrence of pupal diapause in *Cuterebra tenebrosa* (after Baird, 1975). A light:dark ratio of 17:7 is optimal in producing non-diapausing pupae, whereas day lengths of ≤14 h resulted in 100% diapause. Attempts to terminate pupal diapause via temperature or photoperiod manipulations in the *pupal* stage were unsuccessful.

that has only one generation a year (Smith, 1977). He found that pupae from autumn infestations entered a prolonged diapause, whereas those from spring infestations developed more rapidly. Adults from both groups emerged in late summer.

Apart from these well-documented examples of diapause in *Cuterebra* species, only one more species has been shown to have a pupal diapause. Rietschel (1975b, 1980) found that pupae of *Oestromyia leporina* entered diapause. He showed that freezing could interrupt the diapause but, as for some *Cuterebra* species, photoperiod in the larval stage was important, because diapause could be omitted by keeping the host under long day conditions during the last 2–5 days of the third larval instar.

Future studies may reveal that species in other genera also have pupal diapause. Likely candidates are temperate species that have two or more generations per year.

Mortality Factors at the Pupal Stage

When trying to rear oestrid larvae to the adult stage in the laboratory or in the field, survival seldom exceeds 90%, even at optimum temperatures and humidities. The most typical percentage eclosion in experimental rearing is 50–70%. Figure 9.6 shows the results from rearing of the reindeer oestrids *H. tarandi* and *C. trompe* (Nilssen, 1997b). Similar results were obtained by Pfadt *et al.* (1975) for *H. lineatum* and *H. bovis*. The cause of death is unknown, but mortality occurred both early (failure to pupariate or no development after pupariation) and late (in the pupal or adult pharate stages) in the development.

Survival of the pupal stage is of course important for oestrid population dynamics. If mortality approaches 50% for unknown reasons also in nature even at good conditions, we may expect even higher mortality ratios during certain circumstances as specific mortality factors also have been documented (see below).

Predators and Parasitoids

Predation by birds may take place even before the larvae have left the host. North American magpies (*Pica pica hudsonia*) have been observed to pick holes in the backs of cattle while trying to get *Hypoderma* larvae out (Bishopp *et al.*, 1926), and in Sweden, Espmark (1972) observed that magpies (*Pica pica*) pulled out *H. tarandi* larvae from the backs of reindeer. When the larvae have left the host, they are even more vulnerable and many bird species may find a tempting prey. Predators can even find larvae covered by soil (Bishopp *et al.*, 1926), possibly also after pupariation. Colwell (1992) reported magpies and starlings as predators on larvae of *H. bovis* and *H. lineatum*. Sdobnikov (1935) and Grunin (1965) reported that birds (crows (*Corvus corax* and *Corvus corone*), magpies (*P. pica*), long-tailed skuas (*Stercorarius longicaudus*), gulls (*Larus argentatus*) and wagtails (*Motacilla alba*)) searched for larvae (and pupae?) of *H. tarandi* and *C. trompe*. During the author's own efforts to collect larvae of *H. tarandi* and *C. trompe* that were dropping from penned but free-ranging

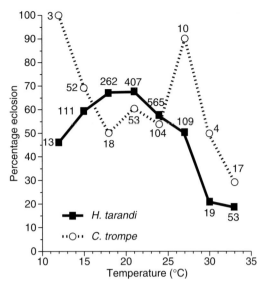

Fig. 9.6. Percentage eclosion as a function of temperature in *Hypoderma tarandi* and *Cephenemyia trompe* (after a table in Nilssen, 1997b). Sample size (larvae set for rearing) is given as numbers near the symbols. For *H. tarandi*, the optimum temperatures seem to be in the range 18–20°C. For *C. trompe*, the pattern is more confusing, possibly because of the smaller sample size.

reindeer, the experiment had to be cancelled because crows (*C. corone*) 'stole' most of our larvae. Rich (1958) found that predatory birds and rodents devoured up to 95% of pupae and these predators were a constant factor throughout his study. Colwell (1992) also stated that rodents are known to feed on mature larvae and pupae of *H. bovis* and *H. lineatum* and suggested them to be a significant mortality factor. In general, however, it is difficult to assess from the above how significant predation is, but it is likely the effect varies with species, season and location. Minár and Samsináková (1977) argued that warble flies have very few natural enemies and parasites, and that pupae are only occasionally eaten by predatory insects and insectivorous birds.

Emerged larvae may also be preyed upon by their own host. Baird (1997) reported that the white-throated woodrats (*Neotoma albigula*) apparently were so irritated by the emerging larvae of *C. austeni* that they killed (and ate?) the larvae, resulting in nearly 30% loss. A reindeer also has been seen eating *H. tarandi* larvae that were newly dropped from its body (A.C. Nilssen, unpublished data). Accidentally, many larvae may also be killed by being stepped upon by their host, in particular if the host animals are in a crowded situation during the dropping period.

According to the review by Scholl (1990) very few reports have been published about parasites (or parasitoids) of cattle grubs. Bishopp *et al.* (1926) reported that *Nasonia brevicornis* (Pteromalidae, Hymenoptera) was recovered from *Hypoderma* spp. pupae in North America, and according to Obitz (1938) (cited in Scholl, 1990) an unidentified ichneumonid was found parasitizing *Hypoderma* spp. pupae in Poland,

and in the former Soviet Union, *Trichopria* sp. (Diapriidae, Hymenoptera) was collected from two *H. bovis* pupae. Interestingly, a *Trichopria* species was also collected from pupae of *D. hominis* in Brazil (Sanavria, 1987, cited in Scholl, 1990). Gansser (1951) (cited in Grunin, 1965) found that pupae of cattle grubs were parasitized by *Alysia* sp. (Braconidae, Hymenoptera).

Thus, only four or five species of hymenopteran parasitoids have so far been found in oestrid larvae/pupae. The apparent absence in other oestrid species may reflect the lack of investigation, but it is likely that parasitoids attack larvae or pupae of other oestrids, especially in areas with high biodiversity and by parasitoids with low host specificity.

Newly eclosed oestrids are especially vulnerable during the period before the wings become functional and thus may easily be targeted by predatory insects. Rocha and Mendes (1996) observed that a newly emerged fly of *D. hominis* was predated by ants. Gregson (1958) also reported that insect predators may eat larvae/pupae and that ants may eat newly emerged *Hypoderma* spp. Colwell (1992) argued that ants feeding on adult flies may cause significant mortality of newly emerged adults of *H. bovis* and *H. lineatum* or adults resting during adverse weather conditions. Solopov (1989b) reported that red ants caused considerable damage to newly emerged adults of *H. tarandi*, and in one experiment ants destroyed 7.3% of the total number.

Temperature and Humidity

For insects in general, the upper lethal temperatures are within the range 40–50°C for short-term exposures (Chapman, 1998), and oestrid pupae exposed to strong sunlight may reach such high temperatures and die. Solopov (1989b) found that if larvae of *H. tarandi* fell on to bare rock at high temperatures and direct sunshine, most larvae (81.6%) died within 3 h of overheating, and if they pupariated, only 3.7–5.0% eclosed. Breyev and Karazeeva (1957) found the upper temperature limit for life functions of *H. tarandi* pupae to be higher than 37°C, because many pupae continued development after the temperature in the substratum had increased to 36.9°C, and he assumed that pupae would probably start to die of heat at 40°C, or possibly even a little higher.

However, temperatures somewhat below the upper limit for extended periods may also cause high mortality. As Fig. 9.6 shows, mortality increased markedly when pupae of *H. tarandi* were kept constantly at temperatures above 30°C. For this species, lowest mortality was observed at 18–22°C, but for species living in warmer areas, optimum temperatures may be higher. In Egypt, Zayed (1992) found that 32°C was optimum for *R. purpureus*, even if 37°C was lethal.

Consequently, high temperatures have both absolute upper limits as well as sublethal effects. The tolerance to heat may be species specific. Pfadt *et al.* (1975) found that the pupal stage of *H. bovis* has less tolerance for high temperatures than does the same stage of *H. lineatum*. This difference may be the reason why the distribution of *H. bovis* does not extend as far south in North America as does that of *H. lineatum* (Weintraub *et al.*, 1961; Pfadt *et al.*, 1975). Breyev (1973) also discussed temperature requirements of the pupae in limiting the distribution of *H. lineatum*.

Low temperatures may also cause death of larvae and pupae. The so-called *supercooling point* is the absolute lower lethal limit: oestrid larvae obviously show freeze-tolerance. Bishopp *et al.* (1926) showed that *Hypoderma* spp. larvae and pupae can withstand rather low temperatures, and could survive minimum temperatures of −15.8°C. Pupae appeared to be more resistant to cold than the larvae. Salt (1944) found supercooling points of pupae of *H. lineatum* to be as low as −25.2°C, and concluded that in southern Alberta there is little chance that early dropped larvae will be frozen and killed. In Canada, Gregson (1958) found that 43% of *H. lineatum* larvae/pupae survived a minimum temperature of −29°C. The supercooling points changed during development, as he found that there was a period of about 3 days (in the prepupal stage) when the cold hardiness is greatly reduced (supercooling point is approximately −15°C). Later in development (pupae and pharate adult stages) the supercooling points reached an average of −28.1°C (see Fig. 9.7).

Larvae of the reindeer oestrids *H. tarandi* and *C. trompe* start departure from the host at the end of April (Nilssen and Haugerud, 1994), a period when there may be low temperatures, especially at night in the northernmost areas. Supercooling points of pupae of these species have not been measured, but Saveljev (1971) found that larvae and pupae of the reindeer warble fly *H. tarandi* could survive 40 days in snow (temperature range 0–12°C) and even 3 days covered by ice and 10 days in water, but Breyev and Breyeva (1946) noted that the pupae of *H. tarandi* did not survive repeated prolonged flooding. Apart from the above-cited works, little information is available on cold hardiness in other oestrid species, but Bennett (1962) reported that pupae of *C. phobifera* could withstand freezing temperatures and argued that low temperature for a time possibly is favourable to development.

As a conclusion, larvae and pupae of most oestrid species, at least those living in temperate and northern areas, seem to survive periods of frost as long as

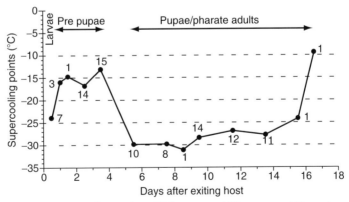

Fig. 9.7. Mean supercooling point of 7 larvae and 91 pupae of *Hypoderma lineatum* according to their ages at 20°C (after Gregson, 1958). The numbers near each point denote the sample size. During the early development of the pupa (the prepupa) there is a period of about 3 days when the cold hardiness is greatly reduced, and it is also reduced near eclosion.

the temperature is above the supercooling point, and larvae dropped on snow will survive even if the development is retarded. Probably, cold temperatures are normally not a serious mortality factor in the pupal stage of oestrids.

When investigating temperature-dependent development, we can calculate the lower thermal limit, or the so-called developmental zero, but this is not directly linked to mortality because the development at this temperature will only be temporarily stopped and will resume as soon as the pupae experience warmer temperatures again. However, if the proportion of the time below the subthermal limit becomes too high, the development will be so lengthy that mortality increases. There may be physiological reasons for this, but another reason is that the larvae and pupae run a higher risk of being attacked by fungi (see below).

Humidity is another factor that is shown to have a strong effect on pupal survival. This mortality factor probably works in two ways. One way is simply that the pupae 'drown', i.e. that they cannot get enough oxygen. This may happen if the larvae are dropped in liquid water or in a substrate saturated with water. The other way, which may be even more important, is that larvae in moist environments are more exposed to fungi and possibly also other microorganisms. There may be more spores of fungi in humid soil, but the fungi may also have better conditions to procreate and grow when water is abundant. Gomoyunova (1973) found that the highest mortality of pupae of *H. bovis* and *H. lineatum* from fungus infections occurred in moist soil when temperatures were low, probably because the long pupal duration increased the risk of fungal infections.

Pfadt (1947) reported that no *H. lineatum* emerged from pupae that were buried 1/4 in. in soil holding >10% water. There are certain areas throughout the USA where *H. lineatum* are scarce or completely absent, e.g. Red River Valley in North Dakota. One hypothesis to explain this is that the type of soil (Fargo clay) is detrimental to the larvae and pupae (Bishopp *et al.*, 1926; Bruce, 1938), but Simco and Lancaster (1964) showed that soil type had no effect on mortality if the soil was properly drained. Also Breyev (1973) found that excessive soil humidity was a factor restricting distribution of *H. lineatum*. Breyev and Karazeeva (1957) reported that *H. tarandi* suffered high mortality in the pupal stage if larvae were dropped on moist places and in places where flooding occurs.

Regarding air humidity, Pfadt (1947) found very little difference in the length of the pupal period and in the survival rate of pupae reared at relative humidities ranging from 0 to 76% RH, but at 100% RH the pupal period was slightly lengthened and mortality of pupa was greatly increased.

Fungi and Microorganisms

Minár and Samsináková (1977) found that pupae of *H. bovis* from which no adults emerged were filled with the mycelium of fungi, identified as the entomophagous fungus *Beauveria tenella* (saprophytic fungi were also found). The larva seemed to be infected by the fungus in the surface layer of soil during pupariation. This fungus occurs very frequently as a parasite of many insect groups, and Minár and Samsináková (1977) suggested that this fungus could be utilized as a natural enemy to control warble flies. Colwell (1992) reported that a fungus has been shown to attack pupae of *H. bovis* and *H. lineatum* in the field, but little is known of its rate of infection. Solopov (1989b) showed that mycosis was connected to humidity. When *H. tarandi* larvae were placed on waterlogged soils, mortality was caused by mycosis. Fungi were found in nearly half of the larvae/pupae that died. In some a white mould was visible on the surface, and when the puparium was removed, he could observe a soft white 'fuzz' (mycelium?) covering the dead pupa. Solopov (1989a) reported that for *C. trompe*, mycosis was the most harmful biological factor for the pupal stage. Kal'vish (1990) isolated the fungus *Tolypocladium niveum* from pupae of the reindeer oestrids *H. tarandi* and *C. trompe* and found that it had an insecticidal effect on adults of *O. ovis*.

Bacteria may also be involved as a mortality factor, and several species have been reported from gut tracts of second- and third-instar cattle grubs (Nogge and Werner, 1970; Norman and Younger, 1979; both cited in Scholl, 1990). According to Scholl (1990), the breeding holes of warble fly larvae in the hide permit entry of bacteria, and symbovine flies feeding at the breathing holes may serve as mechanical transmitters of pathogens. Chamberlain and Scholl (1991) found that when *H. lineatum* failed to pupariate on damp sand, the likely cause of death was bacterial contamination, and administration of antibiotics allowed many to pupariate. The larvae themselves can produce bacteriostatic secretions (Scholl, 1990). Sancho *et al.* (1996) studied the microflora of larvae and pupae of *D. hominis* and found that many species were represented. Interestingly, oestrids may also function as vectors of microorganisms. Vashkevich (1978) found *Brucella* in larvae, pupae and adults of *H. tarandi* that had their origin in reindeer infected with brucellosis. *Brucella* persisted for 6–8 weeks in the pupae without change in virulence.

It is difficult from current knowledge to evaluate the impact of bacteria and fungi on the population dynamics of oestrids, but it is likely that these biological factors may be significant under certain circumstances.

10 Adult Biology

J.R. ANDERSON

University of California, Berkeley, USA

Except for parasites of domestic animals, adult oestrids are rarely collected. Many species, in fact, are known from only a few reared adults. The primary factor contributing to the paucity of specimens in museum collections is low population density (resulting from strong host specificities for wild mammals having populations often limited by available habitat and low infestation rates in obligate hosts). Other factors include: (i) univoltine life cycles of most species; (ii) short life span of adults; (iii) cryptic coloration of many species; and (iv) mating sites of many species being widely scattered and often difficult to access.

Both male and female oestrids have vestigial mouthparts (Zumpt, 1965; Wood, 1987, Chapter 5), and are not known to feed on protein. Nevertheless, in both laboratory and field situations adults of several species are known to ingest water (e.g. *Cuterebra emasculator*, Bennett, 1955; *Cephenemyia apicata*, Catts and Garcia, 1963; *Oestromyia leporina*, Rietschel, 1981), and it has been postulated that ingestion of water and nectar could extend longevity. Ingesting water, however, may not be necessary for flies that eclose with a lifetime supply of fat body reserves. As discussed in the evaluation of the flight capacity of *Hypoderma tarandi* and *Cephenemyia trompe* (Nilssen and Anderson, 1995a), the metabolism of lipids during flight may be a method for these oestrids to obtain water (1 g of fat yields about 1.07 g of water).

All necessary nutrients for basic metabolism, survival, dispersal and reproduction are accumulated during the long larval period in the host. The reproductive organs of males are fully developed at eclosion and they can inseminate females shortly after emerging. Females eclose from puparia with mature eggs or with enough nutrients to complete oogenesis and to survive long enough to larviposit/oviposit. Female oestrids tend to be larger and heavier than males (e.g. Haas and Dicke, 1958; Baird 1971, 1972a; Bennett, 1972; Minár and Breyev, 1982; Nilssen, 1997a; Colwell and Milton, 1998). Larger individuals of either sex generally live longer, are stronger fliers and, in the case of females, have greater fecundity (Forrest, 1987). However, for *H. tarandi* and *C. trompe*, Nilssen (1997a) found that, although male *H. tarandi* were larger than females and male *C. trompe* were smaller than females, longevity was not significantly correlated with size; he also found that larger *H. tarandi* females did not produce more eggs than smaller females.

The life-history tactics and reproductive strategy of oestrids are representative of classical semelparous r-strategists that reproduce once and then die (Force, 1975; Stearns, 1976). Such strategists develop in nutrient-rich environments, and are characterized by rapid sexual maturation, high fecundity, far-flying dispersal potential, population densities usually below carrying capacity and tolerance for stressful environmental conditions. From an ecological perspective the reproductive strategy of oestrids is representative of 'big bang' reproduction (Bell, 1980).

Gonotrophic Development

The oestrids resemble many other Diptera in having meroistic polytrophic ovaries which have 15 nurse cells associated with each developing oocyte (e.g. Cepeda-Palacios and Scholl, 1999). Most oestrids studied (e.g. *Cuterebra fontinella, Oestrus ovis, Cephenemyia apicata, C. jellisoni* and *C. trompe*) initiate development of a single oocyte per ovariole during the pupal period (e.g. Catts, 1964b; Scholl, 1991; Cepeda-Palacios and Scholl, 1999; J.R. Anderson, unpublished data), after which the germarium produces no further oocytes. In fact, the ovaries of *Cephenemyia* species atrophy after the eggs produced have been transferred to the uterus (Catts, 1964b; Anderson and Nilssen, 1996b; J.R. Anderson, unpublished data). *Hypoderma bovis* and *H. lineatum* are unusual in that each ovariole contains two simultaneously developing follicles (Scholl and Weintraub, 1988). The types of ovarian development associated with oestrids preclude the development of subsequent clutches of eggs, as is possible in most other Diptera. Females therefore eclose with all the eggs they are capable of developing. Details of oogenesis and comparisons with other taxa of Diptera are provided in Scholl and Weintraub (1988), Scholl (1991) and Cepeda-Palacios and Scholl (1999).

Initial differentiation and development of the gonads of cattle grubs (*Hypoderma* spp.) begins while larvae are in the second instar (Boulard, 1968). By the late third instar a functioning germarium is absent and each ovariole contains two simultaneously developing follicles (Scholl and Weintraub 1988). Four advanced stages of gonotrophic development occur during the pupal stage after which the female imago emerges with a double compliment of mature eggs. The same process occurs in the reindeer warble fly, *H. tarandi* (Anderson and Nilssen, 1996b). Such *Hypoderma* females can mate immediately after eclosion (Weintraub, 1961; Minár and Breyev, 1982) and need only disperse to a mating site to do so. This gonotrophic scenario eliminates the risks associated with foraging and the time needed by females of other Diptera to find and digest the nutrients necessary for them to produce eggs.

Gonotrophic development in Gasterophilinae and Oestrinae is thought to be like that in *O. ovis*, with females eclosing with fully developed eggs. Except for *D. hominis*, which also emerges with fully developed eggs (Mourier and Banegas, 1970), species of Cuterebrinae studied eclose with oocytes at early stages of development (Scholl, 1991). Eggs do not mature until about 5–8 days post-eclosion

(Haas and Dicke, 1958; Bennett, 1972; Baird, 1972c, 1983, 1997; Colwell and Milton, 1998; Scholl, 1991).

Larviparous oestrids (e.g. *O. ovis*, *Cephenemyia apicata*, *C. trompe*, *C. jellisoni*) emerge with fully developed eggs that are retained in the ovaries until after mating (Catts, 1964b; Anderson and Nilssen, 1996b; Cepeda-Palacios and Scholl, 1999; J.R. Anderson, unpublished data). Eggs are apparently fertilized as they pass through the oviduct into the bilobed uterus (Fig. 10.1). The unique reproductive system of *O. ovis* and *R. purpureus* has been described by Jolly (1847), Dufour (1851) and Cholodkowsky (1908) (cited in Portchinskii, 1913; Townsend, 1938; and Hagan (1951)), and for *C. apicata* by Catts (1964b). The uterus of ovoviparous females serves as an incubation chamber (Hagan 1951). Thus, following an intrauterine egg

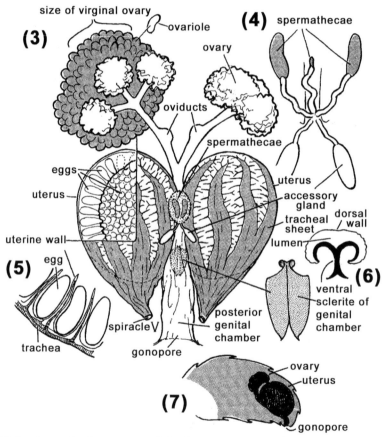

Fig. 10.1. Reproductive system of *Cephenemyia apicata* (after Catts, 1964b); (3) Dorsal view of reproductive system showing comparative sizes of virginal (shaded area at left) and degenerated (at right) ovaries. (4) Detailed dorsal view of spermathecae without surrounding matrix. (5) Diagrammatic cross-section of uterine wall. (6) Detailed dorsal and cross-sectional view of ventral sclerite of genital chamber. (7) Diagrammatic parasagittal view of abdomen showing gross relationships of reproductive organs.

development period of several days, the hatched larvae are maintained in the uterus until a female deposits larvae on an appropriate host. Nearly 10% of larviparous female *C. trompe*, *C. apicata* and *C. jellisoni* examined had 1–2 unhatched eggs in the uterus (Anderson and Nilssen, 1996b; J.R. Anderson, unpublished data). While in the uterus, respiration of the hundreds to thousands of larvae is provided for by an extensive system of trachae that penetrates throughout the uterus (Fig. 10.1).

Mated females first dissected after 10–12 days (*O. ovis*, Bukshtynov, 1978; Cepeda-Palacios and Scholl, 1999), 14 days (*C. stimulator*, Ullrich, 1939) and 18 days (*C. apicata*, Catts, 1964b), had motile larvae in their uteri. However, female *C. apicata* and *C. jellisoni* dissected at just 4–5 days post-copulation after being collected, *in copulo*, at mating sites (J.R. Anderson, unpublished data), had their uteri full of larvae. Also, one of three *C. trompe* females dissected at 72–77 h post-copulation had its uterus full of larvae and no remaining unhatched eggs (Anderson and Nilssen, 1996b). This suggests that eggs in all larviparous oestrids may hatch in about 4–5 days. Females of larviparous species are believed to seek protective sheltering areas during embryonation of their larvae.

After larvae hatch, oestrid females must quickly locate hosts and deposit their larvae because, unlike the phenomenon of adenotrophic viviparity in which adults are adapted to nourish their developing larvae, oviparous oestrid females are not known to provide nourishment for their larvae (Hagan, 1951). One obvious advantage of viviparity is that deposited larvae can immediately invade the host. This eliminates environmental hazards encountered by embryonating larvae within eggs of other oestrid species that are laid on or near the vertebrate host.

Spermatogenesis has not been investigated as often as oogenesis. However, mature spermatozoa were found in male pupae of *D. hominis* (Banegas and Mourier, 1967) and *O. ovis* (Cepeda-Palacios *et al.*, 2001). Males of all species of Gasterophilinae, Hypodermatinae and Cuterebrinae mated under laboratory conditions successfully inseminated females on the first day of emergence, indicating that they also eclosed with mature sperm.

Fecundity

The potential reproductive capacity of a female, as determined by the number of ovarioles per ovary, may be genetically determined. However, the number of ovarioles that produce mature eggs seems governed by the quantity and quality of nutrients obtained from the host in the same manner as host blood sources affect the number of eggs produced by haematophagous flies (Clements, 1992; Crosskey, 1990). Anderson and Nilssen (1996b) noted that it was common to find *Hypoderma tarandi* females with varying numbers of follicles that had not functioned. Examination of reared females associated with individual reindeer revealed that different reindeer produced *H. tarandi* of different mean size (Nilssen, 1997a). Because size differences were not correlated with larval numbers per host, smaller-sized individuals of *H. tarandi* and other oestrids may originate from malnourished or immunologically stressed hosts. Also, because fecundity in oestrids is determined early during larval development, it cannot be increased even if late stage developing larvae receive maximum nourishment.

Most studies of dipteran species have found that larger females of a species produce more eggs than smaller ones. For *H. tarandi*, however, the number of eggs developed was independent of adult size, whereas mean egg weight increased significantly with the weight (size) of newly eclosed females (Nilssen, 1997a). Perhaps larger eggs result in larger, more vigorous larvae that are more successful in invading hosts.

Because both female size and the nutritional state of the parasitized host may have affected the number of eggs recorded in flies examined in different studies, the number of eggs reported for different species is somewhat variable (Table 10.1). But it is clear from all assessments of fecundity that female oestrids produce large numbers of eggs.

Longevity

The longevity and dispersal potential of oestrid adults is governed by the nutrient reserves present at eclosion and the prevailing climatic conditions encountered. For flies eclosing during periods of warm sunny weather, during which both sexes spend a maximum amount of time flying, individuals probably survive for no more than 1–2 weeks.

Although packed with fat reserves, most laboratory-reared oestrids have not lived longer than 12–13 days (Table 10.2). Exceptional survival times (Tables 10.2

Table 10.1. Representative examples of the number of eggs or larvae found in laboratory-reared or virgin female oestrids.[a]

Taxon	Number of eggs	Reference
Hypoderma bovis	551±6	Breyev (1973)
Hypoderma lineatum	530±7	Breyev (1973)
	377–665	Scholl and Weintraub (1988)
Hypoderma tarandi	642±6	Breyev (1973)
	354–772	Nilssen (1997a,b)
	606–754	Anderson and Nilssen (1996b)
Cephenemyia trompe	1280–1520	Anderson and Nilssen (1996b)
	493–1349	Nilssen (1997a,b)
Cephenemyia apicata	959–983	Catts (1964b)
Oestris ovis	337–652	Cobbett and Mitchell (1941)
	572–627	Portchinskii (1913)
Rhinoestrus purpureus	700–800	Zumpt (1965)
Gasterophilus intestinalis	397–770	Dove (1918)
Gasterophilus nasalis	480–518	Dove (1918)
Gasterophilus haemorrhoidalis	134–167	Dove (1918)
Gasterophilus nigricornis	330–350	Chereshnev (in Zumpt, 1965)
Gasterophilus pecorum	1300–2425	Chereshnev (in Zumpt, 1965)

[a] The fecundity of many North American cuterebrid species reviewed by Catts (1982) ranged from 524 to 2511 eggs. More recently, Baird (1997) found that *C. austeni* contained 1321–1731 eggs.

Table 10.2. Survival times of representative laboratory-reared oestrids maintained under various laboratory conditions.

Species	Longevity (days)	Reference
Cephenemyia apicata	12–18	Catts (1964b)
Cephenemyia trompe	4–13	Anderson and Nilssen (1996b)
Cephenemyia trompe	4–44	Nilssen (1997a)
Oestrus ovis	1.5–68	various authors (in Zumpt, 1965)
Cephalopina titillator	4–16	Tzaprun (in Zumpt,1965)
Gasterophilus nasalis	1–12	Dove (1918)
Gasterophilus intestinalis	7–21	Dove (1918)
Gasterophilus haemorrhoidalis	1–7	Dove (1918)
Gasterophilus pecorum	1–4	Chereshnev (in Zumpt, 1965)
Gyrostigma pavesii	3–5	Zumpt (1962)
Cuterebra latifrons	8–11	Catts (1967)
Cuterebra ruficrus	4–13	Baird (1972)
Cuterebra horripilum	1–8	Haas and Dicke (1958)
Cuterebra emasculator	6–8	Bennett (1972)
Dermatobia hominis	1–6	Koone and Banegas (1959)
Przhevalskiana silenus	3–13	Sayin (1977)

and 10.3) were associated with factors such as low temperature and restricted confinement in small containers. Other examples of exceptional longevity include two male *Cuterebra emasculator* that survived for 20 days in darkness at 1.6–10.7°C (Bennett, 1972); an average survival time of 25 days (maximum of 38 days) for *Rhinoestrus purpureus* (Zumpt, 1965); and a maximum survival of 38 days for *Cephalopina titillator* (Grunin, 1957).

Minár and Breyev (1982) found that the laboratory longevity of *Hypoderma bovis* was related to weight loss, as did Nilssen (1997a) for *H. tarandi*, who found that at 5°C, 22°C and 33°C rates of fat loss increased with increasing age. Flies held at 5°C still had most of their fat body remaining at death. Flies held at 22°C and at 33°C had much less fat body remaining than flies held at 5°C, even though they survived for far fewer days than flies held at 5°C. After being confined in vials that prevented flight, flies at all temperatures had comparatively large amounts of fat body remaining at death, so a lack of fat body-derived nutrients was not the cause of death.

Except for female *O. ovis* recaptured up to a week after release (Semenov *et al.*, 1975), the scant data for survival times under natural field conditions have been obtained from the recapture of males caught, marked and released at known mating sites. Although their age at first capture was unknown, marked *Cuterebra latifrons* males were recaptured 10–12 days later (Catts, 1967) and males of *C. fontinella* (as *C. grisea*) survived up to 8 days after marking (Hunter and Webster, 1973a). Utsi (1979) saw *Cephenemyia trompe* males up to 5 days post-marking and Catts (1964b) recaptured marked *Cephenemyia apicata* males up to 6 days post-marking. *Hypoderma*

Table 10.3. The effect of different temperatures on the survival of *Hypoderma* species in the laboratory.

Species	Maintenance temperature (°C)	Survival time (days)	Reference
H. bovis	20	6 (aver.)	Breyev (1973)
H. bovis	5	13 (aver.)	Breyev (1973)
H. bovis (females)	10	5–17	Minár and Breyev (1982)
H. bovis (males)	10	8–22	Minár and Breyev (1982)
H. lineatum	16	1–7	Bishopp *et al.* (1926)
H. lineatum	10	9–17	Bishopp *et al.* (1926)
H. lineatum	14	10	Breyev (1973)
H. lineatum	5	25	Breyev (1973)
H. tarandi	20	8	Breyev (1973)
H. tarandi	5	35	Breyev (1973)
H. tarandi	33	1.2–7	Nilssen (1997a)
H. tarandi	22	3–13	Nilssen (1997a)
H. tarandi	5	3–27	Nilssen (1997a)

lineatum males were resighted at 2–8 days after marking (Catts, *et al.*, 1965) and a *H. tarandi* male was resighted 10 days after being marked (Anderson *et al.*, 1994).

Flight and Dispersal Potential

Females

As a group, oestrids are highly active robust fliers, with some species (*Hypoderma lineatum*, *H. tarandi* and *Cephenemyia trompe*) capable of achieving flight speeds of 28.8–40.0 km/h (Nogge and Staack, 1969; Nachtigall, 1974; Nilssen and Anderson, 1995a). Several species have been reported flying at prevailing wind velocities of 28.8 to 36 km/h (Breyv, 1956; Anderson *et al.*, 1994; Anderson and Nilssen, 1996a) and Baird (1997) reported *Cuterebra austeni* flying at wind velocities as high as 40 to 48 km/h.

For flies flown on a laboratory flight mill (Nilssen and Anderson, 1995a) the longest total flying time for a female *H. tarandi* was 31.5 h, and for a *C. trompe* female it was 11.1 h. The longest flight episodes were 12 h for a *H. tarandi* female and 4 h for a *C. trompe* female. In Russia, marked *H. tarandi* females were recaptured/resighted, within 6–35 h, around a reindeer herd that had moved 13–45 km away from the release point. Some marked females relocated the reindeer herd at a maximum distance of 90 km (Gomoyunov, 1976). At the same time, marked *C. trompe* females were recaptured/resighted around reindeer located 20–45 km from the release point. Based on flight mill studies (Nilssen and Anderson, 1995a), the estimated maximum flight distances were 600–900 km for *H. tarandi* females and 220–330 km for *C. trompe* females. These authors noted that the adaptive significance of such extraordinary flight was related to the migratory behaviour of the vertebrate host, *Rangifer tarandus* (L.). Howard and Conant (1983) hypothesized that nasal bot flies (*Oestrus* and *Gedoelstia* spp.) of migrating wildebeest in Africa might have to fly several hundred kilometres to locate their hosts. At the other extreme, marked-released *O. ovis* associated with flocks of sheep had dispersed only 20–30 km after a week (Semenov *et al.*, 1975).

Flight rapidly depletes the fat body reserves. After laboratory-reared individuals of *H. tarandi* and *C. trompe* had flown on a flight mill for mean times of 8.5 h (SD 7.2 h) and 4.9 h (SD 3.2 h) respectively, they experienced weight losses that were three to four times greater than that of resting flies (Nilssen and Anderson, 1995a). Similarly, Minár and Breyev (1982) found that tethered *H. bovis* females that flew for 1–2 h lost 4.7–7.0% of their initial weight and tethered males flown for 5 h lost 9.6–13.6% of their initial weight. Based on the calorific content of an 85 mg fly, Humphreys and Reynolds (1980) calculated that the energy used in flight by *Gasterophilus intestinalis* would totally consume a fly in only 9 h.

A model illustrating loss of mass as a function of flying and resting (Nilssen and Anderson 1995a) predicted that a 250 mg *H. tarandi* would reach a 'death mass' of 100 mg either by resting for 260 h (at 22°C) or flying only for 74 h, or by other combinations of flying and resting times shown by the model.

Examination of dissected flies revealed that *H. tarandi* and *C. trompe* females that flew for the longest periods of time (5.1–11.7 h) had metabolized most of their fat body (Anderson and Nilssen, 1996b). In most cases such long-flying females were categorized as young individuals (because they had nearly a full complement of eggs or larvae). By contrast, females that flew for less than 5 h but also had very little fat body remaining had only small numbers of eggs or larvae remaining, indicating that they had oviposited or larviposited several times previously. Such individuals were categorized as older flies that probably had metabolized more fat body during more frequent and longer pre-capture flights.

The flight mill studies of Nilssen and Anderson (1995a) revealed that mating greatly altered the flight behaviour of *H. tarandi* and *C. trompe* females. Wild-caught mated *C. trompe* commonly flew continuously for 3–4 h, and *H. tarandi* for 8–12 h. Unmated laboratory-reared females of both species, by contrast, were reluctant to fly; most took continuous flights of less than 30 min. It was postulated that the change in flight behaviour was associated with hormonal changes following mating as reported for other insects (Rankin, 1989).

Males

Flights of male *H. tarandi* on the flight mill typically lasted for only 1–4 min with long periods of rest between flights. Similarly, tethered *Cuterebra latifrons* males did not sustain flight for more than 3 min (Catts, 1967). Such flights were similar to the frequent, but brief, flights of males observed at field mating sites (Catts, 1967; Anderson *et al.*, 1994). The flight pattern of male *C. trompe* on the flight mill was like that of the females and unlike their flight behaviour at mating sites (Utsi, 1979; Downes *et al.*, 1985; Nilssen and Anderson, 1995a; J.R. Anderson and A.C. Nilssen, unpublished data).

At topographical mating sites (next section) male oestrids have been observed to chase conspecifics, as well as all insects near their size, as they pass within view. Thus, oestrid males engage in numerous short, fast flights each day they are present at a mating site (e.g. Anderson *et al.*, 1994). Such frequent flights would rapidly deplete a male's fat body reserve and shorten its longevity. In species utilizing the host as a mating site (next section) both sexes metabolize nutrient reserves as they engage in host-finding flights.

Mating Behaviour

Laboratory studies

Females of *D. hominis* (Banegas and Mourier, 1967), and the few species of oestrids that have been colonized or experimentally mated under laboratory conditions (Table 10.2) will mate within the first or second day after eclosion. Males of all oestrid species studied will successfully mate as early as the first day of their emergence. Under laboratory conditions mating has been induced by tethered flight of both sexes, tumbling both sexes in a confined space, or by contacting females with decapitated males (Weintraub 1961; Catts, 1964a; Rietschel, 1975a; Minár and Breyev, 1982; Baird, 1997; Colwell and Milton, 1998).

Under laboratory conditions female warble flies (*H. bovis, H. lineatum, H. tarandi*) and horse bot flies (*G. intestinals, G. haemorrhoidalis, G. nasalis*) will mate and oviposit within a few hours of eclosing (Dove, 1918; Bishopp *et al.*, 1926; Hearle, 1938; Wells and Knipling, 1938; Gebauer, 1939; Minár and Breyev, 1982; Cogley and Cogley, 2000). Also, reared *G. rhinocerontis* (= *pavesii*) had mated and produced eggs while surviving in a cage for 3–5 days (Zumpt, 1962a).

Females of several species of cuterebrids reared and studied in the laboratory (Table 10.2) mated within 1–2 days post-eclosion (Catts, 1982). However, this may be an artefact of close confinement under laboratory conditions; wild females apparently do not become receptive until about 5 days after emergence. For example, *C. latifrons* females captured at mating sites began laying eggs on the day of capture (Catts, 1967), and under laboratory conditions the earliest oviposition for *C. approximata* and *C. baeri* was 5 days post-eclosion (Catts, 1964a; Colwell and Milton, 1998). Flies that mate on the day of emergence in the laboratory may not do so under natural field conditions – unless mating occurs at the emergence site. Under natural field conditions both sexes may require several days to locate a mating site. This would be especially true for species that utilize the host as a mating site marker (because the host species may have dispersed many kilometres from where mature larvae pupated).

Field studies

Under natural field conditions different oestrid species aggregate and mate either near their free-roaming vertebrate host (Table 10.4) (a mobile attractant), or at a distinctive topographic landmark that is attractive to both sexes (Table 10.5). In either case, because mature larvae may exit hosts anywhere in the host's habitat,

Table 10.4. Oestrid species known to utilize their vertebrate host as the site where the sexes meet and mate.[a]

Species	Vertebrate host	Reference
Gasterophilus intestinalis	Horse	Dove (1918) Wells and Knipling (1938) Rastegaev (1984) Cope and Catts (1991)
Gasterophilus haemorrhoidalis	Horse	Dove (1918) Wells and Knipling (1938) Rastegaev (1984)
Gasterophilus nasalis (= *G. veterinus*)	Horse	Rastegaev (1984)
Rhinoestrus latifrons	Horse	Rastegaev (1984)
Rhinoestrus purpureus	Horse	Rastegaev (1984)
Cephalopina titillator	Camel	Patton (1920)

[a] All species may utilize a dual mating strategy (see Table 10.5).

Table 10.5. Oestrid species that utilize a distinctive topographic landmark as a site where the sexes meet and mate.

Species	Topographic landmark	Reference
Rhinoestrus purpureus	Hilltops/mountain tops	Grunin (1959)
Oestromyia leporina	Rocks on steep mountain slopes	Grunin (1959)
Hypoderma lineatum	Rocky areas & paths along valley creeks; trails along mountain slopes	Gansser (1957); Catts *et al.* (1965)
Hypoderma tarandi	Rocky areas along rivers/streams and road tracks/paths in valleys	Anderson *et al.* (1994)
Cephenemyia jellisoni	Hilltops/mountain tops	Catts (1964b)
Cephenemyia apicata	Hilltops/mountain tops	Catts (1964b)
Cephenemyia trompe	Mountain tops/hilltops	Utsi (1979); Downes *et al.* (1985); Nilssen and Anderson (1995b)
Gasterophilus intestinalis	Tops of hilltop shrubs/trees	Grunin (1959); Catts (1979)
Gasterophilus inermis	Tops of hilltop shrubs/trees	Grunin (1959)
Cephalopina titillator	Tops of hilltop shrubs/trees	Grunin (1959)
Cuterebra austeni[a]	Hilltops/mountain tops	Alcock and Schaefer (1983); Baird (1997)

[a] Reported by Baird (1997) as the species studied by Alcock and Schaefer (1983). For other cuterebrids, Catts (1982) listed *Cuterebra lepivora* and *C. fontinella* as mating at creek basin landmarks; *C. tenebrosa* at cliff face sites, and *C. polita* at stepped slope sites.

dispersal and assemblage of sexually mature adults at the specific host or at a distinctive landmark is an effective strategy to bring together widely scattered males and females of insects having low adult population densities (Anderson, 1974; Thornhill and Alcock, 1983). Because most oestrids are protandrous, large numbers of the earlier eclosing males are active at mating sites several days before females begin to visit the site. This same phenomenon occurs in non-protandrous species, in which females do not visit a mating site until they become receptive at 5–8 days post-eclosion. Important environmental conditions that influence mating behaviour include the time of day, percentage cloud cover, wind velocity and temperature (see 'Environmental Influences').

Mating at host

All oestrids that use the host as a mating site landmark (Table 10.4) parasitize large vertebrates that travel and forage in herds. Rastegaev (1984) found that both sexes of gastrophilid and oestrid parasites of pastured horses near the Caspian Sea flew to the highest points of pastures after emergence.

Shortly thereafter they flew to, and remained around, herds of horses where mating occurred. This enabled female gastrophilids to begin oviposition immediately. Female oestrids however, remained quiescent for 7–10 days before larvipositing on horses. In Rastegaev's studies (1984) males comprised 45% of adult *G. intestinalis* collections around horses, whereas the percentages of *G. haemorrhoidalis*, *G. nasalis*, *Rhinoestrus latifrons* and *R. purpureus* males collected ranged from 53% to 60%.

Although the mating behaviour of *G. intestinalis* has been reported most frequently, and all workers cited in Table 10.4 observed matings around horses, there is some disagreement about the importance of the host as a site where the sexes meet and mate. Several workers have reported collecting males, or observing aggregations of *G. intestinalis* males, at hilltop sites (Brauer, 1863; Walton, 1930; Ullrich, 1939; Grunin, 1959; Catts, 1979; Rastegaev, 1984). This has led to a proposed scenario of an alternative, or dual, mating strategy for this species, with one site being prominent hilltops, and the other site being associated with horses (Catts, 1979; Cope and Catts, 1991). Thus, individuals not successful in encountering mates at hilltop sites may seek out horses as a secondary mating site. Given the limited longevity of adults, this type of mating strategy would ensure that most females in an area would be inseminated in time to disperse their eggs. In contrast, Cogley and Cogley (2000) suggested that *G. intestinalis* mated at horse dung piles, but presented no evidence to support this speculation. The dual mating strategy proposed for *G. intestinalis* also may apply to *G. haemorrhoidalis*, *G. inermis*, *G. nasalis*, *Rhinoestrus latifrons*, *R. purpureus* and *Cephalopina titillator*, as all have been collected at mountaintop/hilltop sites (Table 10.5) (Grunin, 1959; Rastegaev, 1984), as well as at hosts (Table 10.4).

Mating at topographical landmarks

Most species of oestrids studied are known to aggregate and mate at topographical landmarks that usually are distinctive for each species. Mountain tops were the first such site discovered (Brauer, 1863), and presently mountain tops, hilltops and elevated ridges are known mating sites for some species of oestrids in all subfamilies. In a review Grunin (1959) concluded that the following species of oestrids aggregated at the highest points in a locality (e.g. mountaintops/hilltops) for the purpose of mating: *Cephenemyia stimulator, C. ulrichi, C. trompe, C. jellisoni, Pharyngomyia picta, Cephalopina titillator, Rhinoestrus purpureus, Gasterophilus intestinalis, G. inermis, Oestromyia marmotae* and *Hypoderma tarandi* (as *Oedemagena tarandi*). However, recent evidence indicates that *H. tarandi* does not mate at mountaintops/hilltops. Several years of extensive sampling at such sites located in areas of high warble fly abundance revealed no *H. tarandi* males (Utsi, 1979; Anderson *et al.*, 1994; Nilssen and Anderson, 1995b). The mating sites of *H. tarandi*, as well as *H. bovis* and *H. lineatum*, are located at rocky, wind-protected areas along rivers and streams or along paths/trails formed on hillside slopes in valleys (Steck, 1932; Kühl, 1949; Gansser, 1957; Catts *et al.*, 1965; Anderson *et al.*, 1994) Since Grunin's review *Cephenemyia apicata* (Catts, 1964b), *C. auribarbis* (Cepelák *et al.*, 1972), *Cuterebra latifrons* (Catts, 1967) and *C. austeni* (Alcock and Schaefer, 1983; Baird, 1997) have been found to meet and mate at mountaintop/hilltop aggregation sites, and several studies have provided more extensive information about the hilltop mating behaviour of *C. jellisoni, C. trompe* and *G. intestinalis* (Catts, 1964b, 1979; Utsi, 1979; Downes *et al.*, 1985; Nilssen and Anderson, 1995b).

The review by Catts (1982) identifies *Cuterebra lepivora* and *C. fontinella* as mating at distinctive creek basin landmarks, *C. tenebrosa* at cliff face sites, and *C. polita* at stepped slope sites. *Oestromyia leporina* males also aggregate at sites along steep slopes (Grunin, 1959). This species, as well as *Cuterebra latifrons* and *C. polita* (Catts, 1967; Capelle, 1970), mates at sites near the burrows and lairs of their hosts. Several authors have provided both a photographic and verbal description of the mating site used by the species studied (e.g. Catts, 1964b, 1967; Capelle, 1970; Hunter and Webster, 1973a; Meyer and Bock, 1980; Alcock and Schaefer, 1983; Downes *et al.*, 1985; Anderson *et al.*, 1994).

Gansser (1957) reported that in Switzerland *Hypoderma* species used the same mating sites over a span of 18 years, and in northern California, USA, *C. apicata* and *C. jellisoni* have used the same hilltop mating sites over a period of 30 years (J.R. Anderson, unpublished data). It therefore appears that, as long as topographical features remain relatively unaltered, different oestrid species will continue to aggregate at a specific mating site.

Environmental influences

Although there are marked differences among topographical landmark mating sites used by different oestrids, the behaviour and activity of different species at the mating site is very similar, with most in temperate areas arriving at temperatures near 20°C and leaving as the temperature approaches 30°C. On warm, sunny days flies usually are active for only several hours between 0900 and 1400 h, but on cooler, partly cloudy days fly activity is extended. In subarctic areas a few males of *C. trompe* and *H. tarandi* first appeared at mating sites at 6–10°C (Utsi, 1979; Downes, 1985; Anderson, *et al.*, 1994), and a few *H. tarandi* even were sporadically active on days with a cloud cover of 90–95% (Anderson *et al.*, 1994). Males perch on various types of substrata (e.g. ground clearings, rocks, tips of leaves, stems or branches) in a characteristic posture shown for different genera by Grunin (1959), Catts (1964b, 1967), Meyer and Bock (1980), Anderson *et al.* (1994) (see Figs 10.2 and 10.3). Males tend to be most active on bright, sunny days having low to moderate wind velocities. All fly from their perches to pursue conspecifics or other passing insects, or tossed pebbles near their size. Flies generally are not active on dark, cloudy days, at wind velocities above 8 m/s, or during periods of rain or snow (e.g. Catts, 1964b, 1967; Downes *et al.*, 1985; Anderson *et al.*, 1994).

Fig. 10.2. *Hypoderma tarandi* male perched at a mating site in subarctic Norway (photo by A.C. Nilssen).

Fig. 10.3. *Cephenemyia trompe* male launching from a mating site in subarctic Norway (photo by A.C. Nilssen).

Mark–recapture studies (see 'Longevity' section) have revealed that males tend to remain at the mating site where they were captured. Catts (1967) reported that only 3 of 200 marked males moved to an adjacent hilltop.

Each mating site seems to have several preferred perches that are first occupied by arriving males. Thus, when only four males are present at an eight-perch site they frequently occupy the same four perches. Most sites usually accommodate only a few males (<5–10), although some *Cephenemyia* sites may accommodate 25 to 60 males (Utsi, 1979; Nilssen and Anderson, 1995b) and Thomann (1947) estimated a phenomonal aggregation of 'thousands' of *C. stimulator*.

Behaviour at mating sites

Males of some species aggressively exhibit territorial behaviour by defending a choice perching site from other males (e.g. Catts, 1967; Hunter and Webster, 1973a; Meyer and Bock, 1980; Alcock and Schaefer, 1983). In contrast, males of other species (e.g. *C. trompe* and *H. tarandi*) do not defend specific perches (Utsi, 1979; Downes *et al.*, 1985; Anderson *et al.*, 1994; J.R. Anderson unpublished data). In these groups males randomly settle almost simultaneously at various previously occupied perches after male–male pursuit flights, or after the pursuit of a female or another insect species.

During warm, calm, sunny days Anderson *et al.* (1994) reported that *H. tarandi* males perched for only 1–2 min before flying upward to promote cooling. When several males were present at a mating site on such days their frequent thermoregulatory flights and male–male pursuit flights, or pursuit of other insects, occured every few minutes. Under such conditions *H. tarandi* males would engage in about 200 such flights/day if they remained at the mating site during the entire daily activity period (Anderson *et al.*, 1994). At the other extreme, on cold days when few males appear at a mating site, a male may sit at the same perch for 30 min to more than 60 min (Grunin, 1959; Anderson *et al.*, 1994). At these times males will move forward and perch on the tip of one's slowly extended finger, or they can be captured with one's fingers.

The first thermoregulatory behaviour exhibited by *H. tarandi* males on warm sunny days was to select wind-exposed perch sites that absorbed the least radiant energy (Anderson *et al.*, 1994). As the body temperature of perched *H. tarandi* males increased, their behavioural thermoregulatory actions included orientation of the body to the sun's rays, crouching, stilting and flying upward into cooler air. At temperatures above 30°C male cuterebrids left perching/flying areas (Hunter and Webster, 1973a; Meyer and Bock, 1980) and crawled downward into shaded areas. Most studies of oestrids at various aggregation sites have noted that flies are present for only short periods on warm, sunny days, usually leaving within 1–2 h, or less, because they are not able to withstand a longer period of thermal stress.

Gasterophilus intestinalis is the only oestrid known to endothermally increase its thoracic temperature. Under laboratory conditions Humphreys and Reynolds (1980) found that the preferred ambient temperature range for flight of this species was 20–24°C, but that spontaneous take-off did not occur until after the thoracic temperature was raised to 31–32°C. Prior to take-off, flies buzzed loudly as they endothermically warmed the thoracic muscles. During flight, flies maintained nearly a constant thoracic temperature about 10°C above ambient (21.6°C), but when a flying individual was heated with a lamp, its thoracic temperature continued to rise. When warmed by direct sunlight *G. intestinalis*, like other oestrids, may need to cool down rather than warm up.

Intense flying activity may rapidly deplete a male's fat body reserve. The numerous flights taken by males on sunny days may select for high fitness (those possessing superior flight speed, stamina and endurance) allowing maximum time at the mating site (Alcock, 1987; Anderson *et al.*, 1994). The longer a male can remain active at a mating site the greater his chances of mating with one or more females. A female, by contrast, need only visit a male-occupied mating site once in her lifetime, and by arriving later in the day she may enhance the fitness of her progeny by mating with a more vigorous male that has remained active through-out the mating period.

Males perched at mating sites usually fly after and catch females that fly past, after which paired couples land nearby and complete copulation (Fig. 10.4). In addition to pursuing and catching females in flight, *H. tarandi* males would hop on to and grasp females that landed on the ground near them (Anderson *et al.*, 1994). Copulation in most species lasts only a few minutes, but may last for 30 min in some cuterebrids (Catts, 1982). When the sexes disengage, the mated female flies leave.

The lek-like polygynous mating strategy of many oestrids appears to best fit the 'hot spot' model of Bradbury and Gibson (1983). In this model male aggregations result from sequential landings and activities of males at sites preferentially used by females that incite a male pursuit flight. Large, fast-flying males with superior stamina, that may also defend territorial perches, may mate with several females.

Fig. 10.4. Male/female *Cephenemyia trompe in copulo* on the ground at a mating site in subarctic Norway (photo by A.C. Nilssen).

The frequent pursuit of subordinate males by dominant males occupying preferred perching sites at a mating arena is believed to be a behaviour that is beneficial to species that aggregate at special sites. Such behaviour forces the chased males to other nearby mating sites and results in a spaced distribution of males at each of many potential sites. In different geographical areas, for example, there may be many male-occupied mating sites within close proximity (Catts 1964b, 1967; Anderson *et al.*, 1994; Nilssen and Anderson, 1995b). In an area of about 20 km^2 Nilssen and Anderson (1995b) found 18 *C. trompe* mating sites in a 4-h period while flies were active; numbers ranged from 3 to over 60 males per site.

A mating strategy whereby low densities of males are aggregated at numerous topographical mating sites within a limited geographical area is beneficial for males in that, after the initial flight to locate a mating site, most of their fat body reserve can be used to fuel their frequent pursuit flights while at the mating site. Females also benefit by such a strategy because it greatly increases an individual's chance of quickly finding a male-occupied mating site, and thereby retaining most of its energy reserves for host finding flights. Some workers have suggested that oestrids of economic importance could be controlled by insecticide application to the mating sites, but Nilssen and Anderson (1995b) concluded that this would be inappropriate because numerous beneficial insects also frequent hilltops.

Host-finding Behaviour

Although not specifically studied, the host-finding behaviour of mated oestrid females likely involves a ranging pattern of flight (Vale, 1980) until volatile or other types of host cues are detected. As host cues are detected, the specific host-searching behaviour of oestrids is thought to be much like that described for various haematophagous flies (e.g. Sutcliffe, 1986; Allan *et al.*, 1987; Willemse and Takken, 1994) and other myiasis-producing species (Hall, 1995) that use both olfactory and visual stimuli associated with potential hosts.

Unlike most haematophagous flies and other myiasis-producing flies, almost all oestrids are host specific. A generalized model for the host-searching behaviour of most oestrids would include positive orientation to long range, down-wind olfactory cues, followed by orientation to medium range olfactory and visual cues, followed by orientation to close range visual and olfactory stimuli and, perhaps, heat and moisture. For egg-laying species, the final response would involve tactile cues associated with the ovipositor (Karter *et al.*, 1992; Jones, 2000).

In tundra, savanna and steppe habitats inhabited by herds of the large mammalian hosts of oestrids, visual perception of a host's silhouette appears to be an important factor in attracting flies. In such habitats wind would also nearly always be a prominent abiotic factor dispersing odour plumes that could assist in finding such hosts as caribou/reindeer, sheep, wildebeast, horses and zebras.

Cuterebrinae, and one gastrophilid, deposit eggs at off-host sites rather than directly contacting the host. Thus, for cuterebrid species that oviposit on plants and branches along the trails of small hosts like rodents, voles, woodrats and hares, or at host dens/nests, wind-borne odour plumes probably are not a major factor along ground level trails through thickly vegetated, grassy or brushy areas. Instead, such species probably respond primarily to substrate bound, low volatility host odours, such as territorial marking scents, and to visual stimuli associated with trails and entrances or profiles of burrows, dens or woodrat houses (see next section). The large number of chemo-sensilla in the tarsal pads of *Cuterebra* species (Colwell, 1994) may enhance detection of low concentrations of host odours during periodic landings along trails. Responsiveness to substrate bound host karimones would be much like that of certain prey-trailing snakes (Kubie and Halpern, 1979; Chiszar *et al.*, 1990) and hunting dogs that follow freshly laid tracks of game birds and rabbits (Syrotuck, 1972). The common infestation of field dogs with exotic cuterebrid larvae probably occurs as they follow trails of small mammals. Cuterebrid species that oviposit at off-host sites may search for oviposition sites in the evening or early morning hours when host odours along trails and near lair entrances would be at their freshest. Baird (1974, 1997) found that gravid *C. tenebrosa* and *C. austeni* searched for oviposition sites from late afternoon until sunset.

As found for certain haematophagous flies (e.g. Clements, 1992), the hierarchy of behavioural responses involved in host location by oestrids probably includes hormonal changes after mating (Rankin, 1989). Nilssen and Anderson (1995a), for example, reported that the flight behaviour of the caribou/reindeer parasites, *H. tarandi* and *C. trompe*, changed dramatically after females were inseminated. On a laboratory flight mill unmated females were reluctant to fly, whereas mated females generally flew for long periods (e.g. 8–15 h).

Little research has been conducted to determine stimuli that attract oestrids to their hosts. Despite the hundreds of trapping studies conducted in many different geographic areas using such generalized baits as CO_2, l-octen-3-ol (octenol), phenols and acetone, or combinations of such compounds (e.g. Anderson and Hoy, 1972; Murihead-Thompson, 1982; French and Kline, 1989; Nicolas and Sillaus, 1989; Takken and Kline, 1989; Vale, 1993; Green, 1994; Nilssen, 1998; Gibson and Torr, 1999), only *H. tarandi* and a few species of *Cephenemyia* have been caught. Capelle (1970) did not catch *C. polita* in CO_2-baited traps while flies were active near traps. Similarly, CO_2-baited traps in horse pastures (Anderson and Yee, 1995) did not attract and catch *G. intestinalis* at times when females were seen and caught around horses, nor did they catch *O. ovis* when operated near flocks of sheep attacked by flies (J.R. Anderson, unpublished data). CO_2 also did not enhance *O. ovis* larvipositions on a baited sheep or goat model (Cepeda-Palacios and Scholl, 2000a).

For caribou/reindeer oestrids, Anderson and Nilssen (1996a, 1996b) found that CO_2-baited traps functioned as true host mimics in catching only host-seeking females (mated, gravid females of *H. tarandi* and larviparous *C. trompe*). Similar results were obtained for *C. apicata* and *C. jellisoni* of deer (Anderson and Olkowski, 1968; Cogley and Anderson, 1981; Anderson, 2001; J.R. Anderson, unpublished data). Trap catches revealed that even some old *Cephenemyia* females (little fat body remaining and no larvae remaining in the uterus) (Anderson and Nilssen, 1996b; J.R. Anderson, unpublished data) continued to engage in host-seeking behaviour.

Except for the few species associated with cervid hosts, the overall absence of oestrid species captured in, or seen around, CO_2-baited traps indicates that it is likely that most oestrid species are sensing and responding to unique odours and/or visual cues associated with their specific host. Even the oestrids associated with caribou/reindeer respond to host cues other than CO_2. In evaluating electroantennograms, Tommerås *et al.* (1993) found that *C. trompe* specifically sensed components in the interdigital pheromone gland of reindeer and a component from reindeer urine, as well as CO_2 and octenol; *H. tarandi* responded primarily to the same interdigital gland components and CO_2 (Tommeras *et al.*, 1996). Future electroantennogram screenings of odoriferous compounds from other mammals may reveal specific compounds that different oestrid species use to locate their specific host.

In ecosystems that support a variety of mammals that emit CO_2 and a mixture of volatile substances, short-lived, host specific oestrid parasites would waste much time and quickly deplete their energy reserves if they tracked down all potential hosts releasing such common attractants as CO_2 and octenol. Repulsive odours associated with non-host species also would contribute to the efficiency of host seeking activities. In the presence of sheep and sheep odours, for example, CO_2-baited deer models (Anderson, 1989) were almost never attacked by *C. apicata* and *C. jellisoni*. However, when such deer models and white CO_2-baited sheep models were exposed several kilometres away from sheep, both *Cephenemyia* species attacked the models, as well as CO_2-baited cow head models (J.R. Anderson, unpublished data). Like results with deer models (Anderson, 1989), non-CO_2-baited cow head models (Anderson and Yee, 1995) and sheep models, were not attacked.

Oviposition and Larviposition Behaviour

Risk-spreading oviposition/larviposition behaviour

All oestrids studied are characterized by a risk-spreading oviposition/larviposition behaviour whereby females usually deposit only a few eggs or larvae on each of many host animals or at off-host oviposition sites. Such resource partitioning behaviour results in females effectively distributing their progeny among patchily distributed oviposition/larviposition sites. This behaviour reduces competition among a female's progeny, eliminates the possibility of a host being infested with all, or most, of a female's progeny (and then dying before her progeny complete development), and may result in a female distributing her larvae among hosts with different levels of specific immunity. Many workers have commonly observed such risk-spreading oviposition behaviour while working around domesticated animals attacked by *Gasterophilus* and *Hypoderma* species. However, when only a few individual hosts are present, a female may lay many eggs on one animal. Wells and Knipling (1938), for example, saw one *G. intestinalis* lay 301 eggs on one horse in 45 min.

The risk-spreading oviposition behaviour of *H. bovis* was experimentally determined in a *Hypoderma*-free area in former Czechoslovakia by releasing one gravid, laboratory-reared female into each of several pastures and later determining the number of cattle that were infested (Minár and Breyev, 1983). Different females infested 11 of 74, 12 of 112 and 47 of 91 cattle examined from separate pastures. Research with baited deer models whose muzzles and nostrils were treated with insect adhesive that trapped larvae *in situ* revealed that individual

C. apicata and *C. jellisoni* females larviposited only once and then flew away (Anderson, 2001). Examination of dissected *C. trompe* and *H. tarandi* caught in host-mimicking traps (Anderson and Nilssen, 1996b) revealed that these oestrids also were characterized by a risk-spreading reproductive strategy. Similarily, cuterebrid species that oviposit at off-host sites lay small batches of eggs at many different locations. Catts (1982) listed the benefits of such behaviour as: (i) increasing the chance of repeated exposure of an individual host; (ii) permitting a greater proportion of the egg complement to remain unhatched and thus survive longer; (iii) decreasing the possible destruction of progeny by predators or parasitoids; and (iv) increasing the chance of exposure to more hosts.

Oviposition behaviour associated with host-associated visual stimuli

When approaching the vertebrate host, colour appears to be an important visual stimulus for some oestrids. Horak and Boomker (1981), for example, noted that *Strobiloestrus* species oviposited on calves of similar colour to klipspringers, the primary obligate host, and white has long been recognized as being particularly attractive for reindeer oestrids (e.g. Bergman, 1917; Breyev and Karazaeva, 1939 (cited in Breyev, 1950)). Ovipositing horse bot flies, on the other hand, were noted by Dove (1918) as showing no preference for colour, breed or age of horses.

Oestrid species that oviposit on various parts of a host's body visually orient to the host's silhouette. Thus, both *H. lineatum* and *H. bovis* oviposit primarily on the legs and flanks of standing cattle (e.g. Bishopp *et al.*, 1926; Hearle, 1938; Weintraub, 1961; Nogge and Staack, 1969), as does *H. tarandi* on reindeer (Bergman, 1917; Breyev and Savel'ev, 1958; Espmark, 1967), but all three species will oviposit on other parts of the body as well, particularly when *H. lineatum* and *H. tarandi* females lay eggs on reclining animals. The heel fly, *H. lineatum*, usually approaches from the shaded side of the host and lands either on the host or on the ground next to the host. *H. tarandi* behaves similarly, but seems to land on animals more frequently. The most effective oviposition behaviour for both *H. lineatum* and *H. tarandi* is for females to land on the ground and attack recumbent animals. Females approach such animals by raising the abdomen and walking backwards with the ovipositor extended and probing for contact with hairs (op. cit.). When ovipositing, *H. lineatum* lays about 4–12 eggs/hair and *H. tarandi* about 4–10 eggs/hair.

The usual mild reaction of cattle is to kick and move away from ovipositing *H. lineatum*, but the larger *H. tarandi* females provoke a more pronounced reaction in reindeer. When *H. tarandi* females land on reindeer, animals respond first by vigorously shaking their bodies, after which they usually run from the site of attack and then stand motionless in a stiff-legged position as they warily watch for flies. When flies keep landing on reindeer, animals become increasingly agitated; their behaviour culminates in an entire herd responding with periodic bouts of panicked running, after which the animals merge into a closely packed, restless, non-feeding herd.

Contrary to the mostly stealthy oviposition behaviour of *H. lineatum*, *H. bovis* females usually fly to the host at 8–9 m/s, hover momentarily near the rear legs, and then rapidly dart at a leg and oviposit one egg in the astonishing time of just 0.2–0.3 s (Nogge and Staack, 1969). Each time a female strikes downward with her ovipositor she deposits one egg near the base of a hair. The nearly 600 eggs/female *H. bovis* (Table 10.1) equates to 600 separate oviposition attack flights for females successful in laying all of their eggs, whereas an *H. lineatum* female with an equal number of eggs might only engage in 50–150 oviposition attacks. The greater volume of fat body flight fuel in the larger *H. bovis* females probably enables so much oviposition flight activity. Because *H. bovis* usually oviposits on standing or running cattle, its eggs tend to be more widely distributed on the legs and upper body areas of cattle than those of *H. lineatum*. When attacking running animals *H. bovis* females mostly dart downward and oviposit from the shoulder area to the tail head, but they may also occasionally oviposit on the sides and legs (e.g. Weintraub, 1961; Nogge and Staack, 1969). Cattle become greatly frightened when an *H. bovis* female repeatedly attacks the same animal several times in succession as the animal kicks and tries to run from the fly (Bishopp *et al.*, 1926; Hearle, 1938; Weintraub, 1961; Nogge and Staack, 1969). When attacked by *H. tarandi*, cattle also become frightened and try to run away from the flies (Bergman, 1917).

Other *Hypoderma* species, and species in other genera of the large subfamily Hypoderminae that have been studied, also tend to oviposit primarily on the fore and hind legs of their hosts. This includes such species as *Strobiloestrus vanzyli* of lechwe antelope (Howard, 1980), *Przhevalskiana silenus* of goats, sheep and gazelles (Zumpt, 1965; Grunin, 1975; Sayin, 1977) and *P. corinnae* of gazelles (Zumpt, 1965). Another goat parasite, *P. crossi*, differs in attaching eggs to hairs on the back of goats (Soni, 1942). All three species that parasitize elephants attach their eggs in rows at the base of the tusks (Zumpt, 1965). About 2000 eggs of *Cobboldia elephantis* were counted on one tusk of an Indian elephant (Sclater, 1871, in Zumpt, 1965). Of the three species of *Gyrostigma* that parasitize rhinoceroses, *G. rhinocerontis* (= *pavesii*) is attracted to the head where it attaches its eggs to the skin at the base of the ears and around the neck and shoulders (Zumpt, 1965).

Unlike the cattle warble flies, different species of *Gasterophilus* oviposit on different anatomical regions of their equine host. *Gasterophilus intestinalis* oviposits primarily on hairs on the lower inside area of the front legs, but also occasionally on other parts of the legs, the shoulder and mane (e.g. Wells and Knipling, 1938; Hadwen and Cameron, 1918). This species often hovers slowly in one spot as it quickly deposits several eggs before flying to another position or to another horse. *Gasterophilus nasalis* eggs are laid on hairs under the head between the margins of the lower jaw (Dove, 1918), usually one egg to a hair, but rarely as many as 5 (Zumpt, 1965). *Gasterophilus haemorrhoidalis* lays one egg per hair on the short hairs along the edge of the lips, mostly on the lower lip (Wells and Knipling, 1938; Hadwen and Cameron, 1918), and *G. nigricornis* and *G. inermis* females lay one egg at the base of hairs on the cheeks and nose (Zumpt, 1965). All of the latter four species are characterized by a 'hit-and-flee' oviposition behaviour whereby females in full flight collide swiftly with a horse, rapidly deposit an egg and then fly away before repeating the process. Horses protect themselves from *G. nasalis* and *G. haemorrhoidalis* by resting their chin or lips upon the back of an adjacent animal (Figs 10.5 and 10.6).

For *H. bovis* and *H. lineatum* tarsal sensilla may play a role in final host acceptance (Colwell and Berry, 1993), after which ovipositor sensilla determine the egg deposition site. Observations of ovipositing *H. lineatum* and *Strobiloestrus vanzuli*

Fig. 10.5. Horses exhibiting anti-bot fly defensive behaviour used to protect their lips from ovipositing *Gasterophilus haemorrhoidalis* (see also Dove, 1918).

revealed that the extended ovipositor was rapidly thrust in and out between hairs until the flexible tip encountered a hair of the proper diameter to accommodate an egg (Gooding and Weintraub, 1960; Weintraub, 1961; Howard, 1980; Karter *et al.*, 1992; Jones, 2000). In studying the egg attachment organ of various *Hypoderma* and *Gasterophilus* species, Cogley *et al.* (1981) and Cogley and Anderson (1983) found that a key factor involved in egg attachment was separation of the lateral flanges as the ventral attachment organ was pushed on to a hair shaft. These

Fig. 10.6. Horses exhibiting anti-bot fly defensive behaviour used to protect the underside of head/lower jaw from ovipositing *Gasterophilus nasalis* (see also Dove, 1918).

authors found that when gentle pressure was applied along the midpoint of eggs removed from dissected females the lateral flanges opened widely, and immediately closed when the pressure was released. They concluded that, as an egg is deposited naturally, pressure exerted on the larger middle area of the egg when passing through the terminal part of the ovipositor causes the lateral flanges on the basal attachment organ to separate, accommodate a hair, and then close around the hair as pressure terminates when the egg leaves the ovipositor.

The complicated structure of special egg attachment organs (Chapter 6) associated with the eggs of different species (Hadwen and Cameron, 1918; Draber-Monko, 1974) are adapted to fit host hairs of particular sizes (Cogley *et al.*, 1981; Cogley and Anderson, 1983), with adhesive within the attachment groove cementing eggs to host hairs. On lechwe antelope, eggs of *S. vanzyli* are attached to the underside of smaller hairs (diameters 0.032–0.068 mm) (Howard, 1980). The reindeer warble fly, *H. tarandi*, selectively lays its eggs on new hair grown after summer shedding (Bergman, 1932; Breyev and Savel'ev, 1958), and on fine wool and guard hairs (Karter *et al.*, 1992). The latter authors reported that eggs were most often laid at the base of a hair, near the host's skin; eggs were laid on hairs with diameters of 0.017–0.172 mm. *Hypoderma lineatum* also attaches its eggs near the base of hairs and the host's skin (Scholl, 1993). Jones (2000) found that ovipositing *H. lineatum* females selected hairs having a mean diameter of 0.07 ± 0.002 mm, with the more numerous smaller diameter hairs being rejected in favour of the relatively scarce larger diameter hairs.

Larvipostion behaviour associated with host-associated visual stimuli

Larvipositing oestrids that visually orient to contrasting dark cavities include the camel nose bot fly, *Cephalopina titillator*, *Oestrus ovis* of sheep and goats, *Cephenemyia jellisoni* of deer and *Rhinoestrus purpureus* and *R. latifrons* infesting equines; all larviposit into the nostrils of their host (Patton, 1920; Rastegaev, 1984; Cepeda-Palacios and Scholl, 2000a; Anderson, 2001). The presence of larvae of the many other *Rhinoestrus* species in the nasal cavities of such hosts as horse, zebra, giraffe, hippopotamus, springbuck and bush pig (Zumpt, 1965) indicates that all *Rhinoestrus* species larviposit into the host's nostrils. Like *O. ovis*, other species of *Oestrus* also probably larviposit into or near the nostrils of the mammalian host, but *Gedoelstia cristata* and *G. hässleri* were reported to larviposit into the eyes of their antelope hosts in Africa (Basson, 1962a). Females of *Cephenemyia*, *Pharyngomyia*, *Pharyngobolus* and *Kirkioestrus*, whose larvae occur in the nasal cavities, frontal sinuses and pharyngeal region, probably larviposit into the nostrils or on to the lips of their hosts.

Research with animals and baited models has revealed that, in the final phases of attack, *Cephenemyia* species and *O. ovis* first visually orient to the muzzle of the host (Breyev, 1950; Espmark, 1967; Anderson, 1975, 1989, 2001; Cepeda-Palacios and Scholl, 2000a), and that when approaching the muzzle at close range, females visually orient to the nostrils (*O. ovis* and *C. jellisoni*) or the lips (*C. apicata*). For *O. ovis*, movement of sheep and goat models was necessary to induce tethered females to larviposit. From the number of larvae immobilized in insect adhesive on baited models Anderson (2001) found that each larviposition by *C. apicata* contained

Fig. 10.7. Sheep exhibiting the typical anti-bot avoidance behaviour associated with larviposition attacks by *Oestrus ovis*. (Photo by H.V. Daly.)

from 25–68 larvae, and that each larviposition by *C. jellisoni* contained 74–180 larvae. The typical oestrid avoidance behaviour of vertebrate hosts attacked by nasal bot flies (Portchinskii, 1913; Espmark, 1967; Anderson, 1975; Cepeda-Palacios and Scholl, 2000) is to lower the head and place the muzzle and lips on or near the ground (Fig. 10.7).

Prevalence of infestation for oestrids directly attacking their hosts

In spite of various anti-fly behaviours exhibited by various vertebrate hosts, the prevalence of oestrid infestation among non-insecticide treated livestock usually ranges from 75 to 100% (e.g. Meleney *et al.*, 1962; Rogers and Knapp, 1973; Horak, 1977; Rastegaev, 1984; Pandey *et al.*, 1992; Nilssen and Haugerud, 1995). Infestation rates among populations of wild vertebrates are often similar (e.g. Howard and Conant, 1983; McMahon and Bunch, 1989; Perez *et al.*, 1995, 1996).

Behaviour of species ovipositing at off-host terrestrial sites

In the near vicinity of the oviposition site some oestrid females respond to visual stimuli associated with shape and colour. *Gasterophilus pecorum* is an unusual horse bot fly in that females oviposit batches of 10–15 eggs on to certain grasses and leaves that subsequently are eaten by horses (Chereshnev, 1951, in Zumpt, 1965).

All other *Gasterophilus* species oviposit on the host. *Cuterebra buccata* is another oestrid that oviposits on plants eaten by its host (the eastern cottontail rabbit) (Beamer, 1950), and Shannon and Greene (1926) suggested that the howler monkey bot, *Cuterebra baeri*, also oviposited on leaves or fruits eaten by its host. Such bot flies may be attracted by both visual stimuli and odours associated with various plants. When searching for oviposition sites, gravid females of several cuterebrid species visually orient to dark openings associated with entrances to burrows, dens and cliffside crevices or holes. For example, *C. latifrons* entered tunnel openings to wood-rat nests and laid small batches of eggs up to 30 cm inside, as well as on sticks and branches near nest entrances (Catts, 1967). *Cuterebra polita* oviposited on exposed roots of vegetation more than 20 cm inside opened pocket gopher bur-rows (Capelle, 1970), and *C. tenebrosa* oriented to dark cliffside crevice openings that it walked into and laid eggs within 30–40 cm of the opening (Baird, 1974). Gravid *C. tenebrosa* females took short flights from one rock crevice to another as they searched for oviposition sites. Similarly, *Oestromyia leporina* (Hypodermatinae) enters the darkened openings of the burrows and nests of its mouse and pika hosts, but unlike cuterebrids, attaches its eggs to host hairs, usually on the back (Povolny *et al.*, 1960).

Dermatobia hominis oviposition behaviour

The unique oviposition behaviour of *D. hominis* involves the capture of various flies upon which they deposit small batches of eggs (e.g. Guimarães and Papavero, 1999). As females only catch flies in flight, their visual responses to movement are of primary importance in their locating an oviposition site. In this phoretic rela-tionship captured flies, referred to as porters or vectors, ultimately transmit *D. hominis* larvae to different mammalian hosts. However, before this unusual behav-iour plays out, gravid *D. hominis* females usually seek out a mammal and capture flies on or in its vicinity (Bates, 1943; Mourier and Banegas, 1970). By capturing fly species on or near mammals a *D. hominis* female improves the probability that the porter species caught will later visit and land on other mammals when eggs laid on the porter fly contain fully developed larvae. Eggs are glued to one side of a porter's abdomen with their operculum oriented downward, thus enabling the mouth hooks of hatched larvae to immediately contact the mammalian host. At different temperatures *D. hominis* eggs will hatch after 4–15 days (e.g. Koone and Banegas, 1959; Banegas *et al.*, 1967; Mourier and Banegas, 1970), and under nat-ural conditions body heat from the mammalian host is believed to stimulate egg hatching. Guimarães and Papavero (1999) list 54 species of flies found carrying *D. hominis* eggs, including 24 Culicidae, 11 Muscidae, 7 Fanniidae, 5 Calliphoridae, 2 Sarcophagidae, 2 Simuliidae, 1 Tabanidae and 1 Anthomyiidae.

The unique risk-spreading method of distributing eggs among many different captured flies by *D. hominis* has the same benefits described by Catts (1982) for other cuterebrids. As *D. hominis* does not exhibit the strict host specificity of most other oestrids attracted to mammals, traps or models baited with general attrac-tants like CO_2 and octenol might prove useful as tools for learning more about its field behaviour.

Oviposition/larviposition on exotic hosts

Instances of oestrids ovipositing and larvipositing on exotic host species usually seem to involve circumstances in which the exotic host has intermingled with, or been closely associated with, the specific host. *Hypoderma* and *Gasterophilus* species sometimes oviposit on humans that groom or work around horses, cattle and reindeer. Similarly, *C. ulrichii, R. purpureus* and *Gedoelstia* species occasionally larviposit in the eyes of humans associated with reindeer, horses and cattle, as does *P. picta*, the throat bot fly of wild European deer species (Chapter 11). Early-instar *C. trompe* larvae have been recovered from the nasal cavity of dogs (Strom, 1990), and other *Cephenemyia* species have been found infesting llamas in an area inhabited by black-tailed deer (Fowler and Paul-Murphy, 1985; Mattoon *et al.*, 1997). Based on hundreds of published case history reports and notes, *Oestrus ovis* is the most common oestrid known to larviposit in the eyes and nostrils of humans (Chapter 11). This species also has been recovered from the nostrils of dogs (Centurier *et al.*, 1979; Tanwani and Jain, 1986; Lujan *et al.*, 1998).

There are many reports of *H. bovis* and *H. lineatum* ovipositing on and infesting horses (reviewed by Scharff, 1950). Conversely, *G. intestinalis* apparently rarely oviposits on cattle (e.g. Mock, 1973). Other aberrant hosts occasionally infested by *Hypoderma* species are: bison by *H. lineatum* (reviewed by Capelle, 1971); sheep and goats by *H. lineatum* and *H. bovis* (Bishopp *et al.*, 1926; Rastegaev, 1972; Guo and Fu, 1986); cattle and dogs by *H. tarandi* (as *Oedemagena tarandi*, Bergman, 1917; Natvig, 1937, 1939); sheep and horses by *H. diana* (Ross, 1983; Minár, 1987; Hendrickx *et al.*, 1989; Vyslouzil, 1989); and yaks, water buffalo and zebu cattle by *Hypoderma* species (Scholl review, 1993). Species of *Strobiloestrus* that parasitize klipspringers and a few other African antelopes also have been rarely reported from cattle (Horak and Boomker, 1981) and sheep (Brain *et al.*, 1983). Although the above oestrids will oviposit/larviposit on a few aberrant hosts, they do not complete normal development in such hosts. A few mature third-instar larvae of *O. ovis* have been recovered rarely from dogs (Lucientes *et al.*, 1997; Lujan *et al.*, 1998) and humans (Chapter 11), but it is not known whether they could have pupated.

Contrary to oestrids that only sometimes attack aberrant hosts, a few oestrid species exhibit plasticity in host acceptance, which maximizes their encounters with potential hosts in which they can complete development. *O. ovis* is an example of an unusual oestrid that can develop in many different host species, including domestic sheep and goats, many species of wild sheep and goats and ibex. Similarly, *O. aureoargentatus, G. cristata* and *G. hässleri* infest several species of antelopes in Africa, and *Kirkioestrus minutus* infests blue wildebeest, common hartebeest, Lichtenstein's hartebeest and the korrigum (Zumpt, 1965). Another oestrid that is unusual in infesting many host species is *Pharyngomyia picta*, which primarily infests red deer, but also develops in fallow deer, roe deer, Sika deer and elk (Zumpt, 1965).

Cuterebrid infestations of exotic hosts

There are numerous reports of pet cats and dogs becoming infested accidentally with larvae of *Cuterebra* species. These, and the accidental infestations of many different wildlife species, probably occur when the aberrant hosts smell and breathe on cuterebrid eggs laid along host trails and elsewhere. The rare infestation of humans (Chapter 11) probably is associated with their pet cats and dogs. In addition to humans, cats and dogs, Sabrosky (1986) cited records of *Cuterebra* larvae in such aberrant hosts as: cow, sheep, goat, mule, pig, and 11 species of wildlife including red fox; lynx; mink; raccoon; white-tailed deer; and the American woodcock. Several other wildlife species have since been added to the list of aberrant hosts, including various exotic species (e.g. snow leopard, red kangaroo, Bennett's wallaby and Gunther's dik dik) with access to outdoor zoo enclosures (Ryan *et al.*, 1990; Suedmeyer *et al.*, 2000). However, larvae in most aberrant hosts do not develop beyond the second instar (Catts, 1982).

Effect of Climatic Factors on Adult Activity

All species of oestrids typically are most active in engaging in mating and host seeking activities on days typified as being warm, sunny and calm. Maximum activity of most species occurs at ambient temperatures between 20–30°C. Adult activity is reduced on cool, cloudy days and on windy days. During a series of cloudy days having temperatures below 8–10°C, and during inclement weather, most adult oestrids will remain inactive for several consecutive days (Anderson and Nilssen, 1996b). The way in which climatic factors affect different species has been described for cuterebrids by Hunter and Webster (1973), Baird (1974) and Meyer and Bock (1980); for *H. tarandi* and *C. trompe* by Breyev (1956) and Anderson and Nilssen (1996a); for *Hypoderma* species parasitizing cattle by Bishopp *et al.* (1926); for *Gasterophilus* species parasitizing horses by Rastegaev (1984) and Enileeva (1987); and for *O. ovis* by Portchinsky (1913) and Cepeda-Palacios and Scholl (2000a).

Natural Enemies, Mimicry and Protective Coloration

Bergman (1917) reported reindeer biting and eating some *H. tarandi* that landed on them, or on the ground near them, and Anderson (1975) reported deer occasionally biting and killing *C. apicata* females that had landed on them. When sighting *C. trompe* females on the ground in front of them reindeer attack flies by striking at them with their front hoofs (Espmark, 1967).

The adults of many oestrid species apparently are protected from potential predators by virtue of their similarity to venomous bees and bumble bees. Zumpt (1965) noted that all *Cephenemyia*, *Hypoderma* and *Portschinskia* species resembled bumble bees. Nilssen *et al.* (2000) conducted a comparative analysis of colour patterns and size of 11 species of bumble bees and of *H. tarandi* and *C. trompe*. These workers found that three *Bombus* species comprised a Müllerian guild whose members serve as Batesian models for *H. tarandi*, and that three other *Bombus* species comprised a Müllerian guild whose members serve as Batesian models for *C. trompe*. The aposematic colour pattern on these oestrids is believed to provide protection from potential predators. Several of the smaller *Gasterophilus* species are mimics of the honey bee and smaller bees (Zumpt, 1965), and *Gyrostigma* adults resemble certain large African wasps (M.J.R. Hall, London, 2001, personal communication).

Adults of other species of oestrids derive protection from potential predators by virtue of their body texture and disruptive cryptic coloration. Such features enable them to blend into the background on which they are resting and, consequently, be difficult for predators to detect. Portchinskii (1913) commented on how the body texture and coloration of *O. ovis* adults, coupled with countershading changes associated with light from different angles, enabled stationary individuals to blend into various substrata (rocks, soil, wood). When resting on terrestrial surfaces Portchinskii (1915) noted that *Rhinoestrus purpureus* resembled a bird dropping. The tuberculate thoracic texture and speckled earth tone colour patterns observed on examined museum specimens indicates that various other *Oestrus* species, as well as some *Gedoelstia* and *Rhinoestrus* species, would be very difficult to see when alighting and resting on ground, rock or bark surfaces.

11 Larval–Host Parasite Relationships

In order to establish in their host, endoparasitic oestrid larvae penetrate the skin or enter nasal or digestive cavities. They then undertake various routes of migration before leaving their hosts as fully developed third instars for pupation. During these migrations, the larvae accumulate the energy resources to survive during the non-feeding pupal and adult developmental stages (see Chapter 10). The part of their life cycle inside their hosts provides different opportunities for interactions between the host immune system and the host's tissues on which the larvae feed. The clinical manifestations of these myiases vary with the different life cycle stages and the various predilection sites. Even though the pathways of migration are well known for some oestrid species such as *Hypoderma bovis*, *Hypoderma lineatum*, *Dermatobia hominis* or *Oestrus ovis*, and more recently for *Przhevalskiana silenus* (Otranto and Puccini, 2000), only scant information is available about the larval migration of *Hypoderma diana*, for example.

Even though the larval development sites differ from one oestrid subfamily to another, in the same host there may be a common developmental site during one or two larval stages. For example, the two subfamilies Hypodermatinae and Cuterebrinae share similar clinical manifestations for the second and third instars. However, routes of entry differ as does the duration of time spent within the host.

PART A: HYPODERMATINAE HOST–PARASITE INTERACTIONS

C. Boulard, Station de Pathologie Avaire et Parasitologie, France

The most economically important Hypodermatinae were first studied nearly a century ago by Hadwen *et al.* (1915, 1917a,b, 1919) who described the gross pathological lesions induced by penetrating first instars and by hypodermal second- and third-instar *H. bovis* and *H. lineatum*. They also presented preliminary observations on the immune reactions occurring during cattle hypodermosis. Subsequently these same species were extensively studied, particularly with regard to: (i) tissue reactions provoked by the larval penetration and migration through the host connective tissue; (ii) characterization of parasite secretions responsible for lesions in the connective tissue and for host cell recruitment; and (iii) the nature of the immune response.

Immunological and histopathological studies also were conducted on *H. diana* from red deer (Maes and Boulard, 2001), *Hypoderma tarandi* from reindeer (Tashkinov, 1976; Chirico *et al.*, 1987; Monfray and Boulard, 1990; Kearney *et al.*, 1991), *P. silenus* from goats (Cheema, 1977; Puccini *et al.*, 1997) and *Oestromyia leporina*, the field mouse warble fly (Lelouarn and Boulard, 1974; Rietschel, 1975a, 1979).

Histological Studies

During primary infestations, no significant lesions appear in the connective tissue along the first-stage larval (L1) migration pathway. After repeated infestations the migrating larvae appear embedded in large oedema, which may rarely infiltrate the surrounding muscles.

The histopathological effects of migrating *H. lineatum* and *H. bovis* were extensively characterized by Simmons (1937, 1939), Wolfe (1959), Nelson and Weintraub (1972) and Boulard (1975b). Penetrating larvae enter the connective tissue through the skin or hair follicles. Nelson and Weintraub (1972) noted a rapid enzymatic lysis of the calf epidermis produced by larval oral secretions followed by dissolution of the connective tissue. Along the larval pathway there was 'generally no oedema nor was there inflammatory cell infiltration into the tissue or larval trail'. During a subsequent challenge infestation in the same animals, the invading larvae 'were surrounded by oedema and infiltration of inflammatory cells', an observation made earlier by Hadwen (1915).

First instar migration always takes place in the connective tissue, passing through the submucosa of the oesophagus for *H. lineatum*, the epidural connective tissue for *H. bovis*, remaining in the skin cutaneous tissue as in the case for *P. silenus* (Otranto and Puccini, 2000) and *H. tarandi*. The connective tissue surrounding the migrating first instars is always degraded, with collagen and elastin fibres dissociated into fibrils. Fibrin is absent from the infiltrates surrounding larvae as are neutrophils during primary infestations (Boulard, 1975b). Effects of larval passage are noted in the disruption and necrosis of the adjacent connective tissue. Areas behind the passing larvae show massive recruitment of eosinophils. This feature is more pronounced in *P. silenus* than in *O. leporina*.

In previously infested animals, migrating larvae are surrounded by oedema infiltrated mainly with eosinophils and lymphocytes. The number of larvae and the duration of the migration influence the extent of the oedema. However, living first instars are never encapsulated by fibroblasts, whereas dead larvae are found within nodules exhibiting a granuloma structure.

Midguts of both *H. bovis* and *H. lineatum* are filled with a homogeneous viscous material that stains strongly with Periodic Acid Schiff (PAS) (Boulard, 1969). Host connective tissue was never found and blood cells observed very rarely. These observations suggested that the larvae fed on the degraded tissue and cells that were attacked by the digestive enzymes released into the tissue by the larvae.

Second and third instars do not migrate and remain enclosed in a subcutaneous granuloma, producing the familiar swellings known as 'warbles'. There may be an associated oedema in the adjacent connective tissue. The dermis and epidermis of this subcutaneous granuloma is perforated and the posterior spiracles of the larvae face the opening. This granuloma is totally limited by fibrous tissue and the cavity is invaded by an exudate consisting of necrotic cells, fibrous debris, fibrin strands and numerous eosinophils and macrophages. The midguts of second and third instars are packed with these inflammatory cells. Profound modifications of the enzymatic apparatus of the gut between the first instar and the subsequent stages have been noted in conjunction with changes in gut structure (Terada and Ono, 1930; Ono, 1932; Boulard, 1969; Boulard and Garrone, 1978).

Larval Secretory Enzymes

The involvement of larval secretory enzymes in connective tissue degradation has been evoked by all authors describing histopathological changes in tissues of cattle. Early studies of first-instar midguts of *H. lineatum* (Boulard, 1969, 1975b), *H. bovis* (Wolfe, 1959), *H. tarandi* (Kearney *et al.*, 1991) or *O. leporina* (Rietschel, 1975a, 1979) suggested secretions exhibited strong proteolytic activity.

Early characterization of the enzymatic activity showed different proteolytic properties in the L1 midgut versus the warble stages of *H. lineatum* (Ono, 1931). The content of the L1 midgut was named 'hypodermatoxin' and was described as dermolytic. Second and third instars were devoid of such activity (Ono, 1932, 1933). The hypodermatoxin was described as having anticoagulant properties (Ono, 1931), which could explain the absence of fibrin in the histological studies.

An enzyme with strictly specific collagenolytic activity and five other less specific proteolytic enzymes were described from the midgut content of first-instar *H. lineatum* (Boulard, 1970). The collagenase, the first described from a metazoan, was called hypodermin C (HC), and has since been purified by ion exchange chromatography and characterized as a serine protease, having a molecular weight of 25.2 kDa. The enzyme has a specific activity on native collagen, cleaving the triple helix of this molecule at a single region approximately 75% from the *N*-terminus (Boulard and Garrone, 1978; Lecroisey *et al.*, 1979). The amino acid (230 amino acid residues) showed a high degree of sequential homology with the trypsin family serine proteases and was similar to another arthropod serine collagenase secreted by the fiddler crab *Uca pugilator* (Lecroisey *et al.*, 1980, 1987). Modelling of the tertiary structure of this molecule from sequences and X-ray studies indicated a good homology with other trypsins. Crystallographic analyses revealed that this insect collagenase produced a crystal of two molecules associated in an asymmetric unit devoid of any proteolytic activity and that the active site of one molecule of collagenase was trapped by the other molecule of this asymmetric unit (Broutin *et al.*, 1996).

    Initial attempts to isolate mRNA for HC from *H. lineatum* were carried out by
Temeyer and Pruett (1990). Sequencing and gene expression of hypodermins of
H. lineatum were performed by Moiré *et al.* (1994). The sequence deduced from the
cDNA clone indicated that the molecular mass was 28,561 Da, higher than the
protein sequence (see before), and the cDNA clone encoded a 260 amino acid protein. This hypodermin is probably produced as a pre-proenzyme. Analysis of HC
RNA, at the three stages, indicated that the collagenase RNA is expressed only
within the first larval stage. Hybridization of the genomic DNA with an HC probe
indicated large differences between first and third instars; there being three supernumerary fragments that hybridized more intensely in the first stage than in the
third and a specific supernumerary fragment specific for the third stage. These
data suggested that HC might undergo structural genomic changes linked to gene
regulation between first and third instars (Moiré *et al.*, 1994).

    Three other proteolytic enzymes have been isolated and characterized from
the midgut of first-instar *H. lineatum*, i.e. hypodermins A (HA), B (HB) and D (HD).
HA is a serine protease (Boulard, 1970; Boulard and Garrone, 1978; Lecroisey
et al., 1979). The apparent molecular weight of this hypodermin as estimated by
chromatography and SDS polyacrylamide gel was 27.0 kDa (Tong *et al.*, 1981)
with a molecular mass (24,595 Da) derived from the complete amino acid
sequence obtained after gene sequencing of HA (Moiré *et al.*, 1994). Putative glycosylation sites of HA at position 70–72 and 99–101 have been suggested by these
authors and may explain this difference. The catalytic active site contains histidine,
aspartic acid and serine typical of serine proteases. The regions flanking the catalytic site are highly conserved. This enzyme expresses esterase and amidase activity on the synthetic substrates of trypsin (Boulard, 1970; Lecroisey *et al.*, 1979). Its
specificity on other polypeptide substrates differs from trypsin. For instance, contrary to trypsin, HA induces the cleavage of the insulin B chain only at one bond
at the carboxyl side of Arg_{22} and has no activity on the cleavage of Lys_{29}-Ala_{30}
(Tong *et al.*, 1981).

    Analysis of HA protein synthesis and mRNA expression at the different lar-
val stages indicated that both expressions are present only in first instars. Genomic
DNA extracted from first and third instars and hybridization with the same probe
reacted with the same intensity. The genomic fragment of HA, with a 150 bp
longer than expected, evoked the presence of one or more small introns in the
gene and suggests its transcriptional regulation.

    HB is also a trypsin-like protease (Boulard, 1970; Boulard and Garrone,
1978; Lecroisey *et al.*, 1983). Its molecular weight was estimated at 23.0 kDa.
The calculated molecular weight of the mature forms obtained from two clones
of HB and evaluated from their nucleotide sequences indicated a value of
approximately 24.7 kDa and one site of glycosylation for one of these clones.
They encoded proteins of 256 and 260 amino acid residues and, like HA, could
be produced as a pre-proenzyme. These two proteins are closely related with
92% similarity of protein sequence level and 95% of the DNA sequence (Moiré
et al., 1994).

    Enzymatic studies indicated that purified HB exerts esterase and amidase
activities on synthetic substrates of trypsin. A specific catalytic activity on insulin
B chain occurred with a rapid cleavage of Arg_{22}-Gly_{23} and Lys_{29}-Ala_{30} bonds and

to a lesser extent with the Leu_{15}-Tyr_{16} bond. Despite this difference in the catalytic activities, HA and HB share close biological and structural similarities (Lecroisey *et al.*, 1983) and the same mechanisms of expression during the larval stages (Moiré *et al.*, 1994).

The last hypodermin described was by Schwinghammer *et al.* (1988) and it was called hypodermin D (HD). Its biochemical and enzymatic properties indicated that this hypodermin was closely related to HB and shared the same antigenicity. With the description of two cDNA nucleotides coding to HB, it was suggested that this HD was so close to HB that it could be an isoform of HB or an isoenzyme (Moiré *et al.*, 1994).

HC from *H. lineatum* is antigenically similar to proteins secreted by other members of the subfamily (Boulard *et al.*, 1996a). No antigenic cross-reactions were observed between HC and major secretory antigens from specimens from any other oestrid subfamily.

Immune Response

Quantitative investigations

Carpenter *et al.* (1914) were among the first investigators to provide quantitative evidence of an acquired resistance after successive infestations in cattle. The decreasing rate of grubs with successive exposures to *Hypoderma* spp. larvae in natural infestations could sometimes reach total protection in older animals, and more susceptible breeds have been reported (Benakhla *et al.*, 1999). This is now well documented in the long-term work done in different countries, in North America (Knapp *et al.*, 1959), the former USSR (Evstafjev, 1980, 1982) and North Africa (Benakhla *et al.*, 1999). More recently, experimental trickle infestations of naive steers injected with *H. lineatum* L1s were compared with a single infestation and evaluated 8–12 months later by grub survival. In this experiment the repetitive infestations occurring over 1 month induced, as had previously been observed for repetitive reinfestations over several years, a decrease in grub burden compared with the single infestation (Colwell, 2001b). Nevertheless, the antibody response remained very similar in all the animals.

Quantitative studies demonstrated a clear density-dependent effect. When infestations were high (approximately 500–1000 eggs/host), mortality of larvae ranged from 93% to 99% in naive animals. (Weintraub *et al.*, 1961; Breev, 1967). Lower rates of infestation (<20 eggs/host) had larval mortality ranging from 50% to 60% (Breev, 1967).

A mathematical analysis of the results of extensive investigations on more than 12,000 head of cattle naturally infested by *Hypoderma* spp. in Czechoslovakia, Mongolia and the former USSR (Breev, 1967; Breev and Minar, 1976, 1979) showed that the percentage larval mortality increased with the parasite prevalence in the host population. Distribution of larvae within the host population showed a characteristic negative binomial pattern, independent of the location where the investigations were carried out. This was confirmed by research in North America (Lysyk *et al.*, 1991).

The observations and quantitative data accumulated during the last century, led to emerging evidence that the host defence system of cattle is stimulated during infestation and that resistance is developed against the migrating larvae of *Hypoderma* spp. Breev (1967) clearly established that hosts developed substantial protective immune responses during the first 4 months of migration. Breev concluded in this publication: 'it is necessary to concentrate attention on the initial life period of larvae when, due to protective reactions of the host organism, their highest mortality rate is recorded'. He also reaffirmed the lack of reaction during skin penetration in naive animals by invasive larvae, previously observed by Hadwen (1917b), that was generally described in cattle undergoing reinfestation. These results have also been further supported by Gingrich (1980), who considered that resistance was mainly due to internal factors involving cellular components of the immune response that acted mainly on first-stage larvae (Gingrich, 1982).

The Humoral Response

Early research on host humoral responses during natural *Hypoderma* spp. infestations employed primarily antigens from second and third instars. These studies used various approaches to detect and quantify host antibodies: the flocculation test (Peter and Gaehtgens, 1934; Nelson and Knapp, 1961) or agar gel immunoprecipitation (Beaucournu, 1965; Beesley, 1969, 1970; Beesley and Breyev, 1969).

The idea that antibodies were directed against different antigens expressed specifically within each larval stage was ignored for a long time, even though Breev and Obrezha (1969) advanced this idea. A comparison of guinea pig humoral response to injection of different antigenic extracts of *H. tarandi* led to a suggestion that 'antigenic properties of material from first and subsequent instars are likely not identical', thus reinforcing the observations of Ono (1932).

Histological studies describing the connective tissue lesions in the paths of the migrating *H. lineatum* suggested that collagenase was released continuously in the host tissue to facilitate larval migration (Boulard, 1969). Infested cattle and human humoral response to this collagenase (HC) was investigated and the antigenicity of the *Hypoderma* spp. first-instar-specific antigen, HC, was demonstrated in natural and experimental infestations (Boulard, 1970; Boulard *et al.*, 1970; Boulard and Weintraub, 1973). Subsequently, various immunodiagnostics, with HC as an antigen, have been used: gel immunoprecipitation, immunohaemagglutination and direct and competitive ELISA (Boulard, 1970; Boulard *et al.*, 1970; Boulard and Weintraub, 1973; Robertson, 1980; Baldelli *et al.*, 1981; Sinclair and Wassal, 1983; Webster *et al.*, 1997b).

Antibody responses to HC were followed during natural or experimental hypodermosis in cattle, in a rabbit experimental model or in an accidental human infestation (Boulard and Weintraub, 1973; Boulard, 1975a; Boulard and Petithory, 1977; Robertson, 1980; Pruett and Barrett, 1985; Colwell and Baron, 1990; Boulard *et al.*, 1996b; Panadero *et al.*, 1997).

Two types of antibody kinetics have been described. During a low infestation, as was the case in France between 1995 and 2000 before eradication, the early cattle infestations induce a very low antibody response (Fig. 11.1) until 4–5 months post infestation (PI). Then the humoral response starts to increase and reaches a plateau 7–10 months PI. When the infestation intensity is higher (more than 20 warbles) the antibodies can be detected in many animals as early as 6 weeks PI. In any case, the amount of antibody declines as warbles (devoid of this enzyme) develop in the backs of cattle and then disappear in the following 3–4 months after the larvae have exited the warble. Such animals remain seronegative if no other infestation occurs. Maternal anti-*Hypoderma* antibodies that remained for up to 6 months in naive calves (Pruett and Temeyer 1989; Martinez-Gomez *et al.*, 1991) have also been demonstrated.

The HC of *H. lineatum* and *H. bovis* (Pruett *et al.*, 1990) have common epitopes; this unique antigen could be used to diagnose both agents of cattle hypodermosis. Also, immunodiagnosis based on ELISA and using HC has become a useful tool to evaluate the prevalence of cattle hypodermosis in Europe or to monitor the progress of its control in a number of countries including France (Boulard and Argenté, 1993), Greece (Papadopoulos *et al.*, 1997), Morocco (Sahibi *et al.*, 1995), Poland (Cencek, 1995), Portugal (Duque-Araujo *et al.*, 1995), Spain (Reina *et al.*, 1995; Martinez-Moreno *et al.*, 1996), the UK (Webster and Tarry, 1995; Webster *et al.*, 1997a) and North America (Colwell, 2001a).

The dynamics of antibody production were described from milk and found to have the same pattern as in the serum (Boulard and Villejoubert, 1991). The presence of antibodies in milk has made it possible to easily determine hypodermosis prevalence (Charbon *et al.*, 1995; Otranto *et al.*, 2001) with less cost and animal handling.

Immunosurveillance for cattle hypodermosis is now an essential element in eradication programmes in Europe and is performed on the sera collected for other disease monitoring such as brucellosis or leucosis. The capability of this approach to detect infested animals up to 5 months before the first warble appearance makes remedial action possible. In France for instance, this epidemiological tool for hypodermosis survey is increasing every year, and the recommended sampling period is from December to March during the last 4 months of first instar migration. During this period no antibodies resulting from the previous year's infestation remain and if antibodies are detected they are produced with no ambiguity from a current infestation. This method of diagnosis was demonstrated to be much more sensitive than the observations of warbles in the backs of cattle (Vaillant *et al.*, 1997) and is becoming the new tool for qualification of hypodermosis-free departments in France, and has been, since 1985, the only method of diagnosis in the UK (Webster *et al.*, 1997a).

The demonstration of the cross-reactivity of *H. lineatum* HC, with *H. bovis*, *H. tarandi*, *H. diana* and *P. silenus* has opened the opportunity for available immunodiagnosis of reindeer, deer and goat hypodermosis (Monfray and Boulard, 1990; Boulard *et al.*, 1996b). The immune response during goat hypodermosis has been extensively studied in Italy and presents an annual pattern very similar to that of cattle hypodermosis (Puccini *et al.*, 1995). From these data, large immuno-epidemiological studies have been engaged for a goat hypodermosis survey in Italy (Puccini *et al.*, 1995; Otranto *et al.*, 1999) and in Greece (Papadopoulos *et al.*, 1997), and for deer hypodermosis in France (Maes and Boulard, 2000).

Vaccination trials

The early vaccination attempts of naive or previously exposed cattle with crude antigen (Khan *et al.*, 1960) or extracts of first-instar *Hypoderma* spp. (Magat and Boulard, 1970; Pruett *et al.*, 1989; Baron and Colwell, 1991; Chabaudie *et al.*, 1991; Pruett and Stromberg, 1995), produced some promising reductions in grub survival. However, reductions were variable among studies and appeared to be independent of the adjuvants used and the nature of the purified hypodermin (native or recombinant) and of whether the experimental infestation procedure was with natural or experimental challenge. However, in all cases, HA appeared to induce some higher protection even if it was never a total protection. The best results were obtained in HA-vaccinated and experimentally challenged calves, with a 98.5% reduction of the grubs arriving in the back, compared with the number of infective larvae, whereas in the control group 88.5% grub mortality was observed (Pruett *et al.*, 1989). Additional larval mortality in the vaccinates in this study occurred in the second and third instars.

Despite an evident increase in the antibody response following each vaccine trial, the heterogeneity of the resistance to the challenge infestation suggested that antibodies did not appear to be of prime importance in protection. The intensity of the antibody response did not correlate with the number of grubs achieving their biological cycle and did not appear to have any protective value (Pruett and Barrett, 1985).

One of the reasons for this defence failure could be attributed to the proteolytic cleavage of bovine IgG by HA (Pruett, 1993a). However, other adaptations by the migrating larvae to escape the host defence also have been demonstrated, e.g. the presence of an acid mucopolysaccharide surface coat reduces host recognition (Colwell, 1991).

Larval immunomodulation

The fact that invasive larvae fail to induce inflammatory reactions during a primary infection recently received more attention and led to the exploration of the interaction of the larval secretions on the host components participating in inflammation. The complement system is an important component of the innate immune response and constitutes a first line of defence against invading organisms. Activation of complement leads to: (i) a direct attack upon the activating antigenic organism by the generation of a cytolytic attack complex; and (ii) the release of inflammatory mediators that target and recruit cells involved in the early immune response.

The preliminary studies focused on the effects of crude larval secretions, or purified hypodermins, HA, HB and HC, on the cytolytic attack complex activity. Both pathways of complement activation (classical or the alternate) were explored. The powerful activity of HA depleted the complement cytolytic activity via the alternate and the classical pathways, whereas HB interacted mainly on the classical pathway and HC presented no direct interaction on them (Boulard and Bencharif, 1984).

Considering that the third component of complement plays a central role in the initiation of the inflammatory and specific immune response the effects of the three hypodermins on this component were characterized. HA and HB were found to have a dramatic breakdown effect on bovine C_3 (Fig. 11.2). This component is split into numerous peptides that differ totally from physiological activation of C_3 (Boulard, 1989). The total degradation of C_3 at a very low concentration of HA (1 µg/ml of plasma) suggested that in the vicinity of the migrating larvae, the degradation of C_3 could be an important factor that impaired the inflammatory response and immune stimulation. By deflecting the C_3 physiological cleavage and the potential release of chemotactic factors such as C3a and C5a, the migrating larvae of *H. lineatum* may actively circumvent the early host defence system. In fact, the activated bovine complement chemotactic activity for bovine neutrophils appeared to be totally inhibited by HA at very low concentration (Barquet *et al.*, 1993). It was simultaneously demonstrated that HA reduced the expression of the neutrophil adhesion receptor CD18 (Barquet *et al.*, 1993; Moiré *et al.*, 1997). These data corroborated the early histological observations that repeatedly underlined the absence of neutrophils around the migrating larvae (see below).

These preliminary studies have provided some evidence that the invasive strategy of first-instar *Hypoderma* spp. affected both the humoral and the cellular mechanisms involved in the inflammatory reaction (Fig. 11.3). Nevertheless, an exhaustive exploration of other interactions of the invasive larvae to escape the early host defence mechanisms remains to be done.

The cellular response

Immunization trials underscored the fact that in many cases: (i) larval mortality occurred at the end of larval migration; and (ii) it was not primarily antibody-mediated. This suggests that acquired immunity could be supported by a cell-mediated immune response (Pruett *et al.*, 1989). Therefore, antigen-specific or mitogen peripheral blood lymphocyte (PBL) stimulations were carried out, with naive, primary-infested, reinfested or immunized cattle (Baron and Weintraub, 1986; Baron and Colwell, 1991; Fisher *et al.*, 1991; Chabaudie and Boulard, 1992). Among the main conclusions from these studies was that during a primary infestation, especially the 7 months of first-instar migration, naive cattle PBLs were not stimulated to proliferate whatever the stimulating agent, specific antigen or mitogen. On the other hand, reinfested animals after 2 months developed a strong PBL proliferation with either crude first-stage antigen or mitogen stimulation. Moreover, this reaction could be observed as soon as 1 month after infestation in cattle preimmunized with first-stage extracts. Nevertheless, the response could differ when considering purified HA. HA-specific lymphocyte proliferation response from primary or reinfested cattle remained very low. When HA was injected in naive or pre-infested cattle, the lymphocyte proliferative response to mitogens or HA antigen was inhibited (Chabaudie and Boulard, 1992). This immunodepression of the lymphocyte activity lasted 2 weeks after the end of HA injection. Assays with purified HB or HC presented respectively mild or no immunosuppressive activity.

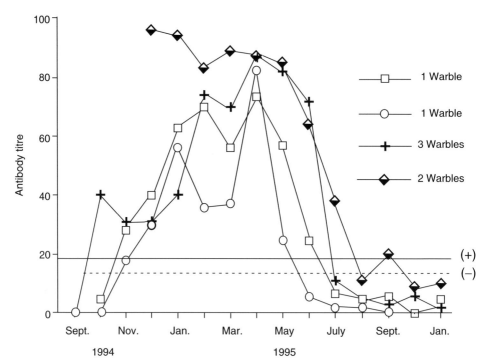

Fig. 11.1. Antibody kinetics in selected animals from a herd of cattle exposed to activity of *Hypoderma bovis* flies. This herd was previously unexposed to fly attack: (a) antibody titres of 4 animals with clinical evidence of hypodermosis.

The mechanism of HA immunomodulation during a primary infestation was approached *in vitro* and different interaction processes have been demonstrated. HA affects the lymphocytes by reducing the production of IL-2 but had no effect on interferon δ production (Nicolas-Gaulard *et al.*, 1995). HA also reduced the expression of different lymphocyte receptors CD2, CD14, CD18 and the α chain of IL-2 receptors (Moiré *et al.*, 1997). HA proliferation inhibition is closely associated with its interaction with macrophages. An HA dose-dependent production of prostaglandin E_2 (PGE_2) by bovine macrophages was characterized. This PGE_2 production could be controlled by indomethacin and it then led to the restoration of lymphocyte proliferation to mitogen (Nicolas-Gaulard *et al.*, 1995).

The following responses indicate a Th2-type response: decreased Il2 production, PGE_2 production (suggesting an Il-4 and Il-10 induction), a dramatic eosinophilia, a negative effect on the macrophages' nitric oxide (NO) release and the induction of IgM and IgG1 antibodies. It must be stressed that this mechanism of immune evasion to *Hypoderma* spp. larvae could downregulate the general Th1 response to unrelated pathogens and make cattle more susceptible to other diseases. For example, increased cases of mastitis have been reported in *Hypoderma* spp.-infested herds in Switzerland (Araujo-Chaveron, 1994).

This direct effect of HA on macrophages, inducing a massive production of PGE_2, could also explain the anaphylactoid shock generated in naive animals after

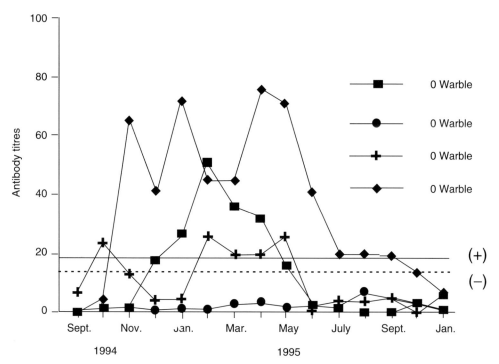

Fig. 11.1. (b) antibody titres of four clinically negative animals (note the two sero-positive animals probably harboured occult infestations in which larvae died prior to moulting to the second instar) (modified from Boulard 1998).

an experimental intravenous injection of HA. This systemic reaction could be totally inhibited by phenylbutazone, a non-steroidal anti-inflammatory drug (Eyre *et al.*, 1980, 1981). The side effects associated with insecticides using the organophosphate (OP) compounds directed against L1 in any heavily infested cattle were not caused by OP toxicity but by HA released from the dying larvae. It is the simultaneous release of the enzymatic gut content, including HA, of the dying larvae that generates the side effects associated with OP treatments (Boulard, 1979; Boulard and Troccon, 1984). Severe haemorrhagic inflammation, with marked vascular changes, has been described around the dead larvae. These large inflammatory reactions induce oesophageal lesions and bloat when larvae are localized in the oesophageal submucosa (*H. lineatum*) or paralysis when they are in the epidural connective tissue (*H. bovis*) (Beesley, 1971). These inflammatory reactions often restricted the progress of national hypodermosis control programmes in different countries until the development of the avermectins. This new class of insecticide presented high efficacy against first instars and a progressive insecticide activity, resulting in a trickle release of the gut content and consequently a progressive slow release of PGE_2.

Despite an increased understanding of the immunopathological activity of HA, other aspects of its activity on the immune system and other biological systems of

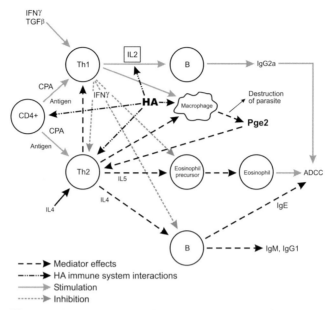

Fig. 11.2. Diagramatic presentation of interactions between hypodermin A (HA) and elements of bovine immune response. ADCC–antibody dependent cell-mediated cytotoxicity; B, 'B' cells; CD4+, Th1–T helper cells mediating humoral immunity; Th2, T helper cells mediating cellular immunity; IL, interleukins; Ig, immunoglobulin; IFNY, interferon gamma.

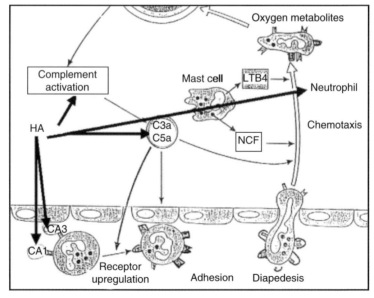

Fig. 11.3. Diagrammatic presentation of interactions between hypodermin A (HA) and elements of the bovine inflammatory system. C3a, complement cascade component; C5a, complement cascade component; LTB4, leukotriene B4.

the host remain to be explored. The activity of the other hypodermins on the immune system have not been deeply studied. Antigens other than HA could participate in the development of acquired resistance to hypodermosis reinfestations. The partial failure of vaccination using HA, HB and/or HC suggests that other molecules or mechanisms could be involved in the control of host defence. Nevertheless, HA plays an important role in the parasite evasion to the host defence mechanisms.

PART B: CUTEREBRINAE HOST–PARASITE INTERACTIONS

E. Lello, Instituto de Biociências, UNESP, Botucatu, SP, Brazil

Information available on interactions between larval cuterebrids and their hosts has been acquired primarily through studies on the economically important species, *D. hominis*. However, some research has focused on other cuterebrids whose life cycles differ from this unique species and which have much stricter host specificities.

First-instar *Cuterebra* spp. enter into the abdominal cavity by penetrating the trachea and oesophagus (Hunter and Webster, 1973b; Gingrich, 1981) and migrate through the peritoneum to the final subdermal site. This migration elicits a very limited host response but does result in occasional symptoms (wheezing, coughing) associated with movement through the trachea and oesophagus (Hunter and Webster, 1973b). This reaction has also been noted in non-normal hosts (e.g. *Lepus californicus* and *Oryctolagus cuniculus* infested with *Cuterebra lepusculi* from *Sylvilagus nuttalli* (Baird, 1983) and with *Neotoma lepida* and *Neotoma cinerea* artificially infested with *Cuterebra austeni* from *Neotoma albigula* (Baird, 1997).

Examination of warble formation around second- and third-instar *Cuterebra* spp. infesting rodents (Bennett, 1955; Payne and Cosgrove, 1966) showed development of a fibrous cyst and accompanying cellular infiltrates that provided nutrient for the rapidly growing second and third instars. Later work reiterated these observations in widely diverse hosts (Cogley, 1991; Colwell and Milton, 1998). In keeping with observations indicating unusual responses to infestation in non-normal hosts, Colwell and Milton (1998) noted that there was a much more intense infiltration of eosinophils and other granulocytes into the 'warble' formed by rabbits artificially infested with the howler monkey bot, *Cuterebra baeri*.

The first studies of immune responses to *Cuterebra buccata* showed that rabbits developed reaginic antibodies as indicated by the development of delayed hypersensitivity and Arthus reactivity (Weisbroth *et al.*, 1973). No indications were given in this study of any impact on larval survival. Acquired resistance was shown to develop in mice experimentally infested with *Cuterebra fontinella*, but was dependent on the site of entry into the host (Gingrich and Barrett, 1976). Mice that were infested via the eyes did not develop resistance to reinfestation. However, reinfestation through the nares induced substantial resistance, particularly when re-exposure was via the same route. Detailed study of rodent antibody responses and larval survival during primary and challenge infestations with *C. fontinella* showed that mice develop IgG antibodies and that an anamnestic response was evident at challenge (Pruett and Barrett, 1983). Acquired resistance was evident on the first challenge infestation, but tended to wane on subsequent challenges. There was an increase in aberrant migratory patterns during the challenge infestations, but this also declined as the number of challenges increased. Larvae were most susceptible to host immune responses during the first-instar migratory phase, as those that survived to reach the subdermal sites usually completed development. These observations suggest that protective immunity is not characteristic of host responses to cuterebrid myiasis. Data from howler monkeys infested with *C. baeri* that showed there was little reduction in prevalence or intensity of infestation associated with older age classes support these contentions (Milton, 1996).

Dermatobia hominis

Unlike other cuterebrid species that infest the host through natural orifices and migrate to subcutaneous tissues (Catts, 1982; Colwell and Milton, 1998; Leite and Williams, 1999), first-instar *D. hominis* pass through intact skin, but do not migrate within the host. They develop in a granuloma below the point of penetration. The histopathology associated with *D. hominis* has been characterized in cattle (Oliveira-Sequeira *et al.*, 1996), in naive and previously immunized rabbits (Lello *et al.*, 1999), in rats (Pereira *et al.*, 2001) and in infested and reinfested mice (Lello and de Rosis, 2003).

Similar tissue changes were observed in all hosts. Twenty-four hours after infestation, the point of larval penetration was characterized by a disruption of the epithelium and the development of an inflammatory exudate accompanied by cellular debris. The adjacent epithelium was thickened and invaded mainly by eosinophils and basophils in cattle and rabbits. In mice and rats the invading cells were predominantly neutrophils. The epithelial cells of the Malpighian layer showed cytoplasmic vacuolization and intercellular oedema, which in some areas grew to vesicles (Fig. 11.4). Normal and degranulated mast cells were present in the papillary dermis, in addition to numerous basophils in rabbits. When larvae reached the dermis, they were surrounded by numerous eosinophils and few basophils (Fig. 11.5). Adjacent to the larvae, blood vessels were congested and the connective tissue showed fibrinous exudation with infiltration by inflammatory cells.

Between 3 and 5 days PI, a fistulous tract extended from the point of entry to the resident larva. Larvae were surrounded by a thin layer of necrotic tissue, numerous eosinophils and few basophils in cattle, but in larger numbers in rabbits. In mice, neutrophils predominated. Fibroblasts proliferated around this inflammatory reaction and along the fistula. Newly formed vessels and mononuclear cells, characteristic of a granuloma, appeared.

Seven days PI, the fistulous tract appeared as a long-necked flask opening at the skin surface and containing the larva in the dilated subcutaneous portion. Most larvae had begun moulting to the second instar and exuvia were present. No inflammatory cells could be observed attached to the larval cuticle, but they were numerous on the cuticle of dead larvae and on the exuviae (Figs 11.6 and 11.7).

In experimentally infested rabbits the inflammatory reaction increased progressively, maintaining the same pattern. Immunization of rabbits, performed with

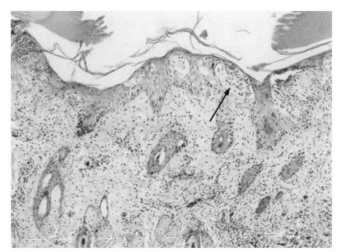

Fig. 11.4. Section of bovine skin adjacent to the penetration point of the *Dermatobia hominis* larva 24 h after infestation showing intradermal microvesicles (arrow). HE 100×. (Courtesy of T.C.G. Oliveira-Sequeira.)

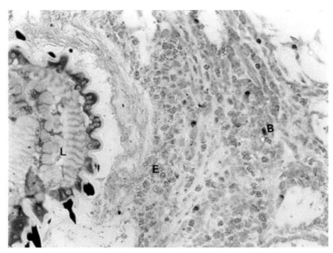

Fig. 11.5. Section of bovine skin 24 h post infestation with *Dermatobia hominis*. The inflammatory reaction around the larva (L), eosinophils (E) and basophils (B). Giemsa 400×. (Courtesy of T.C.G. Oliveira-Sequeira.)

L1, L2 or L3 larval antigens, did not alter the lesion development, regardless of the antigen used. In vaccinates the reaction occurred earlier with more intensity. However, this reinforcement of the local immune response did no damage to the larvae.

Inflammatory responses in mice during primary and challenge infestations are quite different, mainly on day 1 PI. Neutrophils are the predominant infiltrating

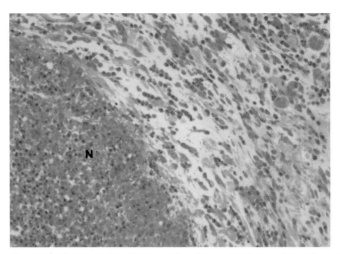

Fig. 11.6. Section of rabbit skin 7 days after *Dermatobia hominis* infestation. Clear limit between the necrosis zone (N) and the granulation tissue with activated fibroblasts and newly formed vessels. HE 600×.

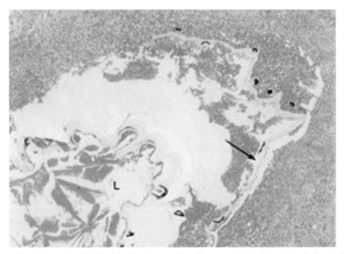

Fig. 11.7. Section of rabbit skin 7 days after *Dermatobia hominis* infestation. Note second-instar larva (L) and the first-instar exuvia (arrow). HE 100×.

cells during the primary infestation. In reinfested mice, neutrophils were more intense and eosinophils became the predominant cell.

The presence of mononuclear cells and the activation of fibroblasts were observed as early as 3 days post-reinfestation and the necrotic halo around the larva was larger when compared with the ones in the primary infestation. Despite being precocious and more intense, the pattern of the reaction evolved similarly to that in the primary infestation. The lesions caused by larvae of *D. hominis* on the epithelium of mice skin were similar to that described by Oliveira-Sequeira *et al.* (1996) in infested cattle and by Allen *et al.* (1977) in *Ixodes holocyclus*-infested cattle. According to Muller *et al.* (1983) these alterations are usual in dermatopathology, and there are no unique characteristics associated with the bot fly infestation. In primary infestations of mice the main inflammatory cells were neutrophils and mast cells, while in cattle and rabbits they were eosinophils and basophils. The absence of basophils in the skin of mice was balanced by the great number of mast cells. The neutrophil reaction to the initial infestation of mice is a non-specific response of innate immunity. The reaction to the second infestation, rich in eosinophils, is associated with acquired immunity.

Oliveira-Sequeira *et al.* (1996) noted the importance of eosinophils in *D. hominis*-infested cattle not only for destroying the parasite, but also for participating in the induction of the encapsulation of the inflammatory reaction, as they stimulate the replication of fibroblasts (Shock *et al.*, 1991) that appear activated in great numbers and in mitosis surrounding the eosinophilic infiltrate.

Immunopathology

The presence of immunoglobulins (Ig) and larval antigens in the host tissue during the first week of an experimental infestation of cattle was described by Oliveira-Sequeira *et al.* (1996). Both IgG and IgM had the same localization and distribution although IgG immunolabelling was always more intense. Twenty-four hours after infestation, Ig were observed scattered in the epithelium, on the interior of the dilated blood and lymphatic vessels and free on the interstitial tissue, externally to the cellular reaction that surrounds the larvae. The same was observed 2, 3 and 7 days after infestation. During this period progressive labelling of plasma cells was observed. Around the viable larvae the Ig were predominantly external to the cellular reaction. Nevertheless, around the dead larvae they were observed within the cellular reaction close to the larvae, on the cuticle and infiltrating the internal organs. The label was also present on the retained and free cuticle of the first instars (Fig. 11.8). Twenty-four hours after infestation, anti-L1 and anti-L2 antibodies labelled the cuticle and internal structures of the first instars. From the second day the larvae were labelled not only by these stage-specific antibodies but also by the penetration point and the fistulous tract. Seven days after infestation during the moulting process, anti-L1 and anti-L2 antibodies labelled the L2 larvae as well as the exuviae of L2 (Fig. 11.9).

Humoral response

The first experimental studies of the humoral response to *D. hominis* were performed using rabbits as a model (Mota *et al.*, 1980; Peraçoli *et al.*, 1980). Immunization with larval extract (L1+L3) induced the production of circulating antibodies, which were detected by immunodiffusion (ID). The kinetics of the humoral response evaluated by ID in rabbits after immunization, immunization and artificial infestation or artificial infestation alone indicated that in immunized animals antibodies were detected by the third week PI. The antibody levels

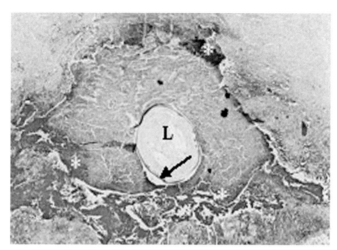

Fig. 11.8. Section of bovine skin 7 days post infestation with *Dermatobia hominis*. Note second-instar larva (L) and the first-instar exuvia (arrow). Immunolabelled with anti-L2 antibodies, 100×. (Courtesy of T.C.G. Oliveira-Sequeira.)

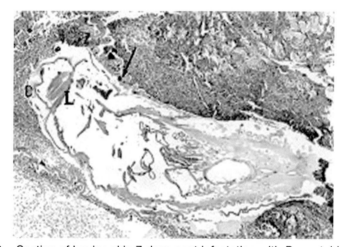

Fig. 11.9. Section of bovine skin 7 days post infestation with *Dermatobia hominis*. Note distribution of immunoglobulins around second-instar larva (L) and the first-instar exuvia (arrow). Immunolabelled with anti-L2 antibodies, 160×. (Courtesy of T.C.G. Oliveira-Sequeira.)

increased in the fourth week and decreased gradually after the eighth week. In artificially infested animals only very low antibody titres were observed 3 and 4 weeks PI. Maximum values, although relatively low, were observed at 5 and 6 weeks, but decreased rapidly thereafter. Immunized and infested rabbits had high levels of antibodies by the fourth week PI, which were maintained up to the 14th week, after which the titres decreased rapidly.

Using ELISA and ID tests, the kinetics and specificity of the antibody response to *D. hominis* antigens were analysed by Lello and Boulard (1990) in experimentally infested rabbits. Three groups of rabbits were utilized; the first was only infested, the second was immunized with second- and third-instar antigens and the third was immunized and then infested 2 weeks later. The ID tests showed the same profile as observed previously in all studied groups by Mota *et al.* (1980) and Peraçoli *et al.* (1980). In rabbits that were only infested, the antibody response against L1 antigen was detected by ELISA as early as the 5 days PI. Subsequently, antibody levels rose sharply and plateaued between 4 and 12 weeks PI. Antibodies to second- and third-instar antigens were detected on 10 days PI, but remained low between 2 and 6 weeks PI. Subsequently, antibody levels rose sharply, reaching a peak by week 14 PI (Fig. 11.10).

Antibodies against second- and third-instar antigens peaked at 4 weeks PI in rabbits immunized and infested at the same time. ELISA continued to detect high levels of antibodies throughout the experiment. Maximum response was observed against third-instar antigens and the lowest against first-instar antigens (Fig. 11.11).

Antibody levels at 1 week PI in rabbits infested 2 weeks PI were comparable with those in rabbits immunized and infested concurrently at 4 weeks PI. The response to first-instar antigen was lower (58%) than in the immunized group at the same time. During the whole infestation the antibodies against L1 and L3 antigens remained at the same level, while those against L2 antigens declined. After the third instars had left the hosts the antibody titres fell dramatically regardless of the technique or the antigen used (Fig. 11.12).

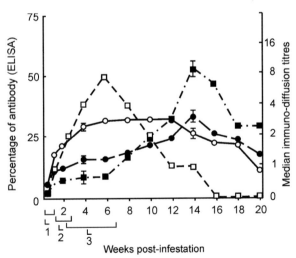

Fig. 11.10. Mean (±SD) anti-*Dermatobia* antibody level in rabbits (*n* = 5) infested with newly emerged L1. Antibody levels were measured by ELISA (left axis) using L1 (solid squares), L2 (solid circles) and L3 (open circles) antigens. Median immuno-diffusion titres (right axis) using combined L2 + L3 antigen is also shown (open squares).

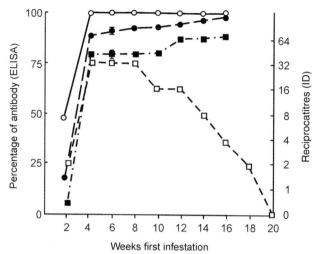

Fig. 11.11. Mean anti-*Dermatobia* antibody level in rabbits (*n* = 5) following immunization with combined L2/L3 antigen. Antibody levels were measured by ELISA (left axis) using L1 (solid squares), L2 (solid circles) and L3 (open circles) antigens. Median immuno-diffusion titres (right axis) using combined L2 + L3 antigen are also shown (open squares).

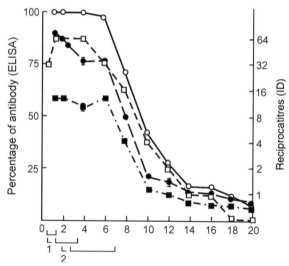

Fig. 11.12. Mean anti-*Dermatobia* antibody level in rabbits (*n* = 5) infested with newly emerged L1 following immunization with combined L2/L3 antigen. Antibody levels were measured by ELISA (left axis) using L1 (solid squares), L2 (solid circles) and L3 (open circles) antigens. Median immuno-diffusion titres (right axis) using combined L2 + L3 antigen are also shown (open squares).

Cellular response

The cell-mediated immune response of rabbits to *D. hominis* larval antigens was evaluated by a leukocyte migrations factor assay (MIF), by Lello and Peraçoli (1993) on groups of rabbits immunized with L2 and L3, immunized and infested 14 weeks later, immunized and infested 4 weeks later or just infested.

The results expressed as migration inhibition index (MII) showed that in immunized rabbits the MII was positive from weeks 4 to 14 PI, becoming negative after 18 weeks. In rabbits immunized and infested at week 14 PI, the MII values that were declining increased rapidly, reaching a peak at 18 weeks PI and then decreasing gradually. Rabbits infested 4 weeks PI, when MII values were high, maintained these levels for 10 weeks, decreasing gradually. Rabbits only infested showed the lowest levels of MII 2 weeks after infestation, which were maintained until the larvae had left the host, 6 weeks after infestation (Fig. 11.13).

The lymphocyte response induced by *D. hominis* was recently examined by de Rosis (2000) in experimentally *D. hominis*-infested mice. One group of 30 mice was prime-infested with four to five larvae each and, five killed each time, 1, 3, 5, 7, 10 and 18 days PI. Another group of 30 mice was equally infested, but the larvae were removed 5 days PI; after 4 weeks the mice were reinfested and killed five each time, 1, 3, 5, 7, 10 and 18 days after reinfestation. Skin biopsies of both groups were prepared for lymphocyte subpopulation studies, defined by monoclonal antibodies Thy+ (total T lymphocytes), CD4+ (helper T lymphocytes), CD8+ (killer T lymphocytes) and CD45 (B lymphocytes). The total of Thy+, CD4+, CD8+ and CD45+ cells in prime and reinfested mice during the 18 days of parasitism is summarized in Table 11.1.

All the four cellular subpopulations in prime and reinfested mice showed the same kinetic with similar rates of growth. The highest increase in number of cells

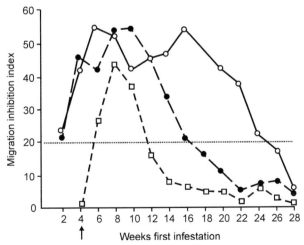

Fig. 11.13. *Dermatobia hominis* antigen-induced leucocyte migration inhibition indices for rabbits immunized (open squares), immunized and infested (open circles) and in rabbits infested without immunization (solid circles). Week of infestation is indicated by the arrow.

Table 11.1. Median number of lymphocyte types (as determined by labelling with specific monoclonal antibodies) in mice infested with *Dermatobia hominis*.

Days post-infestation	Thy+		CD4+		CD8+		CD45	
	First infestation	Challenge	First infestation	Challenge	First infestation	Challenge	First infestation	Challenge
1	13	53	6	38	3	14	2	3
3	25	61	12	40	10	18	7	29
5	128	168	95	99	25	48	42	83
7	133	183	98	115	30	52	55	87
10	140	203	104	130	31	61	56	90
18	179	216	132	143	43	70	61	96

for all lymphocyte subpopulations was observed on day 5 PI in both prime and reinfested mice. Total lymphocytes (Thy+) were counted mainly around the blood vessels near the larvae and on the subepithelium. The number of Thy+ was significantly higher in reinfested mice ($P < 0.01$). During prime infestation the number of CD4+ was significantly higher than CD8+ ($P < 0.01$). The highest value CD4+/CD8+ (3.8) was observed on the 5th day and the lowest on the 3rd day (1.2). During reinfestation the number of CD4+ was also significantly higher than CD8+. The number of CD4+ and CD8+ was significantly higher in reinfested mice than in the prime ones, for all analysed times ($P < 0.01$). The number of lymphocytes B (CD45+) labelled in prime and reinfestation increased gradually with time and was significantly superior in reinfested mice.

In *D. hominis* undergoing a primary infestation and in reinfested mice, CD4+ and CD8+ lymphocytes T are intimately related to the inflammatory reaction pattern. In both groups the inflammatory cells suggested the local participation of cytokines. By studying the cytokine profiles at the site of infestation it is possible to construct a picture of the type of immune response. For a better understanding of host–parasite relationships in this myiasis, a characterization of cytokines would be valuable.

All these data presented lead to the conclusion that despite the intense immune response of the hosts, *D. hominis* probably possesses an efficient mechanism of evasion, by secreting substances capable of annulling the effects of the immune response.

PART C: GASTEROPHILINAE HOST–PARASITE INTERACTIONS

D.D. Colwell, Agriculture and Agri-Food Canada, Canada

There is a distinct paucity of data on the interactions between larval gasterophilids and their hosts. Additionally, there is an almost complete absence of data or observations on any but the genus *Gasterophilus*. This is in contrast to the large volume of research on the control of *Gasterophilus* spp. larvae in horses and other domestic equids.

Thorough descriptions of the pathology associated with early larval migration in the oral cavity of *Gasterophilus intestinalis* have been made (Cogley *et al.*, 1982; Cogley, 1989). Also, pathological changes have been described in the stomach of horses infested with either or both *G. intestinalis* and *Gasterophilus nasalis* (Shefstad, 1978; Cogley and Cogley, 1999; Coles and Pearson, 2000). Principato (1988) described lesions in the oral cavity and throughout the length of the gut in free-ranging horses infested with up to five species (*G. intestinalis*, *G. nasalis*, *Gasterophilus pecorum*, *Gasterophilus inermis* and *Gasterophilus haemorrhoidalis*).

Oral Lesions

First instars of all *Gasterophilus* species, except one, penetrate the host epidermis and migrate within the oral cavity. Several species penetrate at or near the site of oviposition while others, because of the mode of hatching, penetrate the oral mucosa. The one exception is *G. nasalis*, which migrates across the surface to reach the interdental spaces.

A thorough study by Cogley (1989) showed that first-instar *G. intestinalis* migrate no more than 1.5 mm below the lingual surface, leaving distinctive tunnels that appear as tracks on gross examination. Larvae apparently excavate 'air holes' that connect the tunnel to the surface. Tunnels were restricted to the stratum spinosum where adjacent cells were heavily damaged. Haemorrhage into the tunnels was evident.

Presence of first and second instars in the interdental spaces altered the gingival epithelium, often resulting in a separation between the tooth and the adjacent tissue. Anterior portions of the larvae were embedded deep in the submucosa of the periodontal pocket. Ulceration of the gingival was evident with a distinct inflammatory response and accompanying haemorrhage. Cellular infiltrate included large numbers of lymphocytes and plasma cells as well as an increased number of eosinophils.

Lesions in the mucosa of cheeks and soft palate produced by migration of *G. pecorum* and *G. inermis* have a superficial resemblance to those described for *G. intestinalis*. Presumably, similar histopathological changes take place.

Gastric and Intestinal Lesions

Second and third instars are found firmly attached to the mucosa of the stomach or intestine. Species generally exhibit site specificity (see Chapter 4). Attachment of the larvae results in the formation of pit-like erosions in the mucosa of the stomach, duodenum and rectum (Principato, 1988; Cogley and Cogley, 1999). Histologically, the lesions are characterized as a chronic ulcerative gastritis or duodenitis. Villous atrophy marks the site of lesion formation. Also, the mucosa and the lamina propria are thickened, with evidence of an intense fibrosis. Infiltration of eosinophils into the lesion has been noted (Sequeira *et al.*, 2001).

The general consensus is that despite the presence of significant lesions in the oral cavity, the stomach and the intestine, the infestations have little impact on the host. There are occasional reports of perforation of the stomach or intestine resulting in peritonitis (Rooney, 1964; Dart *et al.*, 1987). Massive infestations can lead to blockage of the gastrointestinal tract, but this is also rare.

Antigens and Host Immune Responses

Host immune responses to migration and development of the larvae have received scant attention. One early study has described the digestive enzymes of *G. intestinalis* (Tatchell, 1958). A variety of proteinases and other classes of enzymes were reported from the midgut. They appear to be unrelated to those of other oestrids as Boulard *et al.* (1996a) noted there was no cross-reactivity with antigens of *G. intestinalis* in animals producing antibodies to several oestrids (i.e. *H. bovis*, *H. diana*, *H. tarandi*, *O. ovis*). Crude midgut preparations had anticoagulant properties, but of particular interest was the description of an 'anti-coagulin' isolated from the salivary glands. This intriguing observation has not been pursued, particularly with regard to potential influence on the development of host immune responses. Further characterization of the major immunogens has not been reported; not even by simple SDS–PAGE and Western blot. Development of antibodies is only discussed in the context of sero-diagnosis (Escartin-Peña and Bautista-Garfias, 1993).

Despite having the pathology well characterized there has been no attempt to differentiate between responses to primary infestation and responses to subsequent infestations. An absence of techniques for establishing quantifiable artificial infestations may be a limiting factor in the development of these data.

PART D: OESTRINAE HOST–PARASITE INTERACTIONS

P. Dorchies, G. Tabouret, H. Hoste and P. Jacquiet, Ecole Nationale Vétérinaire de Toulouse, France

Introduction

Oestrosis is a widespread myiasis of small ruminants that severely impairs health and productivity. Many reports provide prevalence data for various regions including both temperate and tropical areas (Pandey and Ouhelli, 1984; Kilani *et al.*, 1986; Belem and Rouille, 1988; Pangui *et al.*, 1988; Yilma and Dorchies, 1991; Hall and Wall, 1995). In hot and dry climates, the nasal discharge caused by oestrosis can become caked with dust, making breathing very difficult. As a consequence, affected sheep and goats breathe through the mouth, which interferes with grazing and rumination. Subsequently, this local nasosinusal infection induces signs of generalized disease, including emaciation, which may sometimes result in death. Moreover, *O. ovis* has sometimes been associated with lung abscesses and interstitial pneumonia, which have been related to inhalation of pus from the sinus (Dorchies *et al.*, 1993).

Despite its high prevalence and the severity of infection, many breeders and veterinary practitioners still remain unaware of the significance of this parasitic disease. Very little research has focused on relationships between larvae and hosts in this important disease.

The pathogenicity of *O. ovis* infection has largely been thought to be the result of mechanical trauma induced by larval hooks and spines. However, recent research has clearly demonstrated that proteolytic activity of excretory/secretory products (ESPs) plus the effect of hypersensitivity phenomena play a critical role in the pathogenesis of resident larvae.

Clinical Manifestations and Lesions

O. ovis is one of the most ubiquitous parasites of both the Old and New World: it can thrive in a wide range of environmental temperatures (Biggs *et al.*, 1998) and hosts. Adult fly activity, numbers of generations per year and regional variations in strike are closely related to the local climate. Clinical manifestations of oestrosis may include long-lasting rhinitis and sinusitis persisting for months (in Mediterranean countries) or the symptoms may have a clear-cut seasonal pattern (areas where climate is temperate or with a hot and dry season as in the Sahel) (Tabouret *et al.*, 2001c). In the latter case, rhinitis is observed during the fly season and sinusitis appears later when the hypobiotic first instars have resumed their development. Few clinical signs are evident during the cold or hot/dry season when larvae are hypobiotic. Subsequently, when larvae resume their development, very severe sinusitis is followed, a few weeks later, by manifestations of fly strike (see below).

Fly Strike

After adult emergence, gravid female flies search for sheep or goats as targets for deposition of their larvae. Fly activity affects sheep behaviour and shepherds observe that their sheep get nervous and gather close together, keeping their noses deep inside the fleece of their neighbours or close to the ground during displacements (see Figs 10.7 and 14.9). Early in the morning and late in the evening, when flies are not active, the sheep resume grazing and are spread all over the pasture. In Ecuador, one can see sheep wearing a leather mask to avoid the larval deposition by flies when a flock is moving from one area to another. Goats seem less sensitive to fly strike perhaps as a result of their browsing habits (Hoste *et al.*, 2001). During this period there is little or no congestion of the nasal septum and turbinates and only a small amount of mucous is observed.

Rhinitis

In the first few weeks following larviposition, nasal discharge and sneezing become more evident and frequent. Sheep are agitated and nasal discharge, which is first serous, becomes sero-mucous, muco-purulent and eventually, in the most severe cases, purulent, and may occasionally be tinged with blood. The amount of nasal discharge is not related to the number of larvae but appears to be related to individual susceptibility and also to interactions with bacteria. Sometimes, severe orf (a pox virus disease) lesions are reactivated by *O. ovis* infection. Severe reactions and the accompanying clinical symptoms may lead to emaciation and death. In addition, these symptoms may affect olfaction in rams, thereby reducing the effectiveness of oestrus detection, which will lead to reduced conception rates (Watson and Radford, 1960). Heavily infested sheep may exhibit neurological symptoms including ataxia, vertigo, nystagmus and amaurosis as well as epistaxis.

As the period of fly activity continues, sheep are repeatedly infested and so antibody levels increase. Because clinical signs and annoyance of sheep are at maximum intensity, breeders usually treat their animals at this time.

Local changes in the mucosae of the upper respiratory tract are independent of the number of larvae present. The changes are characterized by a bright red mucosae with a discrete oedema, but their intensity cannot be correlated to the number of larvae. Usually during this period, similar numbers of each larval instar are found. Furthermore, according to Biggs (1998) any number of larvae above ten is considered to be potentially harmful.

Most of the lesions are localized in the sinus and ethmoidal mucosa. In naive sheep, the epithelial cells of the nasosinus cavities are cylindrical, ciliated and pseudostratified. After natural or artificial infection, hyperplasia and metaplasia develop and the mucociliairy film is abraded. Many cells are positively marked for Ki 67 epitope, indicating a strong cellular proliferation (Nguyen Van Khanh *et al.*, 1998).

Transmission electronic microscopy (TEM) shows the damage associated with the presence of second and third instars (Fig. 11.14). The ultrastructural changes present a gradation, depending on the anatomical sites, the most severe lesions occurring in the sinus. The stratified epithelium is disorganized and the intercellular spaces are enlarged, with epithelial disjunctions. Some of the cells have a rounded shape, presenting signs of cellular degeneration. These ultrastructural changes are likely the result of a combination of mechanical damage associated with the effects of secreted proteases from the larvae. It is also likely that these changes favour the diffusion of antigenic secretory/excretory products through the mucosa to come in close contact with the locally recruited immune cells.

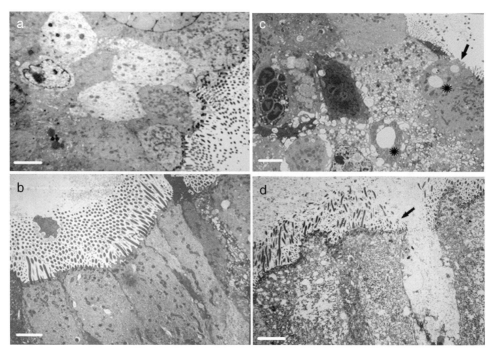

Fig. 11.14. Transmission electron micrographs of sheep nasal mucosal epithelium: (a) nasal septum of uninfested sheep (bar = 5 μm) (b) ethmoid region of uninfested sheep (bar = 5 μm). Note that the epithelium is pseudostratified and the junctions between cells are tight. The cells have a cylindrical shape, with a dense cytoplasm and are apically ciliated. (c) nasal septum of sheep infested with third instar *Oestrus ovis*. (bar = 5 μm) (d) ethmoid region of sheep infested with third instar *Oestrus ovis* (bar = 5 μm). Note the cells appear vacuolated (asterisks) and there is a reduction of the apical ciliature (both indications of degeneration). However, the general organization of epithelium is conserved.

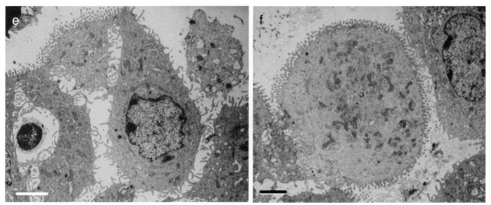

Fig. 11.14. *Continued.* (e) turbinate region of sheep infested with third instar *Oestrus ovis* (bar = 5 μm) (f) sinus region of sheep infested with third instar *Oestrus ovis* (bar = 5 μm). Note the general structure of epithelium is modified. In most cases, the extracellular spaces are enlarged, although the tight junctions remain present. In some cases (f) epithelial cells presented a rounded shape and were totally dissociated from the rest of the tissue and appeared to be dedifferentiating.

Sinusitis

At the end of the fly activity period (in late summer in temperate countries or at the end of the wet season in tropical areas), a high proportion of sheep become infested, many with large numbers of larvae that stop their development as hypobiotic larvae. These hypobiotic first instars will overwinter (or oversummer) and when they resume their development, a sharp increase in nasal discharges and sneezing is usually observed. In hypobiotic larvae, onset of development appears asynchronous as the number of L1 > L2 > L3. As the larvae commence development the nasal discharge is muco-purulent or purulent and the amount discharged increases. The nostrils are not usually directly occluded by the discharge but hay, straw, litter or dust may adhere to the discharge, causing breathing problems. This is particularly evident when sheep are moving. This period is debilitating for sheep, although less severe for goats. Animals lean against walls, trees, racks, etc. and false gid can occur. Expulsion of live third instars via the nostrils can often be observed. After this, sheep breathe more easily for a few days until the next larval expulsion.

Lesions observed during this period are similar to those of the rhinitis period but are more severe. The congestion of mucosae is less significant than during rhinitis. Large amounts of pus and abscesses are commonly found in the sinuses. In some breeds of sheep, neoplasic tumours (adenocarcinoma of the pituitary) 4–5 cm wide can be found. The prevalence of these neoplasic processes is close to 5% in the Lacaune ewes, the local breed from the Roquefort area (Bergeaud *et al.*, 1994).

Lung Lesions

In many cases, oestrosis is worsened by complications associated with infectious bronchopneumonia and pleuropneumonia or pasteurellosis with hyperthermia and coughing. Lung abscesses are also frequent. It may be assumed that they are related to pyogenic focus in the nasosinusal area. In some cases interstitial pneumonia is the most common lesion with interstitial emphysema, pleural adhesions and atelectasis (Dorchies *et al.*, 1993). These atypic lesions associated with oestrosis may be linked to the visna maedi virus for which pathological expression is dependent on a continuous and non-specific antigenic stimulation of the host, as previously described by Dawson (1987).

Similar lesions have been observed in donkeys infected by larvae of *Rhinoestrus usbekistanicus* (Kaboret *et al.*, 1997). The pathology is probably caused by a permanent antigenic stimulation during infection. In both myiases considerable numbers of eosinophils and mast cells are accumulated in the lung parenchyma mainly in the peri-bronchial region. In the absence of any other lung parasite and any other cause of allergic pneumonia, it may be presumed that aspirated larval antigens induce pulmonary sensitization. It has recently been suggested that mast cells could induce lung fibrosis: histamine and serotonin stimulate the growth factor for fibroblasts (FGF) *in vitro* and *in vivo*. *In vitro* co-cultivation of fibroblasts and mast cells showed in the first instance maturation of mast cells, which in return, stimulates fibroblastic growth and collagen synthesis (Tunon De Lara *et al.*, 1996). The same hypothesis could possibly explain the development of interstitial pneumonia in oestrid myiasis both in sheep and in donkeys.

Larval Enzymes

The presence of proteases in the ESPs of *O. ovis* larvae has been clearly identified (Tabouret *et al.*, 2003a). These proteases appear to originate from the gut and are exported on the nasal or the sinusal mucosa. They are mainly trypsin-like serine proteases and participate in extracorporeal pre-digestion of proteins for larval nutrition. No proteolytic activity has been demonstrated in the salivary glands but these organs are the source of the most immunogenic antigens (Innocenti *et al.*, 1995; Tabouret *et al.*, 2001a).

Molecular weights of *O. ovis* larval gut proteases (69, 45, 36, 29, 24 and 20 kDa) are similar to those previously characterized from other agents of myiasis, such as *Chrysomya bezziana* (26, 28, 35, 50–70 kDa) (Muharsini *et al.*, 2000), *Lucilia cuprina* (20, 28, 36, 45 and 54 kDa) (Sandeman *et al.*, 1995) and *H. lineatum* (HC: 26 kDa, Boulard and Garrone, 1978; HA: 32 kDa, Pruett, 1993a). The optimal pH for activity of *O. ovis* proteases is neutral or alkaline and close to the optimal pH values for serine proteases of other Diptera. Inhibitors such as tosyl-L-lysine chloromethyl ketone (TLCK) reduced the activity of larval ESP by 88.1%, indicating tryptin-like activity (Tabouret *et al.*, 2003a). Acid–alkaline phosphatases and oxidase activities have also been detected in ESP.

There are no qualitative changes in the protease profile of ESP among the three instars, indicating that production continues throughout larval development. However, the amount of each protein produced, *in vitro*, by each larval stage is different. The increasing bulk of proteases released on the mucous membrane by L2 and L3 suggests growing nutrient requirements in keeping with the large increase in size as the larvae acquire reserves for the non-parasitic stages that do not feed. Curves of larval growth obtained by Cepeda-Palacios *et al.* (1999) support this theory.

ESP released on the mucosa is able to degrade the components of the extracellular matrix and the lamina of epithelium. Type I (extracellular matrix) and type IV (basal blade) collagens and mucins are degraded into small fragments more easily ingestible by the larvae. The large amount of mucous in the nasal cavities represents a food resource quickly accessible for the larvae. Moreover, digestion of mucin induces a modification of the physicochemical properties of the mucous gel. It may be supposed that L1 can be trapped in mucous covering protecting the epithelium and be exposed to asphyxiation by obstruction of their stigmatic plates or expelled by an immune exclusion (self-cure) as described for gastrointestinal strongyles. Mucolytic activity allows larvae to move more freely in mucous without remaining trapped, avoiding close contact with antibodies and also limiting larval damage.

Disturbance of tissue homeostasis by the larvae and their ESP induces an inflammatory process that favours plasma protein leakage as well as increases in water and electrolyte secretions. This pathophysiological feature is similar to that seen in the gut during gastrointestinal strongylosis. The secreted proteases would thus create an environment rich in varied nutrients, comparable with the organic exudate developed by larval *L. cuprina*. However, leakage of plasma proteins should be a source of plasmatic inhibitors of proteases, such as the α2 macroglobulin, anti-thrombin III or the α1 protease inhibitor, a specific inhibitor of serine proteases. It has been shown *in vitro* that combination of these inhibitors blocks *L. cuprina* proteases and delays larval growth (Bowles *et al.*, 1988). Secretion of α2 macroglobulin, which inhibits collagenases, is maximal during the acute phase of the inflammation. It is thus probable that such inhibitors are released on the nasal mucous membrane as a result of the inflammation induced by the *O. ovis* larvae.

Immunity

Recently, host immune responses to oestrosis have received increasing attention (Dorchies and Yilma, 1996; Otranto, 2001). Research indicates some marked homologies with cellular and humoral immune responses involved against gastrointestinal nematodes, suggesting a Th2 orientation (Urban *et al.*, 1996; Schallig, 2000). This hypothesis is supported by many observations on the larval population regulation and of cellular and humoral reactions.

Regulation of Larval Population

Existence of mechanisms regulating the association between the larval population and the host's immune status are suggested by field observations and as a result of experimental infections in sheep and goats. The mean number of *O. ovis* larvae in adult sheep is usually less than 15 (Horak, 1977, in South Africa; Belem and Rouille, 1988, in Burkina-Faso; Pangui *et al.*, 1988, in Senegal; Yilma and Dorchies, 1991, in France). Kilani *et al.* (1986) and Yilma (1992) unequivocally demonstrated that the larval burdens were higher in lambs than in adult sheep under field conditions where hosts are equally exposed to natural infections. Field surveys show that despite the similarity in prevalence rates of *O. ovis* infections in sheep and goats, the larval burden was found to be lower in goats than in sheep (Tesfaye, 1993). From a survey in slaughterhouses in the south of France, 28.4% of 672 goats were infected with a mean larval burden of 5.3 per goat whereas 43.4% of 631 sheep were infected with a mean larval burden of 10.9 (Dorchies *et al.*, 2000).

An equilibrium between larval populations and the host immune response was clearly demonstrated by Marchenko and Marchenko (1989). On average, 62.9% of the challenge larvae from experimental infections survived in lambs given an immunosuppressive treatment (chlorambucil 1 mg/kg daily). In the same experiment, only 26–36% of larvae survived in untreated control lambs. The last establishment rates are in agreement with data from Yilma and Dorchies (1993). The larval survival rates found at necropsy after 1, 2, 3 and 4 months of a unique experimental infection were 23.5%, 34.5%, 28% and 25.5%, respectively. This limited rate of larval survival after experimental infection may be related to injury of larvae during collection at an abattoir from naturally infected sheep.

Repeated artificial infections of lambs with first instar larvae have been used to simulate natural infections in order to investigate the host immune reactions. In these situations, the mean survival rates of 10–12% suggested the development of some immunity (Yilma and Dorchies, 1993). In Baja California Sur (Mexico), Frugère *et al.* (2000) performed two artificial infections with larvae recovered from gravid flies caught on goats. Establishment rates were 39%: the reason for this high level is perhaps linked with the origin of larvae recovered in gravid females instead of from the heads of sacrificed sheep.

Marchenko and Marchenko (1989) suggested the existence of an immune regulatory mechanism governing the larval population. In lambs preinjected three times at weekly intervals with homogenates of the three larval instars, they obtained a rate of development of 0.4% after challenge with 40 L1. Frugère *et al.* (2000), performing immunizations of lambs with excretory secretory products from *O. ovis* L3, obtained an inhibitory effect on larval growth but did not observe any reduction in larval establishment.

Differences between sheep and goats may be the result of several factors. First, goats, being more sensitive than sheep to the activity of larvipositioning flies, could avoid more efficiently the larvae laid by adult flies through some behaviours. Second, *O. ovis* larvae from sheep are perhaps poorly adapted to goats. After repeated artificial infections with larvae from sheep origin, the survival rates were, respectively, 12.7% (Yilma and Dorchies, 1993) in lambs and 0.9% in kids (Duranton *et al.*, 1996).

Components of the Immune Response

Cellular reactions

Histopathological data suggest that an immediate hypersensitivity phenomenon (type I) is involved in the pathogenesis of oestrosis. The changes resemble those previously described for *L. cuprina* infestations (MacDiarmid *et al.*, 1995). For oestrosis, the first study was performed by Jagannah *et al.* (1989a), who described the presence of eosinophils in the upper respiratory mucosa of *O. ovis*-infected sheep. Recruitment of inflammatory cells (eosinophils, mast cells, globule leucocytes) has been further studied in artificially (Dolbois and Dorchies, 1993) or naturally (Nguyen Van Khanh *et al.*, 1996, 1999a,b) infected animals. There are significant differences between infected and parasite-free sheep both in the cell numbers and in their anatomical (septum, turbinates and sinus) and radial distribution (mucous membrane, interglandular chorion or submucosa). The numbers of serous and mucous mast cells in lambs receiving a single infection were twice those in the uninfected group. In animals subjected to multiple artificial infections the population of serous mast cells increased 11-fold and that of mucous mast cells by 5- to 7-fold compared with control groups. Similarly, the eosinophil counts were 17, 29 and 58 times greater in nasal septum, turbinates and sinus in repeatedly infected groups compared with the control values. These findings suggest that a single infection is tolerated by the host while a massive cellular recruitment characterizes the cellular responses after multiple exposures (Dolbois and Dorchies, 1993). Significant differences between mast cells and eosinophils in naturally infected and parasite-free sheep have been observed in numbers and distribution within the mucous membrane, interglandular chorion or submucosa (Nguyen Van Khanh *et al.*, 1996). Similar observations have been made in camels infected with *Cephalopina titillator* (Viatteau *et al.*, 1999) and in donkeys infected by *Rhinoestrus usbekistanikus* (Kaboret *et al.*, 1997). In naturally infected sheep, the mean number of mast cells is only twice that present in parasite-free animals. This result is remarkably different from that obtained in multiple artificial infections and close to the result with a single artificial infection. Due to histological position of eosinophils and mast cells, it can be presumed that these cells are responsible for limiting parasite larval populations and in sustaining the hypersensitivity phenomenon at the site of tissue damage during infection. *In vitro*, we observed that eosinophils could kill L1 following their degranulation (Duranton *et al.*, 1999a). For mast cells, no observation confirms this hypothesis but the degranulation may be very aggressive for larvae. Sheep mast cell protease (SMCP) has been identified in mucosae of infected lambs, confirming the activity of local mast cells (J. Huntley, unpublished data). An important transepithelial migration of eosinophils has been assessed, indicating that they are attracted in the lumen of the nasal and sinusal cavities to come into contact with the larvae.

More recently, using immunocytochemical methods, it has been shown that lymphocytes, both B and T, in addition to phagocytic mononuclear cells, are numerous in mucosae of the upper respiratory tract of animals infected with *O. ovis* larvae compared with control ones. Cell numbers increase gradually from turbinates to sinus, suggesting that T (CD3+) and B (CD20+) lymphocytes, phagocytic mononuclear cells (CD68+), eosinophils, mast cells and globule leucocytes are mainly recruited where larval moults happen and therefore where L2 and L3 are present (Tabouret *et al.*, 2003a). This cellular profile suggests a Th2-type cytokine production by ovine T lymphocytes but clear evidence to support this hypothesis is still missing in *O. ovis* infection. The recruitment of these cells and the presence of many macrophages suggest that antigenic presentation to T lymphocytes and cellular cooperation are intense. Interestingly, a similar reaction is observed in the skin during *L. cuprina* infection (Bowles *et al.*, 1992). In the same way, a huge mucosal recruitment of eosinophils, mucous mast cells and globule leucocytes is observed during development of resistance to gastrointestinal trichostrongyles (Stevenson *et al.*, 1994; Miller, 1996; Balic *et al.*, 2000; Claerebout and Vercruysse, 2000; Meeusen and Balic, 2000).

Antibodies

ELISA studies performed with serum and mucous collected from nasal cavities demonstrated that infected lambs produced local and systemic IgG and secretory IgA antibodies. Sero-conversion occurs 3 or 4 weeks after artificial infection. Antibodies are transmitted by ewes to their offspring by colostrum. There is no relation between clinical status and blood antibody titres. It is also impossible to demonstrate a clear-cut relationship between the number of parasites and the antibody level.

Based on this antibody response, some serological tests have been described: inhibition of haemagglutination test (Bautista-Garfias *et al.*, 1988; Jagannah *et al.*, 1989a), intradermal reaction (Jagannah *et al.*, 1989a), double diffusion test (Ilchmann and Hiepe, 1985; Bautista-Garfias *et al.*, 1988), ELISA test (Haralampidis, 1987; Yilma, 1992) and Dot ELISA test (Duranton *et al.*, 1995).

For all these diagnostic tests, the first important step is the selection of a reliable antigen. L2 crude antigens are generally used but Tabouret *et al.* (2001a) demonstrated that a 28-kDa protein of the salivary glands' contents could be a more valuable antigen for an ELISA test. Moreover, Goddard *et al.* (1999) used a L1 somatic crude antigen with high sensitivity and specificity.

Immune evasion

An immune evasion by *O. ovis* larvae is suggested by several observations. First, a reduction in lymphocyte reactivity due to somatic larval antigens (Dolbois and Dorchies, 1993) and also to ESP (Tabouret *et al.*, 2001a) has been observed by studies on *in vitro* proliferative responses to mitogens. Following a first infection, an efficient immune reaction is quickly established in relation with an increased lymphocyte proliferation to the two mitogens (phytohaemagglutinin and crude antigen of L1). However, after several reinfections, the lymphocyte proliferative response against crude antigen of L1 decreases whereas no change is observed in the response to phytohaemagglutinin. This suggests a developing immunosuppression possibly involving only the T lymphocytes as the antibody production does not seem impaired.

Second, immunomodulating properties from ESP are also suspected since, like pepsin, they cleave ovine IgG. Only the heavy chains are degraded forming F(ab)′

2 and pFc' fragments. The denaturation of IgG would impair the mechanism of antibody-dependent cell cytotoxicity with eosinophils but this mechanism has not yet been studied for *O. ovis* infection.

Pro-inflammatory properties

The *in vitro* reactivity of monocyte-derived macrophages (MDM) from *O. ovis* artificially infected lambs and kids was determined by measuring the production of nitric oxide (NO) during the course of infestation. In both species, crude antigenic preparations obtained from *O. ovis* first-instar larvae (L1) were found to significantly ($P < 0.01$) inhibit NO production, whereas second-instar extract stimulated it. Furthermore, this NO production by MDM decreased during infestation and was related to blood eosinophilia (Duranton *et al.*, 1999). Moreover, Tabouret *et al.* (2001a) have shown, *in vitro*, that ESP could stimulate NO production by tumoral murine macrophage RAW 264.7. This upregulation is time- and dose-dependent.

Hypersensitive reactions

Type I hypersensitivity has been recently confirmed by intradermal tests with ESP. P. Jacquiet (2003, unpublished data) observed a quick local reaction in less than 30 min after injection. Three hours later, the aspect was modified because the lesion was more congestive and red. In contrast, in control non-infected sheep, there was no reaction at 30 min but the same congestion has been observed at 3 h. The latter phenomenon suggests proinflammatory properties of ESP as earlier observed *in vitro* with ovine and murine macrophages.

Conclusion

In contrast with other parasites, which in host tissue are in close contact with the host immune response and have to counteract this reaction, the larvae of *O. ovis* inhabit the nasal cavities and do not have direct access to nutritive resources except mucin. Several of our observations suggest that larvae produce proinflammatory substances to gain access to plasmatic nutritional products.

Acknowledgements

Many papers reported in Part D of this chapter have been written by the authors in UMR INRA/DGER 1225 with the financial support of Institut National de la Recherche Agronomique and Direction Générale de l'Enseignement et de la Recherche of the French Ministry of Agriculture. The skilful help of the technical team is acknowledged and Mrs C. Duranton-Grisez, Mrs F. Prévot, Mrs M.L. Amigo and Mr J.P. Bergeaud are thanked in particular. Without their daily 'addiction' to *O. ovis*, nothing could and would have been done. We thank them for spending a lot of time and energy on such a marvellous and fascinating parasite.

12 Oestrid Myiasis of Humans

J.R. ANDERSON

University of California, USA

Oestrid myiasis in humans is relatively common, but usually uneventful. Reviews by James (1947), Zumpt (1965) and Hall and Wall (1995) reveal that it has been reported wherever oestrids occur. The only oestrid for which humans serve as one of the natural hosts is *Dermatobia hominis*; however, some species of oestrids in all subfamilies have been implicated in cases of accidental myiasis. Among the species causing accidental myiasis of humans (Table 12.1), those parasitizing domesticated animals (e.g. *Oestrus ovis* and *Hypoderma* and *Gasterophilus* species) are most commonly found involved, undoubtedly because of the frequent contact between humans and the obligatory vertebrate host. According to reports in the literature, humans may be infested with oestrid larvae involved in accidental myiasis by a species directly larvipositing or ovipositing on the victim (e.g. *O. ovis* or *Hypoderma tarandi*) or by a person inadvertently contacting eggs or larvae (*Hypoderma* and *Cuterebra* species). The latter method of transmission almost always results in cutaneous or furuncular infestations, but larvae may also be transmitted more rarely via fingers to the eyes or mouth. Ophthalmomyiasis has been caused by oestrid species in all subfamilies.

Frequency of Human Infestations

Subfamily Cuterebrinae

Because humans serve as one of the many natural hosts for *D. hominis*, human infestations are common throughout its range of distribution in Central and South America. Human infestation with larvae of *D. hominis* has been recognized for several centuries (Guimarães and Papavero, 1999). Infestations with this parasite are so common throughout its native range that most human cases are not reported. Among indigenous people and livestock workers, for example, *D. hominis* infestation is quickly recognized, and the first instars then immediately extracted. Resident infestations show a high prevalence of *D. hominis* larvae in children

Table 12.1. Oestrid species involved in accidental infestations of humans.

Subfamily	Species
Oestrinae	*Oestrus ovis*
	Rhinoestrus purpureus
	Cephenemyia ulrichii
	Gedoelstia hässleri
	Pharyngomyia picta
Hypodermatinae	*Hypoderma bovis*
	Hypoderma lineatum
	Hypoderma tarandi
	Hypoderma diana
Gasterophilinae	*Gasterophilus nasalis*
	Gasterophilus pecorum
	Gasterophilus inermis
	Gasterophilus intestinalis
	Gasterophilus haemorrhoidalis
Cuterebrinae	*Cuterebra* species
	Metacuterebra (=Alouattamyia) baeri

(Sancho, 1988). Throughout its geographic range only human cases involving the most severe pathology (e.g. blindness, death) are usually reported.

Most of the many literature reports of *D. hominis* infestations involve foreign visitors whose infestations were discovered after they returned to their home countries. The bibliography of Guimarães and Papavero (1999) lists references to *D. hominis* infestations in visitors from 18 countries, including 11 from Europe. Increased international travel has increased the number of reports of foreign visitors infested with *D. hominis* to the point where the literature contains an excessive number of redundant case reports. Nevertheless, even in foreign countries, *D. hominis* probably is much more commonly found in people than reported in the literature. For example, many colleagues, like myself, have referred students returning from tropical studies courses to campus health services for removal of larvae without publishing case reports.

In addition to parasitizing humans, *D. hominis* develops in a wide range of mammalian hosts, including native animals such as monkeys, puma, jaguar, agouti and armadillos, as well as in introduced pets (cats, dogs) and livestock (cattle, sheep, goats, equines), and in a few large bird species (Guimarães and Papavero, 1999).

Human infestations with other species of Cuterebrinae have resulted from rare accidental contacts. There are less than 75 literature reports involving species of *Cuterebra*, and only two involving *Metacuterebra (=Alouattamyia) baeri*, whose natural host is the howler monkey. Most cases of human myiasis associated with *Cuterebra* species in North America have occurred in eastern North America (Baird *et al.*, 1989).

Subfamily Oestrinae

Among larvipositing species, *O. ovis* is by far the most common oestrid associated with accidental myiasis in humans (e.g. Krümmel and Brauns, 1956;

Pampiglione, 1958; Zumpt, 1963; Pampiglione *et al.*, 1997). The medical and biological literature contains thousands of case reports from around the world. Most
reports involve people who have been closely associated with sheep and goat hus-
bandry. For cases seen in an eye clinic, the annual incidence of *O. ovis* eye infestations in eastern Libya was estimated at 10/100,000 (Dar *et al.*, 1930). Like the
case for *D. hominis*, most infestations of humans with *O. ovis* probably go unreported.

Although rare in comparison with *O. ovis* infestations, the next most common
member of this subfamily involved in human myiasis is *Rhinoestrus purpureus*, with
most reports associated with people who work with horses in Europe, Russia and
central Asia (e.g. Portchinskii, 1915; Krümmel and Brauns, 1956; Zumpt, 1965;
Rastagaev, 1980, 1982). For other species of Oestrinae there are only one or two
reports of *Pharyngomia picta*, *Cephenemyia ulrichii* and *Gedoelstia hässleri* infesting
humans (Zumpt, 1962b; Bisley, 1972; Mikkola *et al.*, 1982; Sauter and Huber,
1988). Because these and most other species of Oestrinae are seldom-encountered
parasites of wildlife, it is not surprising that attacks on humans are so rare. Because
of widespread human contact with semi-domesticated reindeer and camel herds it
is somewhat surprising that there are no reports of human ophthalmomyiasis associated with *Cephenemyia trompe* or *Cephalopina titillator*. *C. trompe*, for example, has
been recovered from the eyes of reindeer (Rehbinder, 1970, 1977) and the nasal
cavity of dogs (Strom, 1990).

Subfamily Hypodermatinae

James (1947) and Zumpt (1965) noted that the literature contained numerous
reports of human infestation by larvae of *Hypoderma bovis* and *Hypoderma lineatum*. However such reports are not nearly as numerous as those for *O. ovis*, probably because far fewer people are closely associated with cattle than with
sheep/goats, and because of low numbers of cattle grubs due to widespread
eradication efforts during the last quarter century (Chapter 12). The number of
case reports for *Hypoderma*-associated myiasis has declined sharply in modern
times. Zumpt (1965) stated that most larval infestations of humans involved
H. lineatum and that this species was also implicated in most cases of ophthalmomyiasis. However, in 16 publications cited by Scharff (1950) in reviewing
human infestations, larvae of *H. bovis* were listed more frequently than those of
H. lineatum. All of the many cases investigated by Schöyen (1886) in Norway
involved *H. bovis* and more recent larval, and specific serological, identifications
also revealed that *H. bovis* was more commonly involved in human infestations
and in ophthalmomyiasis (e.g. Doby and Deunff, 1982; Kearney *et al.*, 1991).
However, as it has not been experimentally determined whether *H. bovis* has a
greater propensity to infest humans than *H. lineatum*, reports of larvae obtained
from humans may simply reflect their relative abundance in different geographic areas.

Human association with semi-domesticated reindeer undoubtedly accounts
for *H. tarandi* ranking third in frequency of reported cases of hyperdermosis.
Human myiasis associated with other *Hypoderma* species is rare. James (1947),
Sicart *et al.* (1960) and Fidler (1987) each reported a case of human infestation

with *Hypoderma diana*. This species also rarely infests horses and sheep (Dempsey, 1983; Ross, 1983; Minar, 1987; Schumann *et al.*, 1988; Hendrickx *et al.*, 1989; Vyslouzil, 1989).

Subfamily Gasterophilinae

Literature reports reveal that human infestations with various horse bot flies are far less common than cattle warble fly infestations. Infestations with *Gasterophilus intestinalis* and *Gasterophilus pecorum* seem to be most common, with those for *Gasterophilus haemorrhoidalis*, *Gasterophilus nigricornis* and *Gasterophilus inermis* rarely reported. Dove (1937) reviewed early cases in the USA. James (1947) pointed out that early records of *Gasterophilus nasalis* infesting humans were doubtful, and subsequent research by Chereshnev (1954, in Zumpt, 1965) and Rastegaev (1978) revealed that this species was not able to penetrate human skin even when moistened.

Methods of Infestation

Subfamily Cuterebrinae

As discussed in Chapter 10, *D. hominis* infests humans, and all other vertebrate hosts, via active transmission that involves a porter arthropod species on which females deposit small numbers of eggs. First-instar larvae are stimulated to hatch after the porter species lands and remains briefly on the eventual vertebrate host. Details of this unique behaviour are discussed in Chapter 10.

All other reports of human myiasis associated with cuterebrids are believed to result from accidental contact with fly eggs or larvae at natural oviposition sites, or from contact with pet cats or dogs from which larvae are transferred to humans. Given the low population densities of *Cuterebra* populations and female oviposition behaviours, it seems highly unlikely that a female would enter a human dwelling to oviposit. Furthermore, except for Goddard (1997), in cases in which a stinging sensation was associated with a larval infestation, patients did not report the presence of flies. As cuterebrids are not known to larviposit, and their eggs undergo an embryonation period of several days before larvae mature (Chapters 7 and 11), it is illogical to conclude, as Goddard (1997) did, that the stinging sensation of larval penetration is associated with the presence of a fly that may have oviposited on a human shortly before a person felt the 'sting'. It is not known how infants who are only a few weeks old, as well as young children, acquire infestations of *Cuterebra* species, but as the mean longevity of newly hatched larvae of three species of *Cuterebra* varied from 2 to 7 days (Catts, 1982), pets could have transported viable larvae in their fur from field oviposition sites and subsequently transferred them to humans. Catts (1982) noted that the wet and sticky newly hatched larvae readily adhered to objects they contacted. As cats and dogs sometimes have cuterebrid larvae in their moist nares, larvae may be transferred when humans are nuzzled by pets. *Cuterebra* species whose oviposition behaviour has been studied all deposit their eggs at, or near, den or burrow entrances or along host trails, and eggs hatch in response to host warmth and moisture.

Subfamily Oestrinae

As the females of species in this subfamily all larviposit on their vertebrate hosts (Chapter 10), human infestations also result from females abnormally attracted to humans on whom larvae are deposited (Catts, 1982).

Subfamily Hypodermatinae

Nearly all human infestations with *H. bovis* and *H. lineatum* are believed to have resulted from direct contact with cattle, rather than from flies ovipositing on humans. In an exceptional case, Glaser (1913) observed a female deposit an egg on a leg of his wool trousers, after which the hatched larva penetrated through the skin on his thigh. Conjecture associated with many case histories indicates that infestations occur when hands or bare arms or legs contact newly hatched larvae on cattle. The much greater incidence of infestation in children (James, 1947; Bruel *et al.*, 1995) probably results because they are more frequently attired in shorts, and because people sometimes place children on the backs of cattle. Cases of ophthalmomyiasis are believed to result from larvae transferred to the eye via fingers and from larval migration within the host.

As there have been only a few reports of *H. tarandi* infesting humans (Kearney *et al.*, 1991) and only two of *H. diana* doing so (James, 1947; Fidler, 1987), it is not clear how humans become infested with these species. Because *H. diana* is a parasite of wild deer, it seems most likely that a female oviposited on the 1.5–year-old boy found infested or on a pet from which larvae transferred to the boy. Sheep and equines found infested with *H. diana* had shared the same grazing areas with red deer and roe deer (Dempsey, 1983; Ross, 1983; Schumann *et al.*, 1988; Hendrickx *et al.*, 1989; Vyslouzil, 1989). There are a few reports of *H. tarandi* eggs being found on scalp hairs (Bergman, 1919; Natvig, 1939), and Natvig (1939) also reported that eggs of *H. tarandi* (as *Oedemagena tarandi*) were found on hairs on his arm and on a Sami reindeer herder's dog. Chirico *et al.* (1987) reported recovering a larva of *H. tarandi* from the skin of a human.

Subfamily Gasterophilinae

Except for *G. pecorum*, most human infestations with *Gasterophilus* species are thought to occur via newly hatched larvae contracted while grooming, handling or riding a horse. Additionally, Chereshnev (1954 (in Zumpt, 1965)) reported females of *G. haemorrhoidalis* ovipositing on hairs on the back of a human hand and Hearle (1938) reported a female ovipositing on his hat while he was observing oviposition on a horse. Suter *et al.* (1972) reported *G. intestinalis* eggs attached to hairs on the leg of a woman who worked with horses and Cogley and Cogley (2000) reported this species ovipositing on the hair of a child. Humans can become infested with *G. pecorum* by having a moist hand contact with eggs attached to vegetation (Chereshnev, 1954, in Zumpt, 1965).

Chereshnev (1954, in Zumpt, 1965) found that the first instars of several *Gasterophilus* species could penetrate human skin within 3–5 min but that larvae of *G. nasalis* could not do so. Rastegaev (1978) also reported that first-instar *G. nasalis* larvae could not penetrate human skin but that first instars of *G. haemor-rhoidalis*, *G. pecorum*, *G. nigricornis*, *G. inermis* and *G. intestinalis* all quickly penetrated human skin.

In human cases of ophthalmomyiasis and gastrointestinal myiasis (Bisseru, 1967; Faust *et al.*, 1970; Goldsmid and Phelps, 1977; Medownick *et al.*, 1985) humans probably introduced first-instar larvae to the eyes and mouth via their fingers. A case of oral myiasis was associated with a young girl who often kissed a horse (Townsend *et al.*, 1978). A *G. pecorum* infestation in the stomach of a captive male lion (Battisti *et al.*, 1997) probably resulted from a *per os* introduction of eggs.

Anatomical Areas Infested and Related Pathology

Subfamily Cuterebrinae

Because *D. hominis* uses many species of porter arthropods to distribute its eggs, human infestations with *D. hominis* larvae may occur anywhere on the body. The types of myiasis caused by *D. hominis* larvae were listed by Guimãraes and Papavero (1999) as dermal furuncular myiasis, vaginal myiasis, ophthalmomyiasis, rhinomyiasis and cerebral myiasis. Additionally, Bakos and Zanini (1979) reported a *D. hominis* infestation of a woman's tongue. The many case history reports indicate that dermal furuncular myiasis on the head, upper torso, arms and legs is the most frequent type of myiasis encountered. This type of cutaneous myiasis is characterized by a boil-like tumour, or furuncle lesion, with an external opening providing access to the respiratory spiracles of the enclosed larva (James, 1947). In endemic areas where *D. hominis* infestations are quickly recognized, the tiny first instars are considered benign nuisances and immediately removed. The thousands of such infestations that must occur yearly simply go unreported.

When larvae are permitted to grow and enlarge, the non-healing lesion emits a purulent discharge with a fetid odour, and the patient usually experiences itching, local pain, headache and regional lymphadenopathy (Dunn, 1930; James, 1947; Lane *et al.*, 1987; Veraldi *et al.*, 1998). Dunn (1930), who reared larvae to maturity in himself and described associated sensations, noted periodic itching during the 2nd week; by the end of the 3rd week of larval development he was experiencing excruciating pain. Itching would be associated with the immunological response of the host (Grogan *et al.*, 1987). Dunn experienced no pain when mature larvae exited on the 47th and 51st days and then pupated.

The second most common malady associated with *D. hominis* is ophthalmomyiasis (e.g. bibliography of Guimãraes and Papavero, 1999; Goodman *et al.*, 2000; Emborsky and Faden, 2002). The most serious pathology associated with infestations of *D. hominis* is represented by a few cases of cerebral myiasis that have resulted in death (e.g. Dunn, 1934; Céspedes *et al.*, 1962; Rossi and Zucoloto, 1973).

Various *Cuterebra* species associated with human myiasis have mostly infested the eyes, face, neck and chest, but larvae may occur almost anywhere on the body (Baird *et al.*, 1989). Nearly half of the reported cases have been in infants and children less than 13 years old. In humans, larvae may occur in a small, inflamed boil-like tumour with a central breathing pore or they may be associated with a condition commonly called creeping eruption, Hautmaulwurf (in Germany) or volassatik (in Russia) (James, 1947). Creeping eruption results when larvae penetrate the skin and move about in subcutaneous burrows. Intense itching often is associated with this condition. When larvae remain and develop at the portal of entry, furuncular cutaneous myiasis occurs as larvae increase in size (e.g. Schiff, 1993). Larger swellings may be tender or sore and somewhat painful when larvae move around within the tumour. After surgical removal of such larvae the wound quickly heals. In rare cases in which even third-instar larvae have emerged or been surgically removed the wound healed rapidly (Baird *et al.*, 1982). Larvae have been successfully squeezed out by infested people. Blurred vision and various levels of pain were associated with ophthalmomyiasis (Baird *et al.*, 1989) but no cases of blindness were found. Baird *et al.* (1989) noted that all cases of cuterebrid myiasis in North America involved only a single larva.

The human infestations with *Metacuterebra* (=*Alouattamyia*) *baeri* may have involved accidental ingestion of plant material containing eggs; two people had painful throat irritations and expelled larvae via the mouth (Guimarães and Coimbra, 1982) and one person had pulmonary myiasis (Fraiha *et al.*, 1984).

Subfamily Oestrinae

When attracted to humans, the sheep bot fly, *O. ovis*, deposits first instars into the eyes, nostrils, mouth/pharynx or ears (Fraiha *et al.*, 1984). Case reports reveal that the eyes are most often infested. This species is involved in human ophthalmomyiasis much more often than any other oestrid. Zumpt (1965) reported a maximum number of 50 larvae recovered from human infestations, but Cepeda-Palacios and Scholl (2000a) found that tethered *O. ovis* females larviposited only 10.4 ± 6.8 larvae per strike (range 3–24) on sheep and goat models. A review of hundreds of case reports indicates that in most cases less than 10 and mostly only one to two larvae are involved in ophthalmomyiasis of humans. It is uncommon for patients to have larvae in both eyes. For example, Garzozi *et al.* (1989) reported 5 of 27 cases with larvae in both eyes, and in a study in Tunisia all 23 patients had larvae in only one eye (Zayani *et al.*, 1989). The small number of larvae involved in most human cases suggests that such infestations are either primarily associated with old females that have expelled nearly all of their larvae in previous attacks, or that most larvae deposited on humans by larvipositing females missed the intended larvipostion site.

The presence of larvae in people's eyes may cause acute conjunctivitis characterized by inflammation and pain (e.g. Krümmel and Brauns, 1956; Harvey, 1986). People may complain of a burning sensation in infested eyes as well as tear formation, itching and photophobia (James, 1947; Zumpt, 1965); however in most cases the usually mild reactions subside in several days. More severe pathology occurs when larvae penetrate to the posterior or anterior segments of the eye (Kearney *et al.*, 1991), but this rarely occurs when a larva of *O. ovis* is involved (Rakusin, 1970). Pampiglione *et al.* (1997) reported that for most infestations of *O. ovis*, shepherds rely on traditional remedies and rarely visit a physician.

Removal of larvae is curative. Except for one report of a second instar found in the conjunctival sac of a child (Boiko *et al.*, 1978), investigators have noted that the first instars in the eyes of humans soon die, and that symptoms resolve rapidly thereafter (e.g. Krümmel and Brauns, 1956; Pampiglione, 1958; Zumpt, 1963).

In cases of nasolaryngeal myiasis, patients may develop a cough, rhinitis, rhinerrhea, headache, swelling of nasal cavities and nose bleeds before voiding larvae through the nostrils or mouth (e.g. Quesada *et al.*, 1990; Giangaspero *et al.*, 1994; Uriarte and Ell, 1997; Yeruham *et al.*, 1997). James (1947) noted that in oral and nasal myiasis larvae may penetrate the frontal sinuses and cause swelling, severe pain and headache; inflammation of the throat interferes with swallowing. In rare cases, *O. ovis* larvae have completed development to the third instar in humans (Badia and Lund, 1994; Lucientes *et al.*, 1997) and in dogs (Lujan *et al.*, 1998). Early investigators who commented extensively on the pathologies associated with human infestations are Portchinskii (1913), Sergent and Sergent (1952), Pampiglione (1958), Krümmel and Brauns (1956) and Zumpt (1963).

Human infestations with larvae of *R. purpureus*, *P. picta*, *C. ulrichii* and *G. hässleri* all involve the presence of first instars in eyes. Most cases report less than ten larvae, but Portchinskii (1915) reported a maximum of 50 *R. purpureus* larvae in one case. Pathology is similar to that associated with *Oestrus ovis*.

Subfamily Hypodermatinae

The large body of older literature associated with human infestations of *H. bovis* and *H. lineatum* reveals that these species have been most frequently involved in subdermal creeping myiasis in which larvae burrow and wander about in infested individuals. Bishopp *et al.* (1926), James (1947), Scharff (1950) and Zumpt (1965) are among many who reviewed data in early publications. After penetrating the epidermis, larvae behave much like they do in cattle and move around extensively for several months. Larvae have been found almost everywhere on the body, but they most commonly end up in the shoulders, neck and head. In examining many human infestations in Norway, Schöyen (1886) was one of the earliest investigators to determine the upward dispersal of larvae in people, and their frequent appearances in the head, neck and shoulders. Many others have since confirmed this observation, as well as noting that larval movements in people could be rapid and extensive. Maturing larvae usually provoke swellings (warbles) where they stop. The larva that penetrated Glaser's thigh (Schöyen, 1886) eventually was extracted from the base of a lower molar. In some human infestations *H. bovis* and *H. lineatum* have completed their larval development (Smart, 1939; Gansser, 1947 (in Zumpt, 1965); Sigalas and Pautrizel, 1948). Although there have been a few reports of *H. bovis* and *H. lineatum* larvae infesting humans in recent years (e.g. Verger *et al.*, 1975; Fain *et al.*, 1977; Vasil'eva and Yakubovskaya, 1980; Li, 1989; Uttamchandani *et al.*, 1989; Bal and Amitava, 2000; Hu and Wang, 2000), the widespread reduction of cattle warble fly populations associated with area-wide insecticide applications (Chapter 12) has greatly reduced the number of case reports of human infestations in the past quarter century.

Hypodermosis of humans results in the most severe pathology associated with oestrid infestations. Many authors cited by James (1947) and Zumpt (1965) have commented on the severe pain, fever and discomfort that can be associated with more advanced larval infestations, especially warbles associated with third-instar larvae. Sometimes there has been paralysis of the legs (James, 1947), meningitis (Languillat *et al.*, 1976) and intracerebral myiasis (Semenov, 1969; Kalelioglu *et al.*, 1989).

Bishopp *et al.* (1926) described the symptomology and pathology associated with several cases in young children including a fatal case involving a young boy. In another case, from November to March, 14 larvae were removed from a 5-year-old boy (seven of the larvae appeared on the head and face). Human infestations also result in skin allergies and eosinophila (e.g. Boulard and Petithory, 1977; Kearney *et al.*, 1991; Navajas *et al.*, 1998). First instars of *H. bovis* and *H. lineatum*, as well as those of *H. tarandi*, also are sometimes involved in ophthalmomyiasis (e.g. Anderson, 1935; O'Brien and Allen, 1939; Krümmel and Brauns, 1956; Vasil'eva and Yakubovskaya, 1980; Dechant *et al.*, 1981; and many references in Kearney *et al.*, 1991), and the resulting pathology generally is considerably more severe than that associated with *O. ovis*. Eye infestations of *Hypoderma* species often result in severe pain, swelling of the eyelid, nausea, restlessness and occasionally blindness (Edwards *et al.*, 1984; Kearney *et al.*, 1991). In 30 cases of ophthalmomyiasis interna posterior in which a *Hypoderma* larva was recovered, the review by Kearney *et al.* (1991) reported 12 *H. bovis*, 6 *H. tarandi*, 2 *H. lineatum* and 1 *Hypoderma* species.

Subfamily Gasterophilinae

From a review of the early literature by James (1947) and Zumpt (1965) we know that the most frequent human malady caused by various *Gasterophilus* species is subcutaneous creeping eruption resulting from the movement of larvae in tunnels. Several recent reports also describe this condition (e.g. Heath *et al.*, 1968; Rastegaev, 1978; Ma, 1999; Chen, 2001). All species identified in the previous section can penetrate human skin and burrow. This behaviour might appear to be somewhat unusual for species that mature as larvae in the stomach or lower levels of the digestive tract of their natural hosts, but it is consistent with their biology, which involves penetration of the host's oral tissues as immature larvae. Unlike *Hypoderma* species, the movements of *Gasterophilus* species are restricted to subcutaneous tissues and the larvae usually die after several days. In one unusual case, Royce *et al.* (1999) recovered a second-instar *Gasterophilus* larva from an infant. The burrowing activity of the first-instar larvae generally provokes an intense itching reaction, which subsides after larvae die. Larvae also can be removed from burrows without complication. First instars also have been implicated in cases of ophthalmomyiasis (e.g. Anderson, 1935; Medownick *et al.*, 1985; Fukuda *et al.*, 1987) but, as in most cases associated with *O. ovis*, these have been of the benign type. Additionally, *G. intestinalis* has been incriminated several times in cases of gastrointestinal myiasis (Bisseru, 1967; Faust *et al.*, 1970; Danilov, 1973; Goldsmid and Phelps, 1977) and in one case of oral myiasis (Townsend *et al.*, 1978). Ahmed and Miller (1969) reported a pulmonary infestation with a *Gasterophilus* species.

13 Management and Control of Oestrid Flies

P.J. SCHOLL

United States Department of Agriculture, Agricultural Research Service, Midwest Livestock Insects Research Unit, USA

Integrated Pest Management

For several decades the definition of integrated pest management (IPM) has changed and evolved. At its most basic level, IPM was defined as the use of two or more techniques to control arthropod pest populations, and often this was the use of biological plus one more control. My preference is the definition offered by Scholl *et al.* (1990) who defined IPM as: 'An ecological approach to pest management in which all available necessary techniques are consolidated into a unified program so that pest populations can be managed in such a manner that economic damage is avoided and adverse side effects are minimized'. Several other attempts have been made to define this concept, but almost all of them incorporate most or all of the following five considerations:

1. Pest population management, not necessarily eradication.
2. Knowledge of all available and practical control technologies.
3. Awareness of environmental issues and the consequences of misuse.
4. Familiarity with ecological principles.
5. Cost effectiveness, with knowledge of economic thresholds.

Several important global forces are accelerating the shift towards greater use of IPM (Bram, 1994). First, global economics and increasing numbers of corporate mergers have created a dramatic reduction in the number of companies willing to commit both the time and the money necessary for the development of new products. A new drug must have very high potential for economic return in order for these corporations to regain product development costs, especially with systemic drugs that must be approved by the FDA. Second, scientists are finally beginning to understand the consequences of dependence and overuse of a single chemical application in terms of resistance and potential environmental consequences. IPM fulfils the need, especially in developing nations, for strategies that increase profits for producers while at the same time reduce costs to consumers.

The purpose of this chapter is to discuss management of oestrid fly populations. As is discussed in the 'Chemical Control' section, all oestrid flies examined so far are extremely susceptible to control with any of several formulations of macrocyclic lactones, especially the avermectins. But considering the often debated environmental consequences of avermectin use, and realizing that resistance to any chemical probably increases in direct relationship to use, the facetiously suggested one word response to this chapter ('avermectins') for control of oestrid flies will be resisted. Instead, I will attempt to look not only at chemical but also other control strategies that have been or are currently in use to control this interesting group of flies.

Biological Control

This classification of control involves the use of natural enemies of pests. Essentially all animal populations, vertebrate and invertebrate, are reduced in numbers by other forms of life, including predators (both vertebrate and invertebrate), parasites (metazoan arthropods or nematodes) and microbial pathogens (Kilgore and Doutt, 1967; Harwood and James, 1979). The key to successful biological control, therefore, is maximizing the effects of these groups of natural enemies in order to maintain the target organism at lower average densities than would otherwise occur in their absence. Although entire courses are offered in university curricula and numerous textbooks have been written describing the use of organisms for biological control of arthropods, there are few specific references to biological control of oestrid flies.

Pathogens

The groups of organisms that fall into this category are usually defined as being smaller than their hosts. These organisms include viruses and protozoa, bacteria and fungi.

Viruses and protozoa

There are no reports of associations or possible uses of viruses or protozoa in biological control of oestrid flies.

Bacteria

Moroz (1979) described the possible use of bacterial preparations of *Bacillus thuringiensis* for control of *Hypoderma* spp. larvae. This is the only specific reference to the use of bacteria as biological control agents, but there have been numerous descriptions of bacteria associated with oestrid fly larvae, especially the warble-producing Hypodermatinae and Cuterebrinae. Several species of bacteria have been reported from the contents of the subcutaneous furuncles of both *Hypoderma* (Nogge and Werner, 1970) and *Dermatobia hominis* (Sancho *et al.*, 1996), and from the gut tracts of second- and third-stage larvae (Norman and Younger, 1979). *D. hominis* has also been implicated in the association with a bacterial infection

(*Pasteurella granulomatis*) in southern Brazil known locally as 'Lechiguana' (Ribeiro *et al.*, 1989; Riet-Correa *et al.*, 1992). Larvae from both oestrid subfamilies appear to be protected from potentially pathogenic bacteria by production of bacteriostatic secretions (Jettmar, 1953; Landi, 1960; Beesley, 1968; Sancho *et al.*, 1996), and it has been demonstrated that survival of *Hypoderma* spp. larvae collected from hides can be significantly enhanced by immersion in antibiotic saline (Chamberlain, 1989; Chamberlain and Scholl, 1991).

Fungi

There are only a few reports of pathogenic fungi against oestrid flies. Lucet (1914) reported that certain unspecified fungi might have been responsible for the destruction of *Hypoderma bovis* pupae. Minár and Samsinakova (1977) suggested that invasion by the entomogenous fungus, *Beauveria tenella*, may be much more important in mortality of *Hypoderma* spp. than previously thought. They concluded that entomogenous fungi might serve as natural control agents of oestrid flies and even be responsible for reduction in their numbers. The authors further suggested that late third instars leaving the host might pick up spores on the ground as they search for a pupariation site. This would be especially true for *Hypoderma* spp. that, unlike other oestrids, do not burrow into loose soils, but instead crawl along the surface, pupating under leaf litter and loose soils (Bruce, 1938). Both Scholl (1990) and Colwell (1992) suggested that spores from entomogenous fungi might be at least partially responsible for the still unexplained distribution patterns and population densities for the two *Hypoderma* species found in the Americas (*H. bovis* and *Hypoderma lineatum*) since it was demonstrated by Pfadt (1947) that both species could tolerate great variation in soil type, soil moisture and temperature. Finally, Kal'vish (1990) reported that the entomopathogenic fungus *Tolypocladium niveum* collected from pupae of the two reindeer oestrids *Hypoderma tarandi* and *Cephenemyia trompe* might be useful in *Oestrus ovis* control when applied as a suspension of spores.

Parasites/parasitoids

Parasites are normally defined as the smaller member of an association that lives on, or in, and derives all of the benefit at the expense of, the larger member of the association known as the host. Parasitoids are insect parasites (e.g. Hymenoptera and Tachinidae) that kill their host very slowly. In this case the hosts are oestrid flies, especially the larval and pupal stages.

Nematodes

No parasitic nematodes have been reported from the oestrids.

Insects

With the exception of the report of *Trichopria* spp. isolated from puparia of *H. bovis* (Blagoveshchensky, 1970) and *D. hominis* (Sanavria, 1987), the association between parasitic insects and members of the family Oestridae would seem to be mostly anecdotal. Bishopp *et al.* (1926) reported finding several specimens of *Nansonia brevicornis* from pupae kept in screened cages in proximity to parasitized muscid fly pupae. Scharff (1950) indicated that only one case of parasitism had been reported from Europe (Obitz, 1937) by an unidentified ichneumonid wasp. Gansser (1956), in his review of *Hypoderma* spp. biology, concluded that there were no specific biological enemies of cattle grubs, but did, like Obitz, observe attacks on mature larvae by ichneumonids and other wasps. Scholl (1990) suggested that post-parasitic cattle grub larvae and pupae are rarely found in nature, and the lack of numerous observations may be as much due to this sparse distribution as to lack of biological activity. He suggested that parasitic insects might play a larger role in determining distribution patterns especially for *Hypoderma* spp. that, unlike other oestrid fly emergent third instars, do not burrow into the soil. By exiting the warble during the late winter/early spring before there is much biological activity, *Hypoderma* spp. may enhance its survival by discouraging parasitism. Also, although numerous specimens of *Hypoderma* spp. have been collected in South America from imported cattle, there are no verified reports of established populations of this parasite south of northern Mexico. This may be due to the relatively 'rigid' larval development time and due to larvae in infested cattle shipped to South America exiting their hosts at a time of peak biological activity rather than minimal activity such as they would experience in the northern hemisphere. Ants have also been described as playing a role in destruction of oestrid immature stages (Gregson, 1958; Colwell, 1992; Rocha and Mendes, 1996).

Predators

Predators are generally defined as being larger and causing harm to the smaller member of the association known as the prey. In this case the prey are again oestrid larvae and pupae.

With the exception of the citations suggesting that moles (Stegman, 1920) and rodents (Colwell, 1992) may play a part in natural destruction of newly emerged *Hypoderma* larvae and pupae in the soil, most associations between oestrid flies and vertebrate predators have been references to birds. Goodrich (1940) and Bauer (1978) both discussed the observations of starlings attempting to extract *Hypoderma* larvae from warbles along the backs of infested cattle. Bishopp *et al.* (1926) describe the feeding activity of both robins and barnyard poultry on larvae that have dropped to the ground after emergence from their furuncles. Although not documented, it has long been suspected that cattle egrets ingest *Hypoderma* spp. or *D. hominis* larvae unfortunate enough to leave their furuncles during the daytime when the birds are with the cattle. These 'tasty morsels' (to birds at least) led one group of researchers (Leinati *et al.*, 1971) to investigate the possibility of grazing cattle with chickens. They concluded that this practice led to a form of biological control, and to the almost complete elimination of *Hypoderma* infestation in herds where this practice was encouraged.

An observation of possible fenthion toxicity in magpies (*Pica pica*) (Hanson and Howell, 1981) led some to believe that these pest birds were also feeding on grubs in the backs as previously observed for starlings. But a subsequent investigation by Henny *et al.* (1985) indicated that instead of devouring insecticide-killed larvae, the magpies were harvesting hair from the backline of cattle for use in the digestion process. Magpies ingesting enough insecticide-treated hair would eventually become impaired and therefore easy prey for raptors like hawks and eagles. This was hypothesized by the authors as being one of the primary routes of poisoning of predatory birds in the West Coast of the USA. This hypothesis eventually led to the ban of the use of organophosphate systemic treatments for cattle grubs in that region.

Plants

Besides the use of pyrethrum and tobacco leaf extracts to kill furuncular larvae of both *Hypoderma* spp. (Bishopp *et al.*, 1926) and *D. hominis* (Roncalli, 1984a; Guimarães and Papavero, 1999), other plants and plant products have been at times suggested as possible 'natural' controls of *Hypoderma* spp. These include indigo bush, *Amorpha fruticosa* (Brett, 1946); menazon (Goulding, 1962); gardona (Ivey *et al.*, 1968); and devil's shoestring, *Cracca virginiana* (Little, 1931a,b).

Mechanical Control

Mechanical control is probably the most primitive type of pest control, and has been applied in the case of oestrid flies. Mechanical control involves direct or indirect measures, equipment or actions directed specifically against an insect pest in order to destroy it outright, or activities that disrupt the normal physiological activity by other than chemical means. This type of control is sometimes confused with cultural control techniques that tend to be normal or slightly modified agricultural or conservation practices, which are discussed in the 'Cultural Control/Environmental Manipulation' section.

Perhaps the best example of mechanical control is manual extraction of larvae from the furuncles of the families Hypodermatinae and Cuterebrinae. Bishopp *et al.* (1926) and Hearle (1938) devoted several pages in their classic publications to various methods and devices for removal of *Hypoderma* spp. larvae from the backs of cattle hosts, including suction pumps, special forceps and even placement of the mouth of a heated bottle over the breathing aperture of the warble. On several occasions I have observed people inverting a soft drink bottle over the aperture and sharply tapping the bottom of the bottle. If the grub is mature, it is easily extracted by this method. A more recent procedure described by Scholl and Barrett (1986) utilizes hydrogen peroxide infused into the warble using an 'unarmed' hypodermic syringe. In this manner, larvae are removed with minimal damage to them for identification purposes and with minimal discomfort to the animals. Additionally, the antiseptic action of the hydrogen peroxide appears to hasten the healing process and minimize scarring.

Guimarães and Papavero (1966, 1999) reviewed various non-chemical techniques to remove *D. hominis* larvae from their furuncles, especially in humans. The reverse spines and unusual shape makes this species almost impossible to extract manually without damage to parasite and host. For this reason, numerous substances have been described to assist in larval removal including beeswax, adhesive tape, petroleum jelly, gum and even bacon fat (Brewer *et al.*, 1993). All these techniques attempt to prevent respiration; therefore the larvae either die or begin to exit the warble. In either case, they are much easier to remove. Surgical removal of these larvae is the other alternative.

Because second- and third-instar gasterophilids reside in the digestive tract, they are essentially impossible to control except chemically, and therefore mechanical control approaches that help to lessen or avoid infestation have been used historically. Dove (1918) and Knipling and Wells (1935) discussed techniques of washing the eggs attached to the hair on various parts of the body to first stimulate hatching and subsequently remove the larvae. Even rigorous daily currying and hair clipping on a regular basis removed large numbers of potentially infective larvae. Dove (1918) and Hearle (1938) illustrated types of fringed leather or canvas devices placed over the nose and muzzle of the animal to reduce oviposition around the face.

The use of baited traps to reduce adult oestrid populations of *C. trompe* and *H. tarandi* (Anderson and Nilssen, 1996a) and *Cephenemyia* (Anderson and Olkowski, 1968) has been evaluated and may offer a method of reducing populations under those conditions of very limited time and space such as found with reindeer production.

Two of the oldest and least practical mechanical controls are the use of housing and/or screens (Bishopp *et al.*, 1926) to prevent infestation of hosts by oviposition (e.g. Hypodermatinae and Gasterophilinae), larviposition (Oestrinae) and even phoretic infestation (*Dermatobia*). This technique, although probably effective, can be quite expensive, often impractical and probably useful only in reducing populations.

Finally, the most unpleasant mechanical control technique encountered was the practice of trephining of sheep, the surgical perforation of the frontal bones to expose the frontal sinuses and either the removal of the visible larvae or the injection of materials into the sinuses (see Chapter 2). This barbaric procedure was mentioned by Portchinskii (1913), described by Stewart (1932) and repeated by Rogers (1967).

Cultural Control/Environmental Manipulation

Cultural control or environmental manipulation includes the purposeful manipulation of the environment, thereby optimizing animal health while at the same time creating unfavourable conditions for pests. This somewhat broad classification includes not only obvious modifications to the environment such as draining and filling mosquito-producing water, but also activities such as proper agricultural waste management. For some of the most important arthropods affecting livestock (e.g. stable flies, *Stomoxys calcitrans*), source reduction of the number and quality of breeding habitats still remains the most effective means of managing the pest populations. For the oestrid flies, this control classification includes those practices that change cultural practices and/or the environment. Because cultural control methods are often based on modifications in the time and manner of performing necessary operations in production, they are among the cheapest method of insect control. Cultural control methods by themselves may not provide total insect control; nevertheless, they are important in minimizing injury and should be considered whenever possible in integrated control programmes.

Most of the suggestions for cultural control were made in the era before the introduction of efficient chemotherapy, especially the systemic insecticides, or in developing agricultural systems where insecticide use is often not affordable. Suggestions by Bishopp *et al.* (1926) to drain soils to create unfavourable environments or to select soil types that inhibited grub survival, tilling the soil at the time of the year of larval emergence to kill the newly emerged larvae or pupae on the ground and even general sanitation were all considered at the time to be reasonable alternatives. The technique of sanitation and tilling pasture are still recommended and utilized for control of horse parasites as an alternative to pesticide use (Herd, 1986).

Portchinskii (1913) made similar suggestions for *O. ovis* control, including the burning of pastures and tall grass during the season when nose bots were assumed to be exiting the nasal canals. The correct time of the year was made easier by noticing when sheep were sneezing. It was also observed by the author that smoke might enhance sneezing and possibly force more larvae out.

Movement of animals has long been used, when possible, as a control technology for oestrid flies. Alicata (1964) suggested the use of pasture rotation in Hawaii to aid in *Hypoderma* spp. control. This is possible when large tracts of land are available. With all the oestrid flies, the parasitic phase is the larval stage, and if one times the movement of the animals after the larvae exit the host and moves the animals some distance beyond the flight range of the adults, infestation may be avoided. Both Dill (1934) and Hearle (1938) described this method of control in detail in the years before modern insecticide solutions became available for *O. ovis* control. This technique was again described by Hearle (1938) for cattle grubs (*Hypoderma* spp.), and observed frequently by two of this book's authors (Scholl and Colwell) in northern Montana and in southern Alberta, Canada. In this region the timing of grub drop from the animals normally coincides with calving, which typically occurs near the ranch house. Hearle (1938) also suggested that the time of calving was important in this effort. By early summer, before fly emergence from pupae on the soil surface, animals are moved to mountain pastures several miles away from the lower ranges where larvae had dropped.

Chemical Control

The early experience of livestock producers in controlling oestrid flies in all of the commodity groups was frustrating and often involved crude methods and very unpleasant and hazardous application techniques. Reviews of the early chemotherapy, including the first organophosphorus systemics especially for cattle grubs, have been previously published by Bishopp *et al.* (1926), Hearle (1938), Scharff (1950), Gansser (1956), Rogoff (1961), Beesley (1966), Guimarães and Papavero (1966), Khan (1969), Drummond *et al.* (1988), Berkenkamp and Drummond (1990), and Hall and Wall (1995).

The discovery and subsequent introduction of the macrocyclic lactones in the early 1980s dramatically improved control of oestrid flies. The first introduced macrocyclic lactone was ivermectin (Campbell *et al.*, 1983), a potent avermectin with efficacy of 100% against any oestrid larvae tested at recommended dose levels.

In early trials, Benz (1985) and Drummond (1985) described the outstanding efficacy of this material against a very wide range of both internal (endoparasites) and external parasites (ectoparasites), thus the term 'endectocide'. Two additional macrocyclic lactone endectocide compounds have subsequently been introduced with great scientific and commercial success: doramectin, an avermectin (Goudie *et al.*, 1993), and moxidectin (Carter *et al.*, 1987), a milbemycin. Early trials with ivermectin indicated that its efficacy against *Hypoderma* spp. remained high even at titrated doses (Drummond, 1985). Additionally, it was reported that unlike other systemic insecticides, ivermectin could be safely administered as a late-season treatment (Scholl *et al.*, 1985) while the larvae were in their warbles. This permitted the labelling of the avermectin products to include control of not only migrating first instars, but also second and third instars despite some initial reports of adverse reactions to ivermectin in heavily infested animals in France (Preston, 1984).

This very high efficacy has led to dramatically reduced dose levels and the use of these products in microdose form for area-wide control programmes (Minar, 1995). However, because these products are endectocides, their potential efficacy is spread over a wide range of parasitic organisms, both internal and external, all with differing lethal responses. This has led to heated debate over the advisability of using these products in any form other than the labelled dose (Shoop, 1993; Bauer, 1994; Forbes, 1994). It is important to point out that the author is not aware of any case of reduced efficacy of macrocyclic lactones to oestrid flies anywhere that might indicate resistance. However, there are indications that other targeted parasites, especially parasitic nematodes, may be more prone to develop resistance (e.g. Echevarria *et al.*, 1993).

Finally, the other lively debate involving avermectins has been the long-held belief that use of avermectin products has serious environmental consequences, especially to dung-inhabiting insects (Wall and Strong, 1987; Floate *et al.*, 2002). Simultaneously, there is an equally convincing argument that despite the potential for laboratory-demonstrated mortality to arthropods, use patterns of these products permits 'islands' of unaffected individuals and they are often applied in areas where and when few manure inhibitors are present. To date, there are no documented examples of negative impact anywhere despite more than 20 years of intensive use (Wratten and Forbes, 1996).

It would take an entire chapter to list and discuss the literature citations on macrocyclic compounds against parasites, let alone oestrid flies. Therefore I have selected and listed several of the more important citations (Table 13.1) dealing with the use of this group of compounds with the understanding that nearly 100% efficacy against the larvae of various oestrid flies when applied at recommended doses is common among them.

Genetic Control and the Sterile Insect Technique

Knipling (1955) proposed the large-scale release of irradiated sterile males into a wild population in order to achieve eradication of pest populations. The dramatic success of this methodology came to be known as the sterile insect technique (SIT), which was used in the dramatic eradication of the New World screwworm fly

Table 13.1. Citations for efficacy of macrocyclic lactones against oestrid fly larvae.

Subfamily Hypodermatinae

Hypoderma bovis/Hypoderma lineatum

        ivermectin: Benz (1985), Drummond (1985), Leaning (1984), Khan *et al.* (1985).

        eprinomectin: Holste *et al.* (1998).

        doramectin: Hendrickx *et al.* (1993), Lloyd *et al.* (1996).

        moxidectin: Losson and Lonneux (1993), Scholl *et al.* (1992), Boulard *et al.* (1998).

Hypoderma tarandi

        ivermectin: Oksanen *et al.* (1992).

        doramectin: Oksanen (1995).

        moxidectin: Oksanen and Nieminen (1998).

Przhevalskiana silenus

        ivermectin: Tassi *et al.* (1987).

Subfamily Oestrinae

Oestrus ovis

        ivermectin: Drummond (1985), Puccini *et al.* (1983), Roncalli (1984b).

        ivermectin bolus: Rugg *et al.* (1997).

        doramectin: Dorchies *et al.* (2001).

        moxidectin: Dorchies *et al.* (1996), Puccini *et al.* (1994).

Rhinoestrus spp.

        ivermectin: Rastegaev (1988).

Subfamily Gasterophilinae

Gasterophilus spp.

        ivermectin: Bello (1989), Rastegayev (1988).

        moxidectin: Scholl *et al.* (1998), Reinemeyer *et al.* (2000).

Subfamily Cuterebrinae

Cuterebra spp.

        ivermectin: Ostlind *et al.* (1979).

Dermatobia hominis

        abamectin: Cruz *et al.* (1993)

        ivermectin: Maia and Guimarães (1986), Roncalli (1984a), Uribe *et al.* (1989).

        ivermectin: slow release bolus: McMullin *et al.* (1989).

        doramectin: Moya-Borja *et al.* (1993).

(*Cochliomyia hominivorax*) (Diptera: Calliphoridae) from North and Central America (Wyss, 2000). This success led several researchers to suggest SIT for the eradication of oestrid flies, especially *Hypoderma* spp., from North America (Graham and Hourrigan, 1977; Dame, 1984).

The Joint US–Canada Pilot Cattle Grub Project was a 4-year effort (1982–1986) to test this hypothesis and was established along the border of northern Montana, USA and southern Alberta, Canada, targeting *H. bovis* and *H. lineatum* (Weintraub and Scholl, 1984). The goal of this pilot project was to evaluate the feasibility of SIT combined with conventional chemical treatment in a 750 square mile area. In this trial, chemical reduction of the population by itself proved very successful in reducing the local *Hypoderma* spp. population (Scholl *et al.*, 1986). However, with only dependence on larvae collected *in vivo* for subsequent pupal sterilization and release, the final evaluation was not so positive considering both the logistic problems (Kunz *et al.*, 1990) and the economic assessment of the cost per released sterile adult (Klein *et al.*, 1990).

It should be emphasized that despite numerous attempts to rear oestrid flies *in vitro* (Chamberlain, 1964a,b, 1970, 1989; Beesley, 1967; Zeledon and Silva, 1987; Chamberlain and Scholl, 1991), no larvae in any of the four subfamilies of Oestridae have ever been reared *in vitro* from newly hatched first instar to pupae, although some successes have been noted for individual stadia. This barrier has prevented the serious application of SIT for the control of members of this group of flies (Scholl, 1993; Hall and Wall, 1995; Guimarães and Papavero, 1999).

Immunological Control and Vaccines

Since the 1920s attempts have been made to exploit the close association between the oestrid flies (especially *Hypoderma* spp.) and their hosts by injecting crude extracts in an attempt to immunize the animals (Roubaud and Perard, 1924; Peter, 1929). Boulard and Lello describe the subsequent steps in our understanding of this complex process and the resulting potential for development of a vaccine in Chapter 7.

It is appropriate that immunological control and vaccines be discussed last in this chapter. Although enormous strides have been made in our understanding of the potential for vaccines in oestrid fly control (Baron and Weintraub, 1987; Baron and Colwell, 1991; Hall and Wall, 1995), to date there are no vaccines available for oestrid fly control that provide satisfactory results when used alone. Some writers have been optimistic about the use of vaccines for control of ectoparasites of domestic animals (Pruett, 1999; Willadsen, 2001). In my opinion, oestrid fly vaccines have been held at an unrealistically high expectation because of the introduction and outstanding efficacy provided by the avermectins. The expectation of complete eradication made possible by using these drugs has led to the sense that other control technologies are not as effective, and are therefore inadequate. In fact, the development of vaccines over the last two decades makes their use as a component of area-wide control feasible when used as a complement instead of a substitute for other control strategies or methods (Scholl, 1993; Colwell, 2002).

14 A Synopsis of the Biology, Hosts, Distribution, Disease Significance and Management of the Genera

D.D. COLWELL,[1] M.J.R. HALL[2] AND P.J. SCHOLL[3]

[1]Agriculture and Agri-Food Canada, Lethbridge Research Centre, Canada; [2]Natural History Museum, UK; [3]USDA-ARS-MLIRU, USA

This chapter gives a brief overview of the genera within each of the subfamilies, describing briefly the biology, disease, management, hosts and distribution and the management implications. Representative images that illustrate the various stages and host associations are included.

In some instances there has been little or no new published information since Zumpt's synopsis. On the other hand, application of molecular techniques has helped clarify the status of various species within some of the genera. There are several species whose larvae parasitize endangered or rarely encountered hosts. The status of these species is often unknown and new information on their biology and host relationships is not likely to be forthcoming. This situation is most unfortunate given the unique nature of these flies.

The chapter is organized by subfamily and genera are presented in the following order:

Oestrinae

Gedoelstia
Kirkioestrus
Oestrus
Rhinoestrus
Cephenemyia
Pharyngomyia
Pharyngobolus
Tracheomyia
Cephalopina

©CAB International 2006. *The Oestrid Flies: Biology, Host–Parasite Relationships, Impact and Management* (eds D.D. Colwell, M.J.R. Hall and P.J. Scholl)

Hypodermatinae

>*Hypoderma*
>*Strobiloestrus*
>*Portschinskia*
>*Pallasiomyia*
>*Ochotonia*
>*Oestroderma*
>*Oestromyia*
>*Pavlovskiata*
>*Przhevalskiana*

Cuterebrinae

>*Cuterebra*
>*Dermatobia*

Gasterophilinae

>*Gasterophilus*
>*Gyrostigma*
>*Cobboldia*
>*Ruttenia*
>*Neocuterebra*

Gedoelstia Rodhain and Bequaert

Biology

Larvae are parasites of the nasal cavities of members of the subfamily Alcelaphinae (Bovidae) in Africa. They are characterized by a unique first-instar migratory path. First instars ejected by the females into the host's eyes enter the orbit and migrate along the optic nerve where they enter the vascular system and move to the heart. After a brief period they move, still in the vascular system, to the cephalic sub-dura and via the foramen of the cribiform plates into the nasal cavities. Second- and third-instar development takes place in the nasal cavities and sinuses (Basson, 1962a,b; Zumpt, 1965).

Disease, pathogenesis and significance

Little disease is noted in wild ungulate hosts. However, females larviposit on domestic livestock, including sheep, goats and cattle. Larval migration in these hosts, notably sheep and goats, results in one of three severe conditions:

1. An ophthalmic condition with ocular damage evident within 12 h of larval entry (a condition called utipeuloog);
2. An encephalitic condition that is often fatal resulting from larval migrations (usually evident approximately 5 days postinfestation); and
3. A cardiac condition, again the result of larval migration.

Outbreaks are associated with the passing nearby of migrating wildebeest. Heavy losses can be associated with these outbreaks.

Management

Diagnosis depends on the appearance of symptoms, which may be the most cost-effective approach. Presumably, treatment with macrocyclic lactone anti-parasitics will give effective control and alleviate the symptoms. There has been no published work to verify these observations.

Hosts and distribution

Table 14.1. Hosts and distribution (*Gedoelstia*).

Species	Hosts	Distribution	Common names
cristata	*Connoachaetes gnou, Connoachaetes taurinus, Alcelaphus buselaphus, Alcelaphus lichtensteini, Damaliscus korrigum*	Throughout host range, Ethiopia	Larger tuberculous nasal bot fly
hässleri	*Connoachaetes gnou, Connoachaetes taurinus, Alcelaphus buselaphus, Alcelaphus lichtensteini, Damaliscus korrigum, Damaliscus dorcas, Damaliscus lunatus,* domestic livestock	Throughout host range	Lesser tuberculous nasal bot fly

Figures

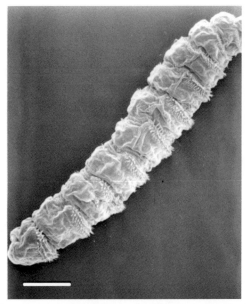

Fig. 14.1. Scanning electron micrograph of a lateral view of first-instar *Gedoelstia* sp. recovered from a bontebok. Note the dorsoventral flattening and the greater abundance of spines on the lateral and ventral surfaces (bar = 115 µm).

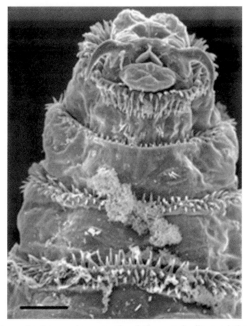

Fig. 14.2. Scanning electron micrograph of the cephalic and thoracic segments of first-instar *Gedoelstia* sp. recovered from a bontebok (bar = 60 µm).

Fig. 14.3. Lateral view of *Gedoelstia cristata* female. Note the characteristic and diagnostic dorsal protuberances on the abdomen (bar = 5 mm).

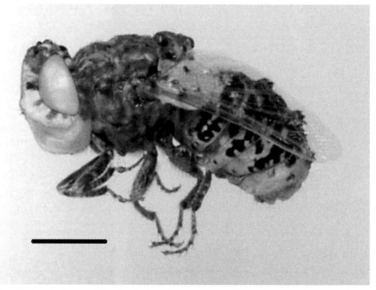

Fig. 14.4. Lateral view of *Gedoelstia cristata* male. Note the smaller eyes in comparison with the female (bar = 5 mm).

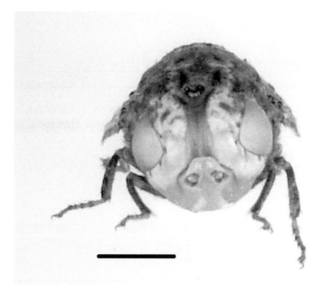

Fig. 14.5. *En face* view of *Gedoelstia cristata* male (bar = 5 mm).

Kirkioestrus Rodhain and Bequaert

Biology

Female *Kirkioestrus minutus* larviposit on or near the host's nares. Larvae are deposited singly or in large groups. First instars develop on the mucosa of the nasal passages. Second instars move to the frontal sinuses where they complete development. Larval development takes approximately 30 days. Mature third instars burrow into the soil where development requires approximately 30 days in the summer (October). The second generation of pupae requires approximately 50 days to develop in mid-winter (July).

Disease, pathogenesis and significance

Symptoms of infestation are apparently mild and no disease is noted in wild ungulate hosts. There is apparently no cross-transmission to domestic livestock despite sharing of the range in some localities.

Management

The absence of significant disease or pathology suggests that treatment is not required in the management of wild host populations. However, it may be presumed that treatment with macrocyclic lactone anti-parasitics will give effective control (see Chapter 13). There has been no published work to verify these suggestions.

Hosts and distribution

Table 14.2. Hosts and distribution (*Kirkioestrus*).

Species	Hosts	Distribution	Common names
minutus	*Connoachaetes taurinus, Alcelaphus buselaphus, Alcelaphus lichtensteini, Damaliscus korrigum*	Throughout Africa	Common hairy nasal bot fly
blanchardi	*Alcelaphus buselaphus, Alcelaphus lichtensteini*	Central Africa	Blanchard's hairy nasal bot fly

Oestrus L.

Biology

Members of this genus are medium to large flies (10–19 mm) with a distinctive appearance and colour, which is said to resemble the excreta of birds, thereby providing some degree of protection from predation by mimicry. Adult flies can deposit live larvae directly on and into the nostrils of their hosts without landing. Only up to about 25 larvae are deposited per 'attack', but females may have up to 500 eggs. *Oestrus ovis* is one of the few species of oestrid for which larvipositional behaviour has been well documented (Cepeda-Palacios and Scholl, 2000a). The larvae of all species develop in the head sinuses and the nasal passages of their hosts (sheep, goats and antelopes), feeding on mucous and epithelial cell materials. First instars have large curved mouth hooks for attachment to the mucous membranes. The larvae can remain within the host for many months, passing the winter of temperate latitudes within the nostrils. However, in optimal conditions they can complete larval development in just 25–35 days. Like all fly larvae they show tremendous growth over their life, for example, those of *O. ovis* increasing in length from 2 to 22 mm and in weight from 0.0002 to 0.52 g (Cepeda-Palacios *et al.*, 1999). When mature the larvae leave via the nostrils, their exit migration often stimulating heavy sneezing of the host in a head-down posture. The larvae burrow into the ground and pupariate, with pupal development taking some 25–30 days in warm weather, longer in cooler periods. Adult flies probably live for no more than 10–20 days in nature, although they can live longer under laboratory conditions. There are two to three generations per year.

Disease, pathogenesis and significance

In terms of economics and both animal and human health, the most important member of this genus is the cosmopolitan species, the sheep nasal bot fly, *O. ovis* (Hall and Wall, 1995). Other species of *Oestrus* parasitize wild animals in Africa or central Asia and there is much less information available concerning all aspects of their biology and pathology (Zumpt, 1965). Adults of *O. ovis* annoy sheep and goats in their persistent larviposition activities, leading to a loss of grazing time and, consequently, animal condition. However, it is development of the larvae within the nasal sinuses that is the most damaging aspect of this species. Infestation levels among animals in a flock vary considerably, depending on the season and geographical location, from 6–50% at lower levels of challenge up to 100% in many areas. It is probable that the pathology of infestation varies with the sheep breed and, possibly, with the geographical strain of fly. Light infestations might be well tolerated and there are usually less than 15 larvae in adult sheep. However, heavy infestations can have serious consequences. Hence, infestation has been reported to cause losses in meat production of 1.1–4.6 kg per animal, 200–500 g of wool and up to 10% of milk (Ilchmann *et al.*, 1986). Infestations are usually accompanied by a mucoid nasal discharge. There is a close immunological relationship between parasite and host and antigenic stimulation of lung tissues can result in interstitial pneumonia (Dorchies *et al.*, 1998).

In addition to the veterinary aspects of *O. ovis*, it can be a medical problem, when larvae are deposited in the eyes or upper respiratory tracts of humans. There are numerous records in the literature of individual cases of infestation of humans by *O. ovis*. In some areas records of such infestations go back over 150 years (Pampiglione *et al.*, 1997). While the nasal passages of humans can be infested, more often it is the eyes that are affected. A study of 30 human cases of opthalmomyiasis showed that the most common symptom was acute conjunctivitis, with 63% also showing lid oedema and 43% a superficial punctate keratopathy, due to movement of the first-instar larva across the cornea (Harvey, 1986). In rare cases invasion of the interior of the eye can cause extensive damage (Rakusin, 1970).

Management

O. ovis-induced nasopharyngeal myiasis is relatively straightforward to manage by the use of oral presentations of closantel (10 mg/kg) or ivermectin (oral dose or subcutaneous injection 0.2 mg/kg), all of which give a 100% therapeutic effect. Slow release capsules of ivermectin given orally can provide a lengthy and effective prophylactic activity.

When bothered by adults of *O. ovis*, sheep frequently flock tightly together with their heads down and nostrils pressed into the fleece of their neighbours (Fig. 14.9). This behaviour might help to reduce the chances of infestation.

Hosts and distribution

Table 14.3. Hosts and distribution (*Oestrus*).[a]

Species	Hosts	Distribution	Common names
ovis	*Ovis ammon, Capra ibex, Ovis canadensis, Odocoileus virginianus* (and many others)	Worldwide (in domestic livestock), no wild hosts in sub-Saharan Africa	Sheep nose bot
variolosus	*Connoachaetes taurinus, Alcelaphus buselaphus, Alcelaphus lichtensteini, Damaliscus korrigum, Damaliscus dorcas, Damaliscus lunatus, Hippotragus niger, Oryx gazella*	Sub-Saharan Africa	
aureoargentatus	*Hippotragus niger, Hippotragus equinus, Damaliscus korrigum, Damaliscus lunatus, Alcelaphus buselaphus, Alcelaphus lichtensteini, Connoachaetes taurinus*	Sub-Saharan Africa	
caucasicus	*Capra caucasica, Capra ibex*	Central Asia	

[a] *Oestrus macdonaldi* and *Oestrus bassoni* are two additional species known from larval stages and from adults, respectively. There is need for critical examination of the relationship of these to *Oestrus variolosus*.

Figures

Fig. 14.6. Lateral view of female *Oestrus aureoargentatus* (bar = 0.25 cm).

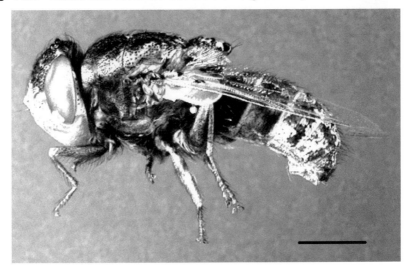

Fig. 14.7. Lateral view of male *Oestrus aureoargentatus* (bar = 0.25 cm).

Fig. 14.8. Lateral view of female *Oestrus variolosus* (bar = 0.25 cm).

Fig. 14.9. Sheep bunched together with heads towards the centre protecting them from larvipositing *Oestrus ovis* (photo courtesy of M.J.R. Hall, taken in Hungary).

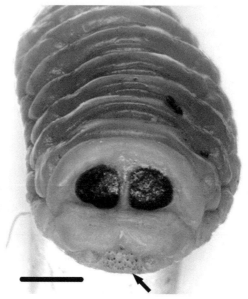

Fig. 14.10. Posterior view of third-instar *Oestrus variolosus* showing spine cluster on terminal abdominal segment (arrow) (bar = 0.5 cm).

Rhinoestrus Brauer

Biology

Most information available for this genus is from studies conducted on *Rhinoestrus purpureus*, which has a significant impact on domestic horses and donkeys. Females larviposit on the nostrils of suitable hosts, depositing 8–40 larvae at each attempt. Total number of larvae ranges from 700 to 800/fly. First instars move to the nasal cavities where they develop for periods ranging from a few weeks to a few months. There is some suggestion that at higher latitudes the first instars exhibit hypobiosis, not completing moult to second instar until the arrival of warmer spring temperatures. Variation in developmental times is seen even within groups of larvae from the same batch. As the first instars mature they move further into the nasal cavities. Second and third instars are found in the same site although some may move into upper pharyngeal regions. Burdens in individual hosts range from 1 to >200 larvae (see below).

Mature third instars exit through the nose and burrow under surface litter. Pupal development times vary slightly among the species, with *R. purpureus* requiring from 15 to 32 days and *Rhinoestrus steyni* requiring approximately 30 days. No data have been reported for the other species infesting members of the genus *Equus*. Pupal development for the species infesting springbuck, *Rhinoestrus antidorcitis* and *Rhinoestrus vanzyli*, requires 49–56 days and 30–50 days, respectively. *Rhinoestrus tshernyshevi* pupal development is amongst the shortest, requiring only 21 days, while *Rhinoestrus nivarleti* from the bushpig requires 28–35 days.

There may be either one or two generations per year, depending on rate of first-instar development. Where two generations occur flies are active from March to mid-June and again from September through October.

Little is known of the adult biology although female flies in captivity live for an average of 25 days, with some living for as long as 38 days. Male flies were observed to live for only 14 days under similar conditions.

Recent studies (Otranto *et al.*, 2004) have revealed that morphological characters used to separate *R. purpureus* from *Rhinoestrus usbekistanicus* are unreliable. In specimens recovered from naturally infested horses from southern Italy these authors report the presence of four morpho-types, two with mixed features representing both species. Molecular evidence in the same work indicated that all morpho-types were identical, representing a single species.

Disease, pathogenesis and significance

Host specificity of the four species reported infesting members of the genus *Equus* is lower than that of species infesting non-perissodactyl hosts. Three species, *R. purpureus*, *Rhinoestrus latifrons* and *R. usbekistanicus*, are capable of developing in a wide range of hosts within the host genus. This is not noted in the members of the genus infesting other non-perissodactyl hosts. The species infesting members of the genus *Equus* cause significant morbidity and mortality, particularly in the domestic horse. In extremely heavy infestations, where burdens exceed 150 larvae per host, the mortality has been reported as high as 82% of affected horses (Akchurin, 1945). In these cases larvae are found not only in the nasal cavities, but also in significant numbers in the pharynx. Larvae are also found in the ethmoid bones that were penetrated, allowing larvae to access the olfactory nerves and portions of the cerebellum.

Pathological changes in the mucosa of the nasal cavities are very similar to those described for *O. ovis* (see Chapter 10). The action of excretory/secretory products, hypersensitivity reactions and mechanical effects of spines and mouth hooks all contribute to these changes.

The close association of humans with equids results in significant occurrence of ophthamomyiasis resulting from the larviposition by females of *R. purpureus*, *R. usbekistanicus* and *R. latifrons*.

Management

Rhinoestrus spp.-induced nasopharyngeal myiasis in equids is easily managed by the use of oral presentations of ivermectin or moxidectin, which give a 100% therapeutic effect. Slow-release capsules of ivermectin taken orally can provide a lengthy and effective prophylactic activity.

Hosts and distribution

Table 14.4. Hosts and distribution (*Rhinoestrus*).

Species	Hosts	Distribution	Common names
purpureus	Horses (*Equus cabalis*) and donkeys	Palaearctic, Ethiopian region, Oriental region, northern Mediterranean (Italy)	Horse nasal bot fly
usbekistanicus	*Equus burchellii*, horses and donkeys	Africa, central Asia, 'Near East', northern Mediterranean (Italy)	Equine lesser nasal bot fly
steyni	*Equus burchellii*, *Equus zebra*	Southern Africa	Zebra larger nasal bot fly
latifrons	Horse		Horse larger nasal bot fly
hippopotami	*Hippopotamus amphibious* (hippopotamus)	Central Africa	Hippopotamus nasal bot fly
nivarleti	*Potamochoerus porcus* (bushpig)	'Congo'	Bushpig nasal bot fly
phacochoeri	*Phacochoerus aethiopicus* (warthog)	'Congo, Cameroons'	Wart hog nasal bot fly
giraffae	*Giraffa camelopardalis*	'Tanganika'	Giraffe nasal bot fly
tshernyshevi	*Ovis ammon*	Central Asia	Argali nasal bot fly
antidorcitis	*Antidorcas marsupialis* (springbuck)	Southern Africa	Springbuck larger nasal bot fly
vanzyli	*Antidorcas marsupialis*	Southern Africa	Springbuck lesser nasal bot fly

Figures

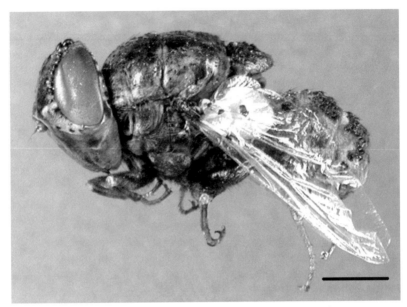

Fig. 14.11. Adult male of *Rhinoestrus purpureus* (bar = 0.25 cm).

Fig. 14.12. Adult female of *Rhinoestrus antidorcitis* (bar = 0.25 cm).

Fig. 14.13. Adult female of *Rhinoestrus vanzyli* (bar = 0.25 cm).

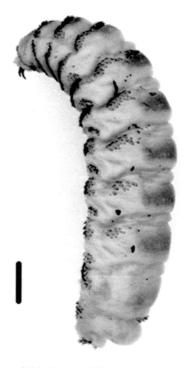

Fig. 14.14. Lateral view of third-instar *Rhinoestrus purpureus* recovered from a naturally infested horse (bar = 0.5 cm).

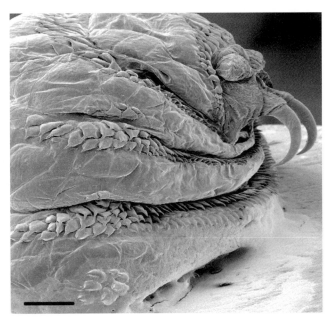

Fig. 14.15. Scanning electron micrograph of the cephalic and thoracic segments of a third-instar *Rhinoestrus usbekistanicus* recovered from a naturally infested horse in Italy (bar = 0.6 mm).

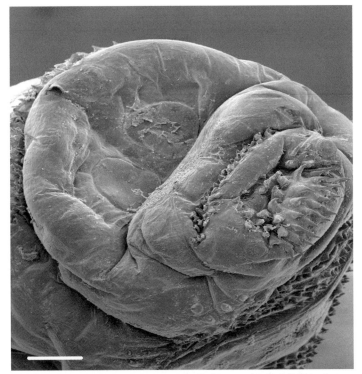

Fig. 14.16. Scanning electron micrograph of the terminal abdominal segments of a third-instar *Rhinoestrus usbekistanicus* recovered from a naturally infested horse in Italy (bar = 0.6 mm).

Cephenemyia Latreille

Biology

Larvae are parasites in the nasal cavities and pharynx of cervid hosts (see Table 14.5). Females hover in front of the host then dart towards the face and eject a packet of larvae on to the host nares (see Chapter 10). First instars migrate into the nasal cavities where they feed on mucous. They remain in the cavities and either moult to the second instar or proceed to the pharyngeal pouches before the first moult. Second and third instars develop in the pharyngeal pouches.

Disease, pathogenesis and significance

Both adult and larval stages have an impact on the host. Larviposition behaviour induces severe reactions in previously infested animals. Animals under attack adopt a rigid posture, which is maintained until the larvae land, at which time the host resumes movement, but is unsteady and frequently shakes its head and sneezes repeatedly. Exit of mature larvae induces severe sneezing attacks.

First instars may become hypobiotic (see Chapter 11d) in order to overwinter. These larvae, deposited in late summer, remain in the deep nasal cavities until spring, when development resumes. White-tailed deer from Alberta, Canada, examined in January, were infested with small first instars (1.5 mm long) resident near the cribiform plates (D.D. Colwell, unpublished data). Occasional larger first instars (3.0 mm long) were found at the same sites, but no second or third instars were noted.

Attachment and feeding activity of the second and third instars results in a bloody nasal discharge. The pharyngeal pouch mucosa becomes inflamed and oedematous. Epidermal sloughing is evident with subsequent development of lesions and necrosis. There are no indications that protective immunity develops.

While adult females larviposit on decoy deer heads (see Chapter 10) the flies are apparently quite host-specific. Larvae are not reported from domestic livestock. In North America the growth of game farming, with concentrations of farmed deer or elk in areas where wild populations are present, provides for the potential of heavy infestations in the farmed animals.

Management

Treatment with macrocyclic lactone anti-parasitics gives effective control (Weber, 1992; Kutzer, 2000; and review by Marley and Conder, 2002). Control of *Cephenemyia* spp. in several host species, including *Cervus elaphus*, *Capeolus capeolus*, *Rangifer rangifer*, *Odocoileus virginianus* and *Llama pacos*, has been high.

Hosts and distribution

Table 14.5. Hosts and distribution (*Cephenemyia*).

Species	Hosts	Distribution	Common names
ulrichii	*Alces alces*	Central and eastern Europe	Elk throat bot fly
auribarbis	*Cervus elaphus, Dama dama*	Europe	Red deer throat bot fly
stimulator	*Capeolus capeolus*	Palaearctic	Roe deer throat bot fly
trompe	*Rangifer rangifer, Odocoileus hemionus*	Holarctic	Reindeer throat bot fly
jellisoni	*Odocoileus hemionus, Odocoileus virginianus, Alces alces americanus, Cervus elaphus canadensis*	Nearctic (western)	
apicata	*Odocoileus hemionus*	Nearctic (western)	
phorbifer	*Odocoileus virginianus, Alces alces americanus*	Nearctic (eastern)	
pratti	*Odocoileus hemionus, Odocoileus virginianus*	Nearctic (southwestern)	

Figures

Fig. 14.17. Male *Cephenemyia trompe* awaiting arrival of passing females. Image courtesy of A. Nilssen (bar = 1 cm).

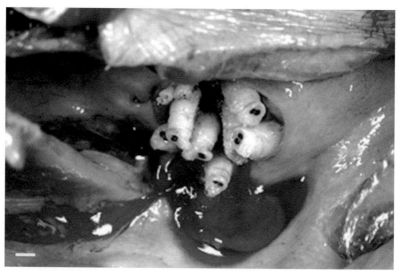

Fig. 14.18. Third-instar *Cephenemyia trompe* attached in the pharynx of a reindeer (*Rangifer tarandus*). Image courtesy of A. Nilssen (bar = 5 mm).

Fig. 14.19. Third instar of *Cephenemyia trompe*, ventral and lateral views (bar = 2 mm).

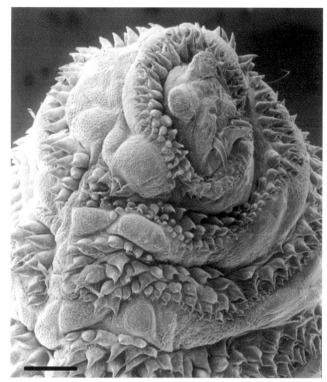

Fig. 14.20. Scanning electron micrograph of ventral/lateral surface of the cephalic and thoracic segments of third-instar *Cephenemyia trompe* (bar = 0.5 mm).

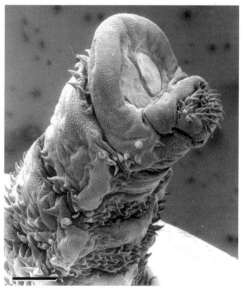

Fig. 14.21. Scanning electron micrograph of posterior abdominal segments of third-instar *Cephenemyia trompe*. Posterior spiracles are evident as are the terminal cluster of spines (bar = 0.6 mm).

Pharyngomyia Schiner

Biology

Females larviposit on to the host nares and first instars migrate into the nasal cavities. Mature first instars migrate to the pharynx, particularly the pharyngeal pouch where they moult and complete the last two larval instars. The later instars are attached to the mucosa of the pharynx and larynx (similar to species of *Cephenemyia*). Larvae are coughed out and pupate in the soil. Pupal development requires from 21 to 40 days. *Pharyngomyia picta* is univoltine, with adults active from June to August in central Europe.

Disease, pathogenesis and significance

There are no descriptions of pathology or effects on the host. It is assumed that the impacts are similar to those described for *Cephenemyia* spp.

Management

Presumably, treatment with macrocyclic lactone anti-parasitics will give effective control (see Chapter 13). There has been no published work to verify these suggestions.

Hosts and distribution

Table 14.6. Hosts and distribution (*Pharyngomyia*).

Species	Hosts	Distribution	Common names
picta	*Cervus elaphus, Cervus nippon, Dama dama, Capreolus capreolus*	Europe, central Asia	Deer throat bot fly
dzerenae[a]	*Gazella gutturosa* 'mongolian gazelle' (probably *Gazella subgutturosa*, goitered gazelle)	Central Asia	Mongolian gazelle throat bot fly

[a] Known only from third instar.

Figures

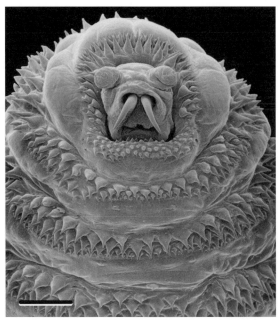

Fig. 14.22. Scanning electron micrograph of the ventral surface of the cephalic and thoracic segments of a third-instar *Pharyngomyia picta* recovered from a red deer in Spain. Note the bases of antennal lobes are widely separated. Note also the large number of spines (bar = 0.6 mm).

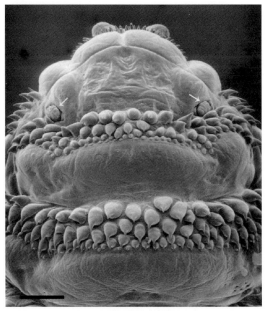

Fig. 14.23. Scanning electron micrograph of the dorsal surface of the cephalic and thoracic segments of a third-instar *Pharyngomyia picta* recovered from a red deer in Spain. Note the presence of anterior spiracles (arrows) (bar = 0.7 mm).

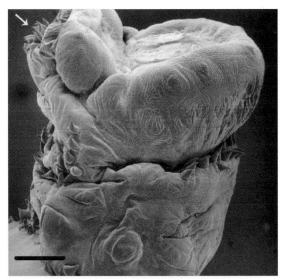

Fig. 14.24. Scanning electron micrograph of the lateral surface of the posterior abdominal segments of a third-instar *Pharyngomyia picta* recovered from a red deer in Spain. Note the presence of the cluster of spines dorsal to the anus (arrow) (bar = 0.7 mm).

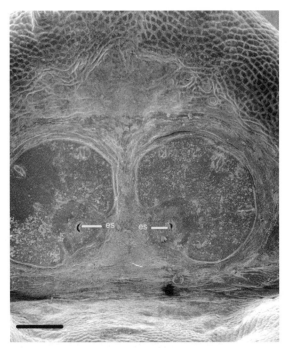

Fig. 14.25. Scanning electron micrograph of the spiracular plates on the terminal abdominal segment of a third-instar *Pharyngomyia picta* recovered from a red deer in Spain (es = ecdysal scar) (bar = 0.3 mm).

Pharyngobolus Brauer

Biology

This is a monospecific genus whose larvae live in the throat of the African elephant. Although adults have been reared there is no information on their biology or behaviour. Second and third instars are attached to the mucosa of the pharynx. Mature larvae detach and are expelled through the trunk.

Disease, pathogenesis and significance

Early observations indicated that the prevalence of infestation was high. However, no pathology or disease has been associated with these larvae.

Management

The absence of significant disease or pathology suggests that treatment is not required in the management of wild host populations. However, it may be presumed that treatment with macrocyclic lactone anti-parasitics will give effective control (see Chapter 13). There has been no published work to verify these suggestions.

Hosts and distribution

Table 14.7. Hosts and distribution (*Pharyngobolus*).

Species	Hosts	Distribution	Common names
africanus	*Loxodonta africanus*	Central Africa (Congo, Uganda, Zimbabwe)	African elephant throat bot fly

Figures

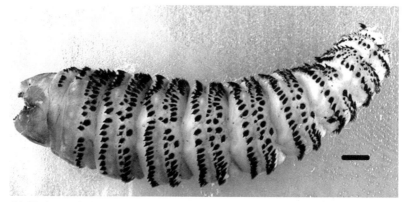

Fig. 14.26. Lateral view of a third-instar *Pharyngobolus africanus* recovered from an African elephant (bar = 2 mm).

Fig. 14.27. Ventral view of the cephalic, thoracic and first abdominal segments of a third-instar *Pharyngobolus africanus* recovered from an African elephant (bar = 2 mm).

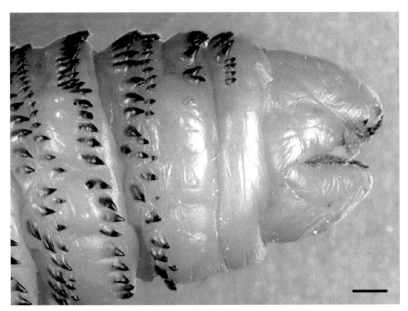

Fig. 14.28. Last abdominal segment of a third-instar *Pharyngobolus africanus* recovered from an African elephant. Note the spiracular plates are concealed within the terminal atrium (bar = 2 mm).

Tracheomyia Townsend

Biology

This monospecific genus is only known from the larval stages found in the trachea where they apparently feed on mucous secretions. Intensity of infestations is low with typically one to two, and occasionally up to six, larvae per host (Zumpt, 1965).

Disease, pathogenesis and significance

There are no descriptions of disease or pathology associated with infestations.

Management

The absence of significant disease or pathology suggests that treatment is not required in the management of wild host populations. However, it may be presumed that treatment with macrocyclic lactone anti-parasitics will give effective control (see Chapter 13). There has been no published work to verify these suggestions.

Hosts and distribution

Table 14.8. Hosts and distribution (*Tracheomyia*).

Species	Hosts	Distribution	Common name
macropi	Red kangaroo (*Macropus rufus*), Cook's kangaroo (*Macropus canguru*), dusky kangaroo (*Macropus robustus*)	Queensland, New South Wales	

Figures

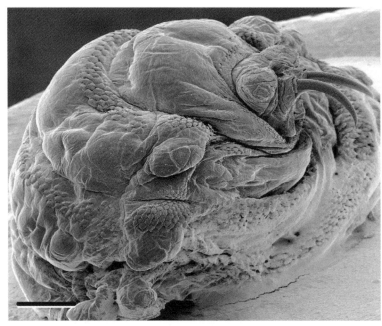

Fig. 14.29. Scanning electron micrograph of the cephalic and thoracic segments of a third-instar *Tracheomyia macropi* recovered from a red kangaroo (bar = 0.6 mm).

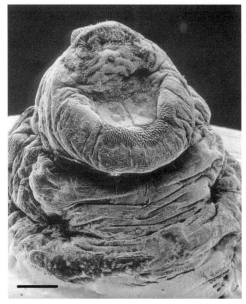

Fig. 14.30. Scanning electron micrograph of the terminal abdominal segments of a third-instar *Tracheomyia macropi* recovered from a red kangaroo (bar = 0.6 mm).

Cephalopina Strand

Biology

Females larviposit on the host nares. Several females may hover near the host's head simultaneously and each may 'strike' several times, depositing small packets of larvae. All instars can be found resident in the nasal cavities, although most second and third instars migrate to the pharynx to complete their development before exiting via the lower nasal meatus or nostrils (Zayed, 1998). Mature larvae migrate from the nasal cavities, usually late in the day and either crawl out of the nares or are expelled by forceful sneezing. Pupal development requires 20–30 days. This species is univoltine in Asian localities, but appears to have two generations in northern Africa. Adults live for 4–38 days with the shorter lifespans being noted in Asia. Each female may have 600–900 eggs. Camel nasal bot fly appears to have spread to every region where camels (*Camelus dromedaries*) have been introduced, including Australia (Spratt, 1984).

Disease, pathogenesis and significance

Adult activity does not apparently elicit a reaction from the hosts, but larval movements within the nasal cavities may induce violent sneezing. In many cases the larvae induce severe lesions which, however, tend to be fairly localized and may not necessarily cause significant symptoms (Hussein *et al.*, 1983). Larvae can induce a considerable discharge of mucous and secondary bacterial infections can result. In a small number of cases (e.g. 3%, Al-Ani *et al.*, 1991) the secondary infections can lead to neurological disorders such as depression, listlessness, walking in circles and stiffness. In Morocco, where rabies is endemic, there is a risk of confusion of *Cephalopina* infestation with the clinical signs of rabies (K. Khallayyoune, personal communication). Occasionally, large numbers of larvae in the nasal cavities may physically block or reduce air flow creating discomfort.

Management

Treatment with macrocyclic lactone anti-parasitics will give effective control (Sharma, 1992, see also review by Marley and Conder, 2002).

Hosts and distribution

Table 14.9. Hosts and distribution (*Cephalopina*).

Species	Hosts	Distribution	Common names
titillator	Camelids (dromedary)	North Africa, Arabian peninsula, Australia	Camel nasal bot fly

Figures

Fig. 14.31. Lateral view of a third-instar *Cephalopina titillator* recovered from a camel (bar = 2 mm).

Fig. 14.32. Dorsal view of a third-instar *Cephalopina titillator* recovered from a camel (bar = 2 mm).

Fig. 14.33. Ventral view of a third-instar *Cephalopina titillator* recovered from a camel (bar = 2 mm).

Fig. 14.34. Scanning electron micrograph of the cephalic and thoracic segments of a third-instar *Cephalopina titillator* recovered from a camel (bar = 0.5 mm).

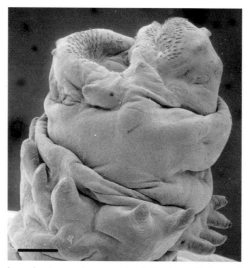

Fig. 14.35. Scanning electron micrograph of the lateral view of the terminal abdominal segments of a third-instar *Cephalopina titillator* recovered from a camel (bar = 0.6 mm).

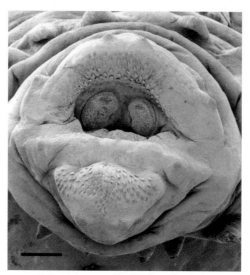

Fig. 14.36. Scanning electron micrograph of the terminal abdominal segment of a third-instar *Cephalopina titillator* recovered from a camel (bar = 0.5 mm).

Hypoderma Latreille

Biology

The six species widely recognized in the genus *Hypoderma*, commonly called heel flies, warble flies or cattle grubs, are restricted to the Holarctic region where they parasitize various Bovidae and Cervidae. The adult flies are either medium to large sized (11–18 mm) bumblebee mimics (*Hypoderma bovis*, *Hypoderma lineatum*,

Hypoderma tarandi) or slightly smaller (11–14 mm) and less hairy flies. Nilssen *et al.* (2000) used robust clustering and multidimensional scaling analyses of the colour patterns of bumblebees and two oestrids to confirm statistically the striking visual resemblance between *H. tarandi* and *Bombus lapponicus*. Most information on oestrid biology comes from studies of the three species of major economic importance, *H. bovis* and *H. lineatum* on cattle and *H. tarandi* on reindeer. Adults are especially active on warm, sunny days. Despite their inability to feed, *H. tarandi* are capable of lengthy flights, an adaptation to pursue their migratory reindeer hosts (Nilssen and Anderson, 1995a). Females of *Hypoderma* lay their eggs on the hairs of the host's lower body and legs, either singly (*H. bovis*) or in batches of up to 15, laid side by side along the same hair shaft (*H. lineatum, H. tarandi*). In *H. lineatum*, at least, females actively select hairs of a particular diameter on which to deposit eggs. Thus, while the mean diameter of randomly selected hairs was 0.04 mm, that of hairs with eggs was 0.07 mm and thicker hairs tended to have more eggs (Jones, 2000). Such selection may be an adaptation to shorten incubation and promote hatching. Karter *et al.* (1992) showed that under controlled conditions in the laboratory, *H. tarandi* oviposits close to the base of newly grown hairs and that newly hatched larvae show a positive thermotaxis. They concluded that these behaviours also promote near-maximum hatchability, a short incubation period and high transmission efficiency. A specialized basal attachment organ with a prominent clasper is found on the eggs, enabling them to lock firmly around the hair shaft and attachment is further enhanced by an adhesive material (Cogley *et al.*, 1981).

After hatching, the first instars penetrate the skin and undergo a migration through the body tissues to overwintering sites. Routes of migration and sites for overwintering differ between species of *Hypoderma*. Larvae of *H. bovis* migrate along nerves to the spinal cord and overwinter within the epidural fat around the thoracic and lumbar vertebrae. Larvae of *H. lineatum* migrate along connective tissues and between the fascial planes of muscles to overwinter within the submucosa of the oesophagus. Migration routes of *H. lineatum* are characterized by yellow or greenish gelatinous, oedematous areas with an overwhelming eosinophil infiltration (Panciera *et al.*, 1993). The larvae resume migration the following spring and move to the back where a small, subcutaneous swelling develops and the growing larva cuts a hole to the skin surface. The larva orientates in the swelling, the 'warble', so that its posterior respiratory spiracles are placed at the opening of the hole, enabling the larva to breathe as it grows, feeding on cellular debris and exudates. Larvae exit their warbles after 30–60 days, falling to the ground where they pupariate. After eclosion, the adults live for only a few days. Those of *H. bovis* live on average for 3–4 days and the females are reproductively well adapted to this short, non-feeding lifestyle; they emerge from the puparium with all their eggs fully developed, and the capacity to immediately mate and then oviposit on cattle (Scholl and Weintraub, 1988). Mating takes place away from the host, at aggregation points where females are intercepted in flight or when they land close to a male (Anderson *et al.*, 1994).

Disease, pathogenesis and significance

The species of major economic importance are *H. bovis* and *H. lineatum*. Their impact is due to the breathing holes of the warbles, which are enlarged during development and significantly degrade the value of the hide, resulting in serious losses to the cattle industry of the Holarctic region (Scholl, 1993). Cattle grubs were estimated to cause annual losses (excluding control) of over US$600 million in the USA (Drummond *et al.*, 1981). More recent data from northern China indicate annual losses to hides of US$15,000,000 (Yin *et al.*, 2003). Hides of reindeer are similarly affected by *H. tarandi*, with records from Alaska of up to 2000 warble scars per hide (Washburn *et al.*, 1980). In the Magadan region of the far east of the former USSR, 23% of deer hides were downgraded due to warble fly damage (Shumilov and Nepoklonov, 1983). This reflects the high prevalence of infestations by all species of *Hypoderma* in endemic areas. For cattle warbles, prevalence in Europe ranges from 0 to over 80%, depending on the levels of control (Colebrook and Wall, 2004). In China the infestation rate in yaks is generally 80–90% but can rise to 100% (Yin *et al.*, 2003). In northern Norway, Folstad *et al.* (1991) recorded 99.9% prevalence of infection of reindeer with *H. tarandi*. Prevalences of *Hypoderma diana* and *Hypoderma actaeon* in deer are typically over 50% and up to 90–100% (Colebrook and Wall, 2004).

Compared with losses to the hide and packing industry, other production losses from hypodermosis are probably small, e.g. losses due to decreased weight gain and milk production, but these are more difficult to quantify. The persistence of the females, in particular those of *H. bovis*, in laying from 300 to 800 eggs induces a dramatic escape response in cattle, termed 'gadding'. This behaviour is thought to be a potential cause of injury, spontaneous abortion and reduced milk production but, again, the losses are difficult to quantify. Although the escape responses are less dramatic in reindeer when *H. tarandi* are egg-laying, *H. tarandi* (with *Cephenemyia trompe*) have been shown to cause a major disturbance to reindeer in the summer in Norway, having a negative impact on their grazing behaviour with a probable negative impact on nutrition and autumn carcass weights: surprisingly, the reindeer did not compensate for decreased daylight grazing by increased grazing at night, when insect harassment was low (Colman *et al.*, 2003). In Canada also, these two oestrid species have a significant effect on caribou activity budgets as they increase the time spent standing and moving and decrease the time spent feeding and resting (Downes *et al.*, 1986).

Migrating first instars of *Hypoderma* secrete a suite of three serine proteases. Two of these act to inhibit host immunity in animals infested for the first time, by hindering the inflammatory response around the larva and inducing low lymphocyte proliferation (Otranto, 2001). However, resistance to *Hypoderma* does develop in infested animals, such that larval survival is reduced on subsequent infestations. This immune response has been the stimulus for a great deal of research into a possible vaccine against *Hypoderma* species, but the major benefits of the immunological studies to date are tools for detection of infestation by serology. These are now used routinely, not only to examine blood samples of individual animals (Tarry *et al.*, 1992) but also to examine pooled milk samples from dairy herds.

Occasionally, mammals other than the natural hosts can be infested by larvae of *Hypoderma*, including humans (see Chapter 12). Hypodermosis in man most frequently features skin allergies accompanied by blood eosinophil differential counts varying from subnormal to 60% above the normal (Boulard and Petithory, 1977).

A rare case of hypereosinophilia of a 2-year-old boy was reported by Navajas *et al.* (1998), with fever, muscle pain and heart, CNS and skin involvement. The condition resolved completely when several larvae of *H. bovis* were removed from scalp nodules. The severity of human infections varies with the site of the larvae, from a 'creeping myiasis' caused by subdermal migrations (Uttamchandani *et al.*, 1989) to ophthalmomyiasis interna resulting in visual loss (Edwards *et al.*, 1984; Kearney *et al.*, 1991), to rare intracerebral myiasis (Kalelioglu *et al.*, 1989), which has also been reported in calves (Caracappa *et al.*, 1996).

Management

Control of hypodermosis in the individual animal or within a herd is nowadays a relatively straightforward matter. Following the widespread use of systemic organophosphates against migrating larvae from the 1950s, in the early 1980s ivermectin was used and found to be highly effective against larvae in warbles as well as those migrating through the body. Subsequently, moxidectin (200 µg/kg against *H. lineatum*) and doramectin (200 µg/kg subcutaneous against *H. bovis*) were shown to be equally effective in cattle. A topical formulation of ivermectin was 100% effective against migrating first-instar *H. lineatum* for 3 weeks following treatment (Colwell and Jacobsen, 2002). The advantages of killing the larvae as soon as possible are that the adverse reactions that can result from death of large numbers of larvae are reduced. Disadvantages of the use of ivermectin are the lengthy preslaughter withdrawal period and a restriction on use in lactating animals (Jackson, 1989). Macrocyclic lactones such as eprinomectin, which does not have a withdrawal period for lactating cattle, could be the drug of choice for dairy animals.

Ivermectin was 100% effective against *H. tarandi* in reindeer, whether given by injection, orally or as a pour-on. However, oral preparations were less effective against nematodes and therefore this formulation was not recommended when given as part of a broad-spectrum antiparasite treatment. Doramectin was similarly effective, but moxidectin was less effective (93%) (Oksanen, 1999).

Despite the relative ease of control using macrocyclic lactones, even in herds that have consistent therapeutic treatment for hypodermosis, annual reinvasion of flies from outside the treated area produces a continual need for treatments (Colwell, 2000, 2002). Area-wide control is difficult to achieve for similar reasons. Hence, even on a small island with limited numbers of animals, all treated with ivermectin, eradication was not achieved, probably due to reinvasion by flies from the mainland, 2–3 km away (Nilssen *et al.*, 2002). However, eradication of hypodermosis has been achieved in the UK and in Ireland. The campaign was launched in 1978, at which time there were an estimated 4 million infected cattle. Control was based on a combination of the voluntary use of pour-on organophosphorous treatments, with compulsory treatment of cattle showing *Hypoderma* larvae in the spring, plus appropriate movement restrictions. It was immediately successful, with a decrease from the original infestation levels of about 40% to less than 1% within 4 years. From 1982 it became compulsory to treat a whole herd in which *Hypoderma* was found (Tarry, 1986) and the programme was so effective that by the end of the decade it was only possible to detect infestations by serological analysis of hosts. By spring 1991, all blood samples from 227,000 cattle were sero-negative (Tarry *et al.*, 1992). Similarly, all samples were negative in the annual survey of 2002–2003, which tested 200,769 animals in 5189 herds (Swallow, 2003). Other European countries are adopting control campaigns that embrace the concepts of simultaneous treatments of live-stock combined with legislation to ensure treatments are compulsory and follow-up epidemiological surveys to ensure continued effectiveness of the control campaign. A good example of this is France, which is almost free of hypoder-mosis (Boulard, 2002).

The sterile insect technique (SIT) has been considered as a method for con-trol/eradication of *Hypoderma*. The major constraint is the difficulty in produc-ing sterile flies due to the lack of an *in vitro* rearing method. In spite of these difficulties, a trial release of sterile males of *H. bovis* and *H. lineatum* was successful in eliminating both species, one species each from two zones of 625 km² along the US–Canadian border (Kunz *et al.*, 1990). The trial followed earlier insecti-cidal applications to cattle to reduce the wild populations to a low level. This integrated approach is an attractive option for the future, because the combined treatment attacks both immature and adult stages of the parasites in a comple-mentary manner. However, before large-scale integrated control programmes using SIT can be initiated, *in vitro* rearing techniques would have to be devel-oped. The responses of *H. tarandi*, and presumably other species, to host odours (Tommerås *et al.*, 1996) suggests that trapping (Anderson and Nilssen, 1996a) could also play a part in future management techniques for *Hypoderma*.

Hosts and distribution

Table 14.10. Hosts and distribution (*Hypoderma*).

Species	Hosts	Distribution	Common names
actaeon	*Cervus elaphus*	Palaearctic	Red deer warble fly
bovis	*Bos taurus*	Holarctic	(Larger) cattle warble fly
diana	*Cervus elaphus, Capreolus capreolus, Ovis orientalis*	Palaearctic	Roe deer warble fly
lineatum	*Bos taurus*	Holarctic	(Lesser) cattle warble fly
tarandi	*Rangifer tarandi*	Holarctic	Reindeer warble fly
sinensis*	*Bos grunniens* (yak)	Palaearctic	

* This species is known from three adult specimens deposited by Pleske in the Zoological Museum of St. Petersburg and was synonymized with *Hypoderma lineatum* by Grunin who examined the specimens. A distinct species is known from yaks and recent molecular evidence indicates that the specimens used by Pleske are identical to the larvae from yaks and therefore *H. sinensis* has been revived (Otranto *et al.*, 2004).

Figures

Fig. 14.37. *Hypoderma bovis* male aggregation site in southern Alberta, Canada. J. Weintraub indicating location of males along dry stream bed.

Fig. 14.38. *Hypoderma bovis* female resting.

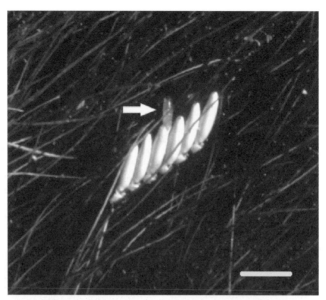

Fig. 14.39. Eggs of *Hypoderma lineatum*. Note the emerging larva (arrow).

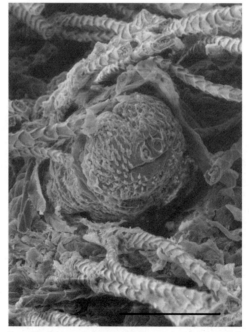

Fig. 14.40. Scanning electron micrograph of newly hatched *Hypoderma lineatum* penetrating host skin. Only the posterior abdominal segments, with posterior spiracles evident, are visible (bar = 90 µm).

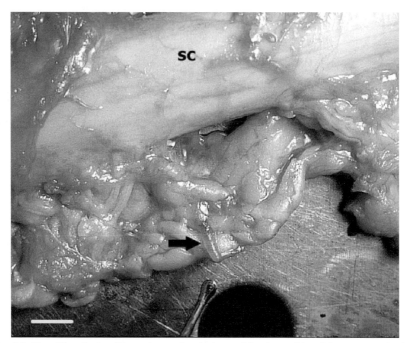

Fig. 14.41. First-instar *Hypoderma bovis* (arrow) in the epidural fat; sc = spinal chord (bar = 1.0 cm).

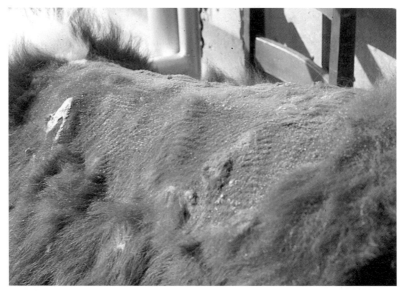

Fig. 14.42. Back of bovine showing numerous 'warbles' containing third instars. The long winter hair-coat has been shaved to reveal the warbles. Note purulent exudate from one 'warble'.

Strobiloestrus Brauer

Biology

Eggs are attached to hair shafts using an attachment organ similar to that of *Hypoderma* (see Howard, 1980, for *Strobiloestrus vanzyli*). First instars presumably penetrate the skin and undergo a subdermal migration although this has not been clearly established. Subdermal first instars tend to be found on the dorsal thoracic and lumbar regions.

Disease, pathogenesis and significance

There are no descriptions of disease associated with infestations, despite indications that the intensity can be very high. The overall impact on the hosts has not been described.

Management

None has been shown to be necessary as the impact appears limited and there are no other known hosts that are affected. The absence of significant disease or pathology suggests that treatment is not required in the management of wild host populations. However, it may be presumed that treatment with macrocyclic lactone anti-parasitics will give effective control (see Chapter 13). There has been no published work to verify these suggestions.

Hosts and distribution

Table 14.11. Hosts and distribution (*Strobiloestrus*).

Species	Hosts	Distribution	Common names
clarkii	*Oreotragus oreotragus* (klipspringer), *Raphicerus campestris* (steinbok), *Pelea capreolus* (grey rhebok), *Redunca fulvorufula, Redunca arundinum* (reedbuck), *Tragelaphus strepsiceros, Capra hirsicus* (domestic goat)	South Africa	Clark's antelope warble fly
ericksoni	*Kobus leche* (Lechwe)	'Southern Congo'	Northern lechwe warble fly
vanzyli	*Kobus leche* (Lechwe)	'Northern Rhodesia'	Southern lechwe warble fly

Figures

Fig. 14.43. Adult male *Strobiloestrus clarkii* (bar = 0.5 cm).

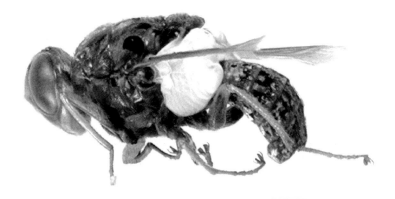

Fig. 14.44. Adult male *Strobiloestrus vanzyli* (bar = 0.5 cm).

Fig. 14.45.   Ventral view of a second-instar *Strobiloestrus vanzyli* (bar = 2 mm).

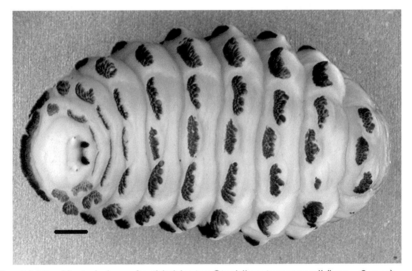

Fig. 14.46.   Ventral view of a third-instar *Strobiloestrus vanzyli* (bar = 2 mm).

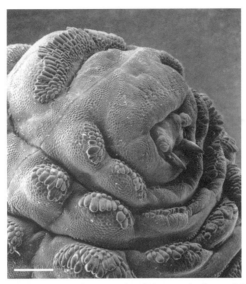

Fig. 14.47. Scanning electron micrograph of the cephalic and thoracic segments of a third-instar *Strobiloestrus vanzyli* (bar = 0.9 mm).

Fig. 14.48. Scanning electron micrograph of the cephalic and thoracic segments of a second-instar *Strobiloestrus vanzyli* (bar = 0.75 mm).

Portschinskia Semenov

Larvae of this genus are subcutaneous parasites of rodents and lagomorphs. The adults are large with a striking resemblance to bumblebees, which is suggestive of mimicry (see Chapters 6 and 10).

Biology

Details are known for only *Portschinskia magnifica*, which infests the field vole, *Apodemus speciosus*. Voles generally have only a single larva that is found within a warble on the ventral surface usually near the tail. Larval development requires approximately 2 months and there is a distinct pupal diapause for overwintering.

Disease, pathogenesis and significance

Zumpt (1965) indicates that infested mice are sluggish, and easily approached, thus making them susceptible to predation. He also indicates that mice with 'large' numbers of larvae often die. No further descriptions of pathology or disease have been made.

Management

None has been shown to be necessary as the impact appears limited and there are no other known hosts that are affected. The absence of significant disease or pathology suggests that treatment is not required in the management of wild host populations. However, it may be presumed that treatment with macrocyclic lactone anti-parasitics will give effective control (see Chapter 13). There has been no published work to verify these suggestions.

Hosts and distribution

Table 14.12. Hosts and distribution (*Portschinskia*).

Species	Hosts	Distribution
magnifica	*Apodemus speciosus*	Asia
loewi	*Ochotona alpina*	Central Asia
neugebaueri	Unknown	Swiss Alps
himalayana	Unknown	Garhwal district, India
gigas	Unknown	Amur River
bombiformis	Unknown	Szechuan, China
przewalskyi	Unknown	Zinchai, China

Figures

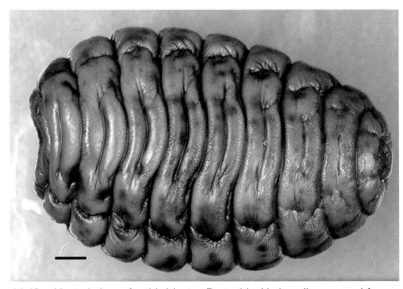

Fig. 14.49. Ventral view of a third-instar *Portschinskia loewii* recovered from *Ochotona alpina* (bar = 2 mm).

Fig. 14.50. *En face* view of cephalic, thoracic and first abdominal segments of a third-instar *Portschinskia loewii* recovered from *Ochotona alpina* (bar = 2 mm).

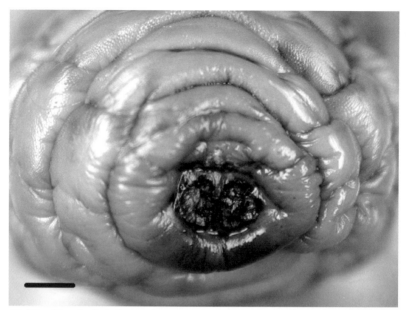

Fig. 14.51. Terminal abdominal segment of a third-instar *Portschinskia loewii* recovered from *Ochotona alpina*. Note the typical spiracular plates (bar = 2 mm).

Pallasiomyia Rubtzov

This is a monospecific genus known only from second and third instars that are found as subcutaneous parasites. The main host, the saiga antelope (*Saiga tatarica*), is an endangered species (Milner-Gulland *et al.*, 2003).

Biology

The single species is known from early descriptions of second and third instars recovered from the type host. Specimens in the collection at the Natural History Museum (London) recorded as originating from the Tibetan gazelle are unverified. There is absolutely no information available on the biology of this species.

Disease, pathogenesis and significance

Impact on the hosts is largely unknown. Apparently heavy infestations affect the behaviour of the host, causing them to have less fear of predators (e.g. hunters). Zumpt (1965) also noted that hides of heavily infested antelopes were not usable and that the meat was unpalatable. Given the recent heavy exploitation of the host species for meat and hides (Bekenov *et al.*, 1998) there is a surprising absence of data on occurrence of the parasite and its effects on the host.

Since the main host is an endangered species (Milner-Gulland *et al.*, 2003; Huffman, 2004) there may be a need to investigate the impact of this parasite. Current precipitous decline, related to overhunting and selective poaching of males, which has apparently skewed sex ratios (Milner-Gulland *et al.*, 2003), may be exacerbated by parasitism. However, such a dramatic decline in primary host numbers will also have likely led to a decline in parasite numbers.

Management

Treatment with macrocyclic lactone anti-parasitics would presumably result in effective control (see Chapter 13). There has been no published work to verify these suggestions.

Hosts and distribution

Table 14.13. Hosts and distribution (*Pallasiomyia*).

Species	Hosts	Distribution	Common name
antilopum	*Saiga tatarica* (Saiga antelope), *Procapra picticaudata* (Tibetan gazelle) (NHM collection)	Kazakhstan, Mongolia, Tibet	Saiga warble fly

Figures

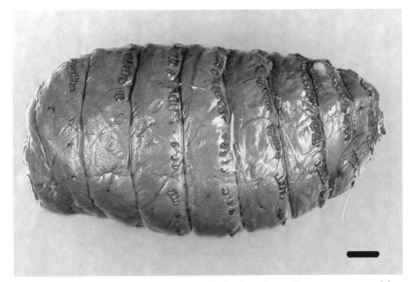

Fig. 14.52. Lateral view of a third-instar *Pallasiomyia antilopum* recovered from *Procapra picticaudata* (Tibetan gazelle) (NHM collection) (bar = 2 mm).

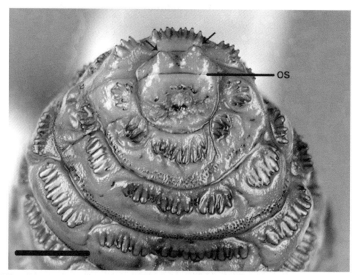

Fig. 14.53. Ventral view of the cephalic and thoracic segments of a third instar *Pallasiomyia antilopum* recovered from *Procapra picticaudata* (Tibetan gazelle) (NHM collection). Note the characteristic fleshy, conical protuberances (arrows) dorsal to the opercular scar (os) (bar = 2 mm).

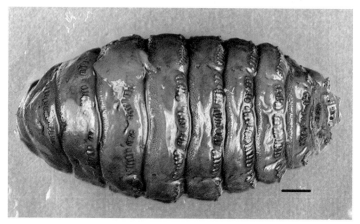

Fig. 14.54. Ventral view of a third instar *Pallasiomyia antilopum* recovered from *Procapra picticaudata* (Tibetan gazelle) (NHM collection) (bar = 2 mm).

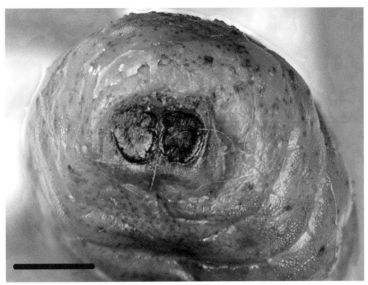

Fig. 14.55. Posterior abdominal segments of a third instar *Pallasiomyia antilopum* recovered from *Procapra picticaudata* (Tibetan gazelle) (NHM collection) showing the posterior spiracular plates (bar = 2 mm).

Ochotonia Grunin

Biology

This monospecific genus is known only from third instars collected as subcutaneous parasites of lagomorphs, in particular the pika. No information is available on the biology of this species.

Disease, pathogenesis and significance

There is no information available.

Management

In the absence of information on the biology and host–parasite interactions of this species there is no requirement for management of this infestation.

Hosts and distribution

Table 14.14. Hosts and distribution (*Ochotonia*).[a]

Species	Hosts	Distribution
lindneri	*Ochotona* spp.	Asia

[a] Known from a single third instar.

Oestroderma Portschinsky

Biology

This is a monospecific genus, whose larvae are subcutaneous parasites of the pika. Females attach eggs to the host hair using an attachment organ similar to that known for *Hypoderma*. No further details of the life cycle are known.

Disease, pathogenesis and significance

There is no information on the host–parasite interaction of this genus.

Management

The absence of information on the host–parasite interaction leaves open the question of managing the parasite.

Hosts and distribution

Table 14.15. Hosts and distribution (*Oestroderma*).

Species	Hosts	Distribution
potanini	*Ochotona alpina* (Altai pika)	Central Asia

Oestromyia Brauer

Biology

Eggs are individually attached to the base of host hair with an attachment organ typical for the subfamily. Eggs of this genus are characterized by a terminal umbrella-like elaboration not associated with the attachment and whose function is unknown (Grunin, 1975; Rietschel, 1980). Eggs of *Oestromyia marmotae* are found on the ventral surface and on the legs of marmots (Grunin, 1975). Larvae hatch within

48 h of deposition and quickly penetrate the skin. Young rodents are occasionally infested with larvae of *Oestomyia leporina* that have migrated from their mothers (Rietschel, 1975a). First instars penetrate the upper layers of the skin and migrate for 3–4 days. The migration is subcutaneous, in connective tissue, often terminating near the posterior regions of the host. Subsequently, the larvae digest a respiratory opening in the skin and moult to the second instar. A host reaction or 'warble' forms around the second instar. Larval development requires 26–40 days.

Hosts infested with *O. leporina* tend to have single larvae (Volf *et al.*, 1990) while reports for marmots infested with *O. marmotae* harboured 7–21 larvae.

O. leporina has a single generation per year and pupae undergo a diapause, which can be artificially interrupted by exposure of the pupae to freezing temperatures (Rietschel, 1980).

Disease, pathogenesis and significance

There has been little indication of significant pathology or disease associated with infestation.

Management

None has been shown to be necessary as the impact appears limited and there are no other known hosts that are affected. The absence of significant disease or pathology suggests that treatment is not required in the management of wild host populations. However, it may be presumed that treatment with macrocyclic lactone anti-parasitics will give effective control (see Chapter 13). There has been no published work to verify these suggestions.

Hosts and distribution

Table 14.16. Hosts and distribution (*Oestromyia*).[a]

Species	Hosts	Distribution
leporina	(Europe) *Microtus arvalis, Microtus agrestis, Pitymus subterraneus, Arvicola terrestris, Ondatra zibethica* (Asia) *Ochotona alpina, Ochotona daurica, Microtus oeconomus, Microtus gregalis, Citellus undulatus*	Palaearctic
marmotae	*Marmota caudate*	Central Asia
prodigiosa	*Ochotona daurica, Ochotona pallasii*	Central Asia

[a] Two other species have been described (*O. koslowi, O. scrobiculigera*), but only scant details are available.

Pavlovskiata Grunin

Biology

This genus is monospecific. Larvae are subcutaneous parasites of the goitered gazelle in central Asia. Females, active in September, attach eggs to hair of both forelegs and hind legs, usually close to the hoof. There is no information on biology of the larval stages except for notes by Zumpt (1965) that indicate third instars are present in subcutaneous cysts in July and August. Infestations can be large, with Grunin (1962) reporting 132 larvae on one host.

Disease, pathogenesis and significance

There is no information on the host responses beyond the development of subcutaneous cysts. Similarly, there have been no studies to describe disease associated with infestations of this species. The goitered gazelle is the only known host. The host's conservation status is 'near threatened' (Huffman, 2004) and until more information is available on the impact of this parasitism it is difficult to assess its importance in the management of the species.

Management

Requirement for management is unknown. The absence of described disease or pathology and the absence of other known hosts suggest that treatment is not required in the management of wild host populations. However, it may be presumed that treatment with macrocyclic lactone anti-parasitics will give effective control (see Chapter 13). There has been no published work to verify these suggestions.

Hosts and distribution

Table 14.17. Hosts and distribution (*Pavlovskiata*).

Species	Hosts	Distribution	Common names
subgutturosae	*Gazella subgutturosae*	Central Asia	Pavlovsky's gazelle warble fly

Przhevalskiana Grunin

Biology

Larvae are subcutaneous parasites of goats, sheep and gazelles across a wide area of the Palaearctic. Females attach eggs, often in groups, to hair shafts on the back and flanks of the goats. First instars penetrate the skin using enzymes similar to those described for *Hypoderma* spp. Disagreement about larval migrations within the host were put to rest by Otranto and Puccini (2000) who established that larvae develop subcutaneously at the site of penetration. There was no evidence for significant migration as seen with the cattle species of *Hypoderma*. First instars are apparently anaerobic, living within a small cyst that has no opening to the exterior. First instars grow quickly to about 5 mm in length, but growth slows during the period from June to November when they moult to second instar. Second instars open a small hole to the outside, which enlarges as the larvae grow. Second- and third-instar growth is rapid and dramatic, with the third instars nearly doubling in length in a month (Otranto and Puccini, 2000).

There are indications that goats develop protective or at least partially protective immunity following a primary infestation (Otranto and Puccini, 2000). This is evidenced by the presence of dead first instars in older goats. However, more detailed research is required as no significant differences in larval burdens were evident in goats between 1–3 and 4–6 years of age (Otranto and Puccini, 2000).

In southern Italy flies are active from April through June. Larvae have been found in the subcutaneous tissues from May through the following February. Larvae exit the host and pupate in the soil, with the pupal development requiring 30–60 days.

Recent molecular studies have demonstrated that *Przhevalskiana silenus*, *Przhevalskiana crossii* and *Przhevalskiana aegagri* are a single species (Puccini and Otranto, 2000; Otranto and Traversa, 2004). This suggests a degree of morphological variation that may be associated with host effects.

Disease, pathogenesis and significance

The host response around each larva is a typical granulomatous cyst seen with other members of the subfamily. No other pathological or physiological changes in the host have been described.

There are reports of economic impacts on milk and meat production (see Puccini and Otranto, 2000), but there is generally very little well-documented research.

Management

Infestations can be controlled through the use of either organophosphate systemic insecticides or macrocyclic lactone anti-parasitics. Timing of treatment to coincide with the end of fly activity will allow optimal control and limit the economic consequences of the infestation.

Hosts and distribution

Table 14.18. Hosts and distribution (*Przhevalskiana*).

Species	Hosts	Distribution
silenus	*Gazella dorcus*, *Gazella granti*, *Ovis ammon*, domestic sheep (*Ovis aries*), domestic goats (*Capra hircus*)	Middle East, Mediterranean, central Asia, north and central Africa
silenus shugnanica	*Ovis ammon*	Pamir (central Asia)
corinnae	*Gazella dorcus*, *Gazella subgutturosa*	North Africa, central Asia

Two additional species, *P. aenigmatica* from the 'Mongolian gazelle' (probably *Gazella subguttorosa*) and *P. orongonis* from the rare Tibetan antelope or Chiru (*Pantholops hodgsonii*) have been described, but are known only from second and third instars.

Figures

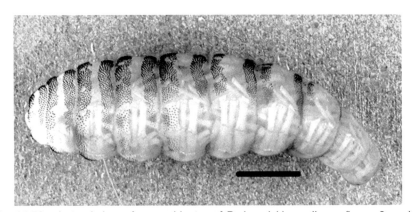

Fig. 14.56. Lateral view of second instar of *Przhevalskiana silenus* (bar = 2 mm).

Fig. 14.57. Dorsal view of second instar of *Przhevalskiana silenus* (bar = 2 mm).

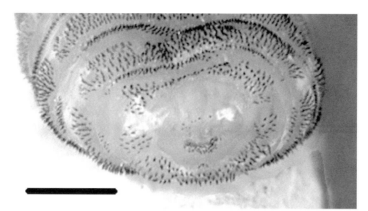

Fig. 14.58. *En face* view of second instar of *Przhevalskiana silenus* (bar = 2 mm).

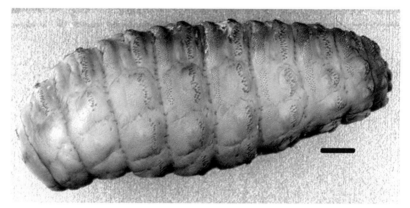

Fig. 14.59. Lateral view of third instar of *Przhevalskiana silenus* (bar = 2 mm).

Fig. 14.60. *En face* view of third instar of *Przhevalskiana silenus* (bar = 2 mm).

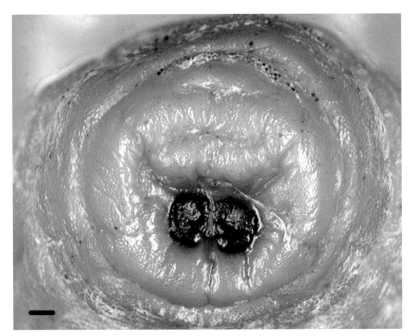

Fig. 14.61. Terminal abdominal segment of third-instar *Przhevalskiana silenus* showing paired posterior spiracular plates (bar = 2 mm).

Cuterebra Clark

This genus contains a large number of species whose larvae parasitize rodents, lagomorphs, marsupials and, in one case, monkeys, of the New World. Recently, several other genera (e.g. *Alouattamyia, Metacuterebra, Andiocuterebra*, etc.) have been synonymized with *Cuterebra*, thereby increasing the number of species (see Chapter 3, but also see comments in Chapter 6, which give a dissenting opinion on the genus *Andiocuterebra*). Extensive treatment of the Nearctic species is given in Sabrosky (1986) and of the Neotropical species in Guimarães and Papavero (1999). Therefore, we have limited treatment of this genus to the following summary of the biology.

Biology

Mated females oviposit in a locality frequented by potential hosts. Location of oviposition sites may have a role in the general agreement that members of this genus are highly host-specific (see Colwell and Milton, 1998). High numbers of tarsal sensillae suggest that females are stimulated to oviposit by the presence of low concentrations of host-associated compounds. (Females have higher numbers of sensillae than males.) The eggs are firmly attached to the substrate with an adhesive material. Eggs require 4–10 days to embryonate (Catts, 1982). Hatch occurs in response to sudden temperature increase associated with a nearby potential host. Not all eggs in a clutch will hatch synchronously.

Newly hatched larvae are active, searching for a host and once on the host, for a suitable entry site. Moist body openings (e.g. nares, mouth, eyes) are the portal of entry, the larvae being unable to penetrate unbroken skin. Following entry into the host, larvae undergo a brief migration, probably within the body cavity, of 3–5 days after which they arrive at subcutaneous locations where a granulomatous cyst forms around the developing larvae. Location of the cysts tends to be species-specific (e.g. around the throat of howler monkeys infested with *Cuterebra baeri*, see Milton, 1996). Larvae complete development within a period of 20–40 days and then exit the cyst through the breathing hole that opens just as the cyst is forming. Pupariation takes place under the surface litter or deep in the soil. Species in northern latitudes often exhibit a pupal diapause that aids in survival through winter.

Male flies exhibit classic lekking behaviour, gathering at prominent ecological features where they wait for passing females (see Chapter 10). The features are highly species-specific and there are cases where a feature is subdivided between two species. Consistent use of the same feature for extended periods has been reported.

Disease, pathogenesis and significance

Infestations produce no significant pathology or disease symptoms other than the characteristic cyst. There is some debate regarding overall impact on the host, with some authors suggesting that infestations affect survival and others suggesting no impact on host survival or productivity. Two instances where impact on host survival is strongly suggested are the infestation of Ord's kangaroo rat (*Dipodomys ordii*) and the howler monkey (*Alouatta palliate*). In the first case infestation of young kangaroo rats by *Cuterebra polita* affects overwintering survival because of depletion of nutrient reserves (Gummer *et al.*, 1990). In the second case survival of juvenile monkeys with infestations of *C. baeri*, on Barro Colorado Island, Panama, is dramatically reduced during the rainy season when nutrient availability is limited (Milton, 1996).

Management

The absence of significant disease or pathology suggests that treatment is not required in the management of wild host populations. However, it may be presumed that treatment with macrocyclic lactone anti-parasitics will give effective control (see Chapter 13). There has been no published work to verify these suggestions.

Hosts and distribution

Table 14.19.   Hosts and distribution (*Cuterebra*).

Species	Hosts	Distribution
fassleri (= *Andinocuterebra fassleri*)*	Unknown	Unknown
semiater (= *Pseudogametes semiater*)	Unknown	Brazil
hermanni (= *Pseudogametes hermanni*)	Unknown	Brazil
dasypoda (= *Rogenhofera dasypoda*)	Unknown	Brazil
gilvopilosa (= *Rogenhofera gilvopilosa*)	Unknown	Peru
grandis (= *Rogenhofera grandis*)	Rodents, *Sciurus aestrans*, *Akodon azarae*, *Akodon molinae*, *Reithrodon physodes*	Argentina
lopesi (= *Rogenhofera lopesi*)	Unknown	Brazil
trigonophora (= *Rogenhofera trigonophora*)	Unknown	Brazil
almeidai (= *Metacuterebra almeidai*)	Unknown	Brazil
apicalis (= *Metacuterebra apicalis*)	Rodents, *Holochilus brasiliensis*, *Oligoryzmys flavescens*, *Oryzomys capito*, *Pseudooryzoms wavrini*, *Zygodontomys brevicauda*; Marsupials, *Marmosa mitis isthimica*, *Metachirus nudicaudatus*	Neotropics

Continued

Table 14.19. *Continued.* Hosts and distribution (*Cuterebra*).

Species	Hosts	Distribution
baeri (= *Metacuterebra baeri*)	Howler monkey (*Alouatta palliata*, *Alouatta belsebul*), *Aotus trivirgatus*	Costa Rica, Panama, Brazil, Peru, Guyana
cayennensis (= *Metacuterebra cayennensis*)	Marsupials, *Caluromys* spp, *Didelphis* sp.	Nicaragua, Panama, Surinam, Brazil
detrudator (= *Metacuterebra detrudator*)	Rodents, *Oryzomys capito*, *Proechimys* spp.; marsupials, *Marmosa murina*	Panama, Trinidad, Peru, Ecuador, Brazil
funebris (= *Metacuterebra funebris*)	Rodents, *Makalata armata*	Trinidad, Brazil
infulata (= *Metacuterebra infulata*)	Unknown	Brazil
megastoma (= *Metacuterebra megastoma*)	Unknown	Brazil
patagona (= *Metacuterebra patagona*)	Unknown	Argentina
pessoai (= *Metacuterebra pessoai*)	Marsupials, *Caluromys* spp	Trinidad, Brazil
rufiventris (= *Metacuterebra rufiventris*)	Rodents, *Nectomys squamipes*, *Oryzomys* sp	Peru, Ecuador, Brazil
simulans (= *Metacuterebra simulans*)	Marsupials, *Caluromys* spp., *Didelphis marsupialis*	Peru, Surinam, Brazil
townsendi (= *Metacuterebra townsendi*)	Marsupials, *Marmosa microtarsus*	Brazil
praegrandis	Unknown	Peru, Brazil
sabrosky	Unknown	Brazil
emasculator	*Tamias striatus*, *Sciurus carolinensis*	North America, east of the Great Plains
sterilator	Striped goper, *Spermophilus tridecemlineatus*	North America, USA, Illinois, Minnesota
fasciata	*Tamias striatus*	Central USA

Continued

Table 14.19. *Continued.* Hosts and distribution (*Cuterebra*).

Species	Hosts	Distribution
fontinella	Mice, *Peromyscus leucopus*, *Peromyscus maniculatus*, *Microtus* spp.	North America
cuniculi	Lagomorphs, *Sylvilagus palustris*	Southeastern USA
abdominalis	Lagomorphs, *Sylvilagus floridanus*	East and central USA
maculosa	Unknown	Panama, Guatemala
mirabilis	Lagomorphs, *Lepus californicus*	Southwestern USA
princeps	Lagomorphs, *Lepus alleni*	Mexico, southwestern USA
ruficrus	Lagomorphs, *Lepus californicus*	Southwestern USA
cochisei	Unknown	Arizona, USA
lepusculi	Lagomorphs, *Sylvilagus nuttalli*	Western USA
buccata	Lagomorphs, *Sylvilagus floridanus*	USA
albata	Lagomorphs, *Sylvilagus floridanus*	Arizona, California, USA
lepivora	Lagomorphs, *Sylvilagus audubonii*, *Sylvilagus nuttallii*	Western USA, primarily California
jellisoni	Lagomorphs, *Lepus californicus*	Western USA
postica	Unknown	Mexico
polita	Pocket gopher, *Thomomys talpoides*; Kangaroo rat, *Dipodomys ordii*	Western USA, western Canada
clarkii	Mice, *Peromyscus pectoralis*	Mexico
atrox	Unknown	Southwestern USA, Mexico
terrisona	Unknown	Central, southern Mexico, Guatemala
americana	Pack rat, *Neotoma floridana*	Southeast USA
bajensis	Pack rat, *Neotoma lepida*	California, USA

Continued

Table 14.19. *Continued.* Hosts and distribution (*Cuterebra*).

Species	Hosts	Distribution
enderleini	Pack rat, *Neotoma micropus*	Texas, USA
histrio	Pack rat, *Neotoma mexicana*	Mexico
indistincta	Unknown	Southwest USA
latifrons	Pack rat, *Neotoma fuscipes*	California, Oregon, USA
albipilosa	Unknown	Mexico, southern California, Arizona, USA
approximata	Mice, *Peromyscus maniculatus*	Western North America
arizonnae	Unknown	Mexico, Arizona, USA
austeni	Pack rat, *Neotoma spp.*	Southwest USA
neomexicana	Mice, *Peromyscus* spp.	Southwest USA
tenebrosa	*Neotoma* spp.	Western North America
tenebriformis	*Neotoma* sp.	Southwest USA

Figures

Fig. 14.62. Adult female of *Cuterebra baeri*. The fly was reared from a third instar recovered from a naturally infested howler monkey (*Alouatta palliata*) on Barro Colorado Island, Panama (bar = 0.25 cm).

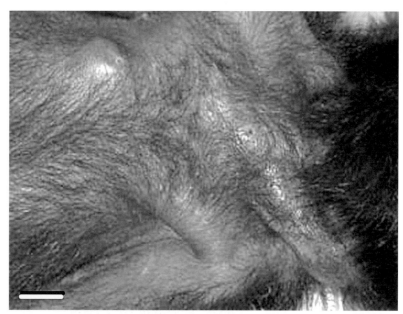

Fig. 14.63. Naturally infested howler monkey (*Alouatta palliata*) from Barro
Colorado Island, Panama. Note the small breathing holes made by the larvae and
the small size of the cyst, which are indicative of early developmental stage,
probably second instar (bar = 1 cm).

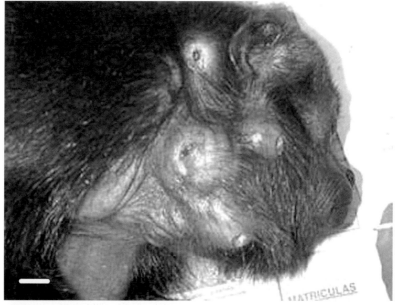

Fig. 14.64. Naturally infested howler monkey (*Alouatta palliata*) from Barro
Colorado Island, Panama. Note the large size of the cysts containing larvae and the
large breathing hole, which are indicative of third-instar development. Note also that
the cysts are located predominantly around the neck of the host (bar = 1 cm).

Fig. 14.65. Mature third-instar *Cuterebra baeri* recovered from a naturally infested howler monkey (*Alouatta palliata*) on Barro Colorado Island, Panama (bar = 1 cm).

Dermatobia L.

This monospecific genus is of great economic significance in the neotropics where it affects livestock production and also has zoonotic potential. Extensive literature on the biology and management of this parasite is amply summarized in Guimarães and Papavero (1999). Therefore, we have provided only a short synopsis of this genus.

Biology

Mated females capture zoophilous insects on the wing and attach up to 30 eggs to the captured insect's abdomen. The process is repeated until the female expends her egg complement, which may reach up to 800. Up to 1 week is required for the eggs to embryonate, after which they hatch in response to sudden increases in temperature associated with the porter insect landing on a potential host. A variety of insects are known to act as porters (see Guimarães and Papavero, 1999), but all are zoophilic, diurnal, moderate in size (flies that are too large or too small cannot be captured by the female *Dermatobia*) and moderately active (female *Dermatobia* are attracted and stimulated to pursue actively moving objects of the appropriate size).

Newly hatched larvae are able to penetrate directly into the skin where they complete their development without further migration. Distribution on the host varies dramatically as it is affected by the behaviour of the porter fly. Development requires 40–60 days in cattle, but is less in smaller hosts. Larval development takes place in a characteristic cyst or furuncle.

There appears to be little host specificity, with infestations reported from a large variety of domestic and wild mammals, including dogs, cats (wild and domestic), cattle, sheep, goats, various equines and humans (see Chapter 12). Mature larvae exit the host through the breathing hole of the cyst and quickly burrow beneath the surface, where pupariation takes place. Adults emerge 35–60 days later.

Adult males exhibit the lekking behaviour or hilltopping characteristic of many oestrids (Guimarães, 1966). Mating can occur within a few hours after the flies have emerged.

Disease, pathogenesis and significance

Development of the larvae has a major economic impact on domestic livestock and has resulted in a great deal of research on the biology and control of the insect and on the pathology of the infestations. In addition to the damage to hides there are indications of reduced weight gain and milk production. Generalized immunosuppression has also been described. Hosts develop limited immunity to reinfestation (see Chapter 11d).

Management

Current management of *Dermatobia hominis* in domestic livestock relies largely on the use of synthetic parasiticides (see Chapter 13).

Hosts and distribution

Table 14.20. Hosts and distribution (*Dermatobia*).

Species	Hosts	Distribution	Common names
hominis	Numerous mammals including cattle (*Bos taurus* and *Bos indicus*), domestic sheep, goats, equines, humans and some large birds	Tropical Central and South America	Human bot fly, numerous local names (see Guimarães and Papavero, 1999)

Figures

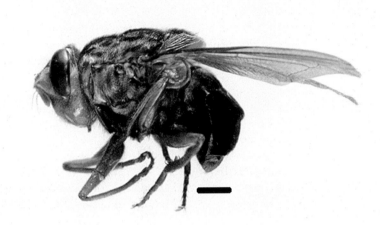

Fig. 14.66. Female *Dermatobia hominis* (bar = 2 mm).

Fig. 14.67. Muscoid flies with cluster of *Dermatobia hominis* eggs (bar = 2 mm) (photo by P.J. Scholl).

Fig. 14.68. Eggs' cluster on muscoid fly abdomen showing several newly emerged first instars (arrow) (bar = 2 mm) (photo by P.J. Scholl).

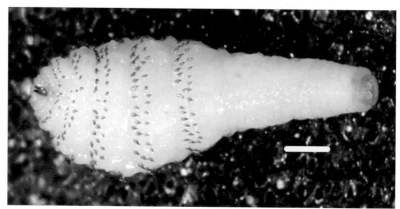

Fig. 14.69. Second-instar *Dermatobia hominis* (bar = 2 mm) (photo by P.J. Scholl).

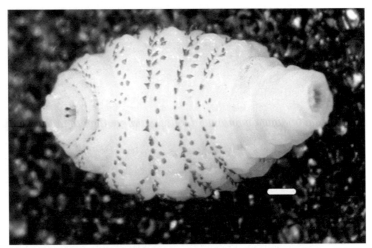

Fig. 14.70. Third-instar *Dermatobia hominis* (bar = 2 mm) (photo by P.J. Scholl).

Fig. 14.71. Pre-pupa of *Dermatobia hominis*: note the everted anterior spiracles (white arrow) (bar = 2 mm) (photo by P.J. Scholl).

Fig. 14.72.   Bovine host heavily infested with *Dermatobia hominis* (photo by P.J. Scholl).

Gasterophilus Leach

Biology

The nine species of *Gasterophilus* are all parasites of equids and, like their hosts, they were originally restricted to the Palaearctic and Afrotropical regions. Nowadays, representatives of the genus can be found worldwide. The larvae of all species develop inside the digestive tract of their hosts. Adults are medium to large sized flies; most have a body length in the range of 9–12 mm, but some are up to 15–16 mm long (*Gasterophilus intestinalis*, *Gasterophilus pecorum* and *Gasterophilus ternicinctus*). They tend to be hairy bee-like flies, but their body coloration can be quite variable. The females of all but one species, *G. pecorum*, lay their eggs directly on to the hosts, attaching their eggs to hairs, in a manner similar to females of *Hypoderma*, by means of an attachment organ situated basally or ventrally on the egg and an adhesive (Cogley and Anderson, 1983). Females have the potential to produce between about 160 (*Gasterophilus haemorrhoidalis*) and 2500 (*G. pecorum*) eggs. The production of so many eggs is probably a specific adaptation to the egg-laying habit of *G. pecorum*, because this is the single species that lays its eggs away from the host, attaching them to the leaves and stems of plants, mainly grasses in meadows and pasture. Larvae develop inside the eggs within 5–8 days, but can then remain dormant but viable within the egg case for many months, until ingested by the host with a mouthful of plant material. The females of species that oviposit directly on to their hosts choose hairs at particular oviposition sites (Table 14.21). The potentially most fecund species that lays on the host is *G. intestinalis*, producing up to 1100 eggs. Interestingly, it is the only species that does not lay its eggs in the head region and perhaps, again, larger numbers of eggs are an adaptation to laying eggs at sites where the chances of the larvae being ingested are least. However, Cope and Catts (1991) suggest that >90% of eggs of *G. intestinalis* can be ingested through self- and mutual-grooming. Like *G. pecorum*, unhatched larvae within the eggs of *G. intestinalis* can remain viable for several weeks, depending on the temperature. Larval development after hatching is similar for all species, involving some period of development in the buccal cavity and/or the digestive tract before being passed out with the faeces, but there are important differences, as indicated below for those species for which information is available (mainly from Zumpt, 1965).

G. haemorrhoidalis: First-instar larvae penetrate lips and undergo a subcutaneous migration into the mouth (not cheeks or tongue). Second instars are found in the stomach and duodenum. Third instars detach before maturity and pass to the rectum where they reattach near to the anus to complete development.

G. inermis: First-instar larvae penetrate the external cheek surface and migrate subcutaneously towards the corner of the mouth. Entering the mouth, they continue their movement under the mucous membranes of the cheek. Second- and third-instar larvae have been collected from the rectum.

G. intestinalis: The larvae are taken into the mouth when the host licks an egg-laden patch of hair. In the mouth, the larvae penetrate the dorsal mucosa of the anterior end of the tongue and gradually migrate to the posterior end. They leave the tongue and move to interdental spaces, mostly in the upper jaw, where they moult to the second instar. During their time in the interdental spaces the tracheal system is developed and there is an increase in haemoglobin content, in preparation for the growth phase within the low-oxygen-tension environment of the stomach (Cogley *et al.*, 1982). Larger second instars leave the interdental spaces and migrate to the root of the tongue, pharynx and epiglottis. They move to the stomach within a few days and the third instars complete development grouped around the interface of the glandular and non-glandular epithelia, the margo plicatus (up to 75% of clusters localized on the non-glandular mucosa near the margo plicatus) (Agneessens *et al.*, 1998). The strongly curved mouth hooks, with tile-like ridges, that may function as barbs and large depressor muscles are a good example of the adaptation of *Gasterophilus* larvae to its environment, to aid attachment in the digestive tract (Erzinçlioglu, 1990).

G. nasalis: On hatching, larvae migrate to the lips and penetrate into the gums, around and between the teeth. The first instar lasts for about 18–24 days. A few days after moulting to the second instar, larvae move to the proximal duodenum (Reinemeyer *et al.*, 2000) and attach to the wall, moulting to third stage and completing development within about 11 months of hatching.

G. nigricornis: Larvae migrate subcutaneously to the corner of the mouth and then into the mucous membrane of the cheek, like *G. inermis*. Larvae moult to the second instar after 20–30 days, move to the inner surface of the cheek and pass down the digestive tract to the duodenum, where they reattach. The wall of the duodenum forms swellings around the larvae, which develop within them to the third instar within 60–90 days. Third instars leave the swellings and reattach more superficially to the mucous membrane to complete development.

G. pecorum: Ingestion stimulates rapid hatching followed by penetration into the mucous membranes of the lips, gums, cheeks, tongue and hard palate. The larvae burrow towards the root of the tongue and the soft palate where they may remain for up to 10 months, moulting to the third instar. At that stage they usually move to the stomach to complete development. If swallowed as first instars, they may settle and develop within the walls of the pharynx, oesophagus or stomach.

Usually only one generation of bot flies develops each year, the adults appearing in the summer after the larvae have passed the previous autumn, winter and spring within the protected environment of the host. On exiting the host, the larvae pupate within the faecal pile or in the soil nearby. The pupal period lasts 2–4 weeks depending on species, but may be extended up to 2–3 months in cool conditions. Adult flies live for a few days at most and the females are ready to mate as soon as they emerge. Hilltop aggregations have been suggested as a mechanism for bringing males and females of *G. intestinalis* together (Catts, 1979) and they have also been observed mating around horses (Cope and Catts, 1991), but Cogley and Cogley (2000) proposed that faecal piles of the host were the most likely mating site, especially in locations without obvious hills.

There are few accounts of the adult behaviour of *Gasterophilus* species and most relate to *G. intestinalis* around horses (Cope and Catts, 1991; Cogley and Cogley, 2000). They demonstrate the remarkable tenacity of ovipositing females, which lay eggs on walking and trotting horses in addition to laying them on standing horses. If the horse gallops away due to the harassment of the fly, the females pursue at a distance or even alongside the flank until the horse stops, when the flies immediately resume oviposition. Females of *G. intestinalis* can oviposit their total load of 900–1100 eggs within 1 h, a rate that is in keeping with their short life span, effectively no more than 1 day (Cope and Catts, 1991).

Disease, pathogenesis and significance

In general the larvae of *Gasterophilus* species are well tolerated and do not cause significant symptoms unless they are in large numbers. However, the high value of many horses, the recurrent expense of treatments and the potential for self-injury by horses harassed by ovipositing flies make *G. intestinalis* a major economic pest of equines in North America (Cope and Catts, 1991). Infestation rates can be very high. In Morocco, 98% of donkeys were infested with *G. intestinalis* and 95% with *G. nasalis* and the mean monthly burden of second and third instars of both species varied from 75 to 311, with a maximum of 715 (Pandey *et al.*, 1992).

Erosions, ulcers, nodular growths and stomach perforation were recorded as a result of these infestations and, elsewhere, glossitis, peridontitis, gastric abscesses, ruptures, peritonitis, general debilitation in heavy infestations and even rectal prolapse (Daoud *et al.*, 1989) have been reported. Zumpt (1965) also observed rectal prolapse in zebra heavily infested in the anus with larvae of *G. haemorrhoidalis*. The degree of gastric ulceration can vary with the species, e.g. ulcers produced by *G. intestinalis* are more distinctive than those produced by *G. nasalis*; ulceration by both species leads to proliferation of fibrotic tissues below the ulcer (Cogley and Cogley, 1999). Such tissue production is important because it maintains the strength and thickness of the wall of the digestive tract. A case has been recorded where deep penetration of the colon of a horse by a larva of *G. intestinalis* caused perforation of the colon and leakage of the intestinal contents, leading to septic peritonitis (Lapointe *et al.*, 2003). Infestation with first instars of *G. nasalis* can lead to necrosis of tissues at the infestation sites in the mouth and the resulting pus pockets can extend into the tooth sockets. Pandey *et al.* (1992) concluded that treatment against bots is justified, especially if linked with treatments for helminth parasites to increase the performance of equines during periods of agricultural operations. In Europe the most frequently recorded species is *G. intestinalis*, with a common prevalence in the order of 40–60%, sometimes up to 94% in southern areas or down to 12% in northern areas (Colebrook and Wall, 2004). The numbers of larvae recovered from horses varies greatly, but the means are generally in the range of 20–40 per animal. Hosts infested with more than 100 larvae of *G. intestinalis* are rare, but numbers can rise to >300 (Lyon *et al.*, 2000). In addition to *G. intestinalis*, in England (Lyon *et al.*, 2000) and France, cases of *G. nasalis* infestation are infrequently found, but other species are rare in western, central and northern Europe. By contrast, in southern Europe there is much greater abundance and diversity of *Gasterophilus*, with prevalences of 93.8% for *G. intestinalis*, 76.5% for *G. nasalis*, 71.0% for *G. inermis*, 39.3% for *G. pecorum* and 10.8% for *G. haemorrhoidalis* recorded in Italy (Principato, 1989).

When the rare opportunity to survey wild equids has arisen, high prevalence rates similar to or greater than those in domesticated equids have been found. For example, prevalence rates in 35 Burchell's zebra (*Equus burchellii*) in the Kruger National Park, South Africa, were 100% for *G. terninctus*, 100% for *G. nasalis*, 97.1% for *G. haemorrhoidalis*, 94.3% for *G. pecorum*, 91.4% for *G. meridionalis* and 77.1% for *G. inermis* (Horak *et al.*, 1984). Interestingly, adult females of all of these species except *G. pecorum* were observed while ovipositing on Burchell's zebra that had just been shot in a study in Zambia (Howard, 1981). Mean larval burdens of 50–300 third instars of each species per zebra were recorded. In spite of these high burdens, no significant pathological effects could be attributed to the presence of these bot fly larvae. Howard (1981) suggested that the bot flies should be considered as wildlife just as much as their hosts and that their extinction in local populations should be considered to be just as alarming as the gradual disappearance of larger and more familiar wildlife species – a forward thinking perspective.

Gasterophilus species can have a negative impact on hosts in ways other than just the internal damage due to larval feeding and attachment. The harassment of horses by ovipositing *G. intestinalis* can lead to self-injury. Hair loss above the subcutaneous migratory tracks of larvae of *G. inermis* gives rise to a condition known as 'summer dermatitis of the cheek of horses' (Zumpt, 1965). Even the appearance of large numbers of eggs laid on horses can be considered aesthetically displeasing, but they are relatively easy to remove with fine-toothed combs.

The larvae of *Gasterophilus* species are rarely found in the digestive tracts of scavengers or carnivores that have probably acquired them through feeding on infested carcasses (whiteback griffon vulture, *Gyps africanus*, Cooper and Houston, 1972; lion, *Panthero leo*, Battisti *et al.*, 1997). The larvae are either found attached to the wall of the digestive tract or free-living, and the wall usually shows signs of attachment lesions.

Humans can become infested with larvae of *Gasterophilus* species, which migrate just below the skin, leaving a distinct trail ('creeping myiasis') and causing severe itching (Zumpt, 1965; Hall and Smith, 1993). Larvae usually die as first instars, but they rarely develop to second instars (*G. haemorrhoidalis*; Royce *et al.*, 1999). Experimental studies on first-stage larvae of *G. haemorrhoidalis*, *G. pecorum*, *G. nigricornis* and *G. inermis* have shown that larvae readily penetrate unbroken human skin in a few minutes but that those of *G. intestinalis* and *G. nasalis* do not do so, even after 2 h. However, human infestations with *G. intestinalis* have been recorded.

Management

Since the 1980s macrocyclic lactones have dominated the treatments for gasterophilosis. Ivermectin paste administered orally at 0.2 mg/kg was shown to be highly effective (94–100%) against *G. intestinalis* (Klei *et al.*, 2001). A single dose of moxidectin 2% equine oral gel (0.4 mg/kg body weight) was 100% effective against *G. nasalis* and 97.6% effective against *G. intestinalis* when measured 14 days post-treatment (Reinemeyer *et al.*, 2000).

Although not practical for control in most circumstances, placing horses inside stalls such that at least their lower bodies are out of sight appears to prevent oviposition by *Gasterophilus* (Cogley and Cogley, 2000).

Serological techniques exist for the diagnosis of *Gasterophilus* (Escartin-Peña and Bautista-Garfias, 1993). In addition, endoscopic techniques can be used to identify larvae in hosts without the need to euthenase them (Reinemeyer *et al.*, 2000).

Hosts and distribution

Table 14.21. Oviposition and egg-hatching data for species of *Gasterophilus* (details for *G. lativentris*, *meridionalis* and *ternicinctus* are unknown).

Species	Number of eggs	Oviposition site	Embryonic period (in days)	Stimulation to hatch
haemorrhoidalis	50–200	Lips, mainly the upper lips	2	Moisture from host
inermis	320–360	Cheeks	Unknown	No external stimulation

Continued

Table 14.21. *Continued.* Oviposition and egg-hatching data for species of *Gasterophilus* (details for *G. lativentris, meridionalis* and *ternicinctus* are unknown).

Species	Number of eggs	Oviposition site	Embryonic period (in days)	Stimulation to hatch
intestinalis	400–1000	Mainly lower forelegs, also back and flanks	5	Application of moisture and friction from rubbing and licking of the host
nasalis	300–500	Under the chin in the groove between the halves of the lower jaw	5–10	No external stimulation
nigricornis	330–350	On cheeks, rarely on nasal region	3–9	Unknown
pecorum	1300–2500	Leaves and stems of plants, mainly grasses	5–8	Ingestion by host

Table 14.22. Hosts and distribution (*Gasterophilus*).

Species	Hosts	Distribution	Common names
nasalis	Horse, donkey, *Equus burchellii*	Holarctic, Ethiopia	Linnaeus' horse bot fly
lativentris	Known only from adult		
haemorrhoidalis	Horse, donkey, *Equus burchellii*, *Equus zebra*	Palaearctic, South Africa	Rectal horse bot fly
inermis	Horse, *Equus burchellii*	Palaearctic, Africa	Unarmed horse bot fly
intestinalis	Horse, donkey	Worldwide, with hosts	Armed horse bot fly
ternicinctus	*Equus burchellii*	Sub-Saharan Africa	
meridionalis	*Equus burchellii*	Sub-Saharan Africa	Non-spotted zebra bot fly
migricornis	Horse, donkey	Palaearctic (Asian portions)	Broad-bellied horse bot fly
pecorum	Horse, donkey, *Equus burchellii*	Palaearctic, Africa	Dark-winged horse bot fly

Figures

Fig. 14.73. Ovipositing *Gasterophilus intestinalis*. Note eggs already attached to host hair (white arrows) (bar = 0.5 cm).

Fig. 14.74. Third-instar *Gasterophilus intestinalis* (bar = 0.5 cm). (Lateral view, cephalic segment to the right.)

Fig. 14.75. Third-instar *Gasterophilus nasalis* (bar = 0.5 cm). (Lateral view, cephalic segment to the right.)

Fig. 14.76. Third-instar *Gasterophilus pecorum* (bar = 0.5 cm). (Lateral view, cephalic segment to the right.)

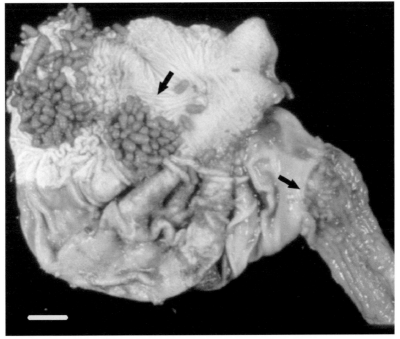

Fig. 14.77. Equine gastrointestinal tract showing dual infestation of *Gasterophilus intestinalis* (black arrow on left) and *Gasterophilus nasalis* (black arrow on right) (bar = 5 cm).

Gyrostigma Brauer

Biology

Adults are large conspicuous flies with dramatic red coloration. They bear a strong resemblance to pompiliid wasps. Larvae are parasites in the stomach of rhinoceroses. Most of the information on this genus comes from observations on *Gyrostigma rhinocerontis*, but presumably applies to the other species. Eggs are attached to the skin of the host through the use of attachment organs that resemble those of *Gasterophilus pecorum* (see Cogley, 1990). No information is available on development times, hatching stimulus or migration routes of newly hatched first instars. Second and third instars are found in the host stomach, firmly attached to the mucosa with the aid of robust mouth hooks (see Chapter 8). Mature third instars pass out with the faeces and pupate in the soil. Pupal development requires about 6 weeks.

In South Africa *G. rhinocerontis* has two generations per year with flies active in March–April and October–December. From the limited information available *Gyrostigma conjugens* appears to be univoltine.

Disease, pathogenesis and significance

Although there are no descriptions of disease associated with these maggots the impact on the gastric mucosa is similar to that described for *Gasterophilus* spp. which attach and feed in similar locations. Occasionally larvae of *Gyrostigma* spp. are unknowingly imported by zoos with their rhinoceros hosts (Warnecke and Göltenboth, 1977; Brum *et al.*, 1996). For example, the Asiatic species *G. sumatrensis* is known only from third instars recovered from hosts in zoological gardens in Germany. An additional example is from England where several larvae of *G. rhinocerontis* were recovered in early October 1995 from the faeces of a southern white rhino (*Ceratotherium simum simum*) that had been imported from South Africa in Aguust. Adult flies emerged in early November, in keeping with the pattern observed in South Africa (M.J.R. hall, unpublihsed)

Management

The absence of significant disease or pathology suggests that treatment is not required in the management of wild host populations. However, it may be presumed that treatment with macrocyclic lactone anti-parasitics will give effective control (see Chapter 13). There has been no published work to verify these suggestions.

Hosts and distribution

Table 14.23.   Hosts and distribution (*Gyrostigma*).

Species	Hosts	Distribution	Common names
rhinocerontis (= *pavesii*)	*Diceros bicornis* (black rhinoceros), *Ceratotherium simum* (white rhinoceros)	Sub-Saharan Africa	Pavesi's Rhinoceros bot fly
conjugens	*Diceros bicornis*	Sub-Saharan Africa (east)	Enderlein's rhinoceros bot fly
sumatrensis	*Dicerorhinus sumatrensis* (Asiatic two-horned rhinoceros)	Sumatra	Asiatic rhinoceros bot fly

Figures

Fig. 14.78.   Adult *Gyrostigma rhinocerontis*, live, perched on the hand of Dr C. Dewhirst (photo courtesy of C. Dewhirst).

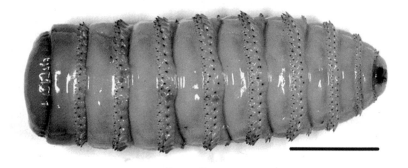

Fig. 14.79.   Ventral view of third-instar *Gyrostigma rhinocerontis* (bar = 1 cm).

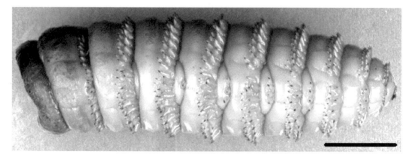

Fig. 14.80.   Lateral view of third-instar *Gyrostigma rhinocerontis* (bar = 1 cm).

Fig. 14.81.   View of the cephalic and thoracic segments of third-instar *Gyrostigma rhinocerontis* (bar = 0.5 cm).

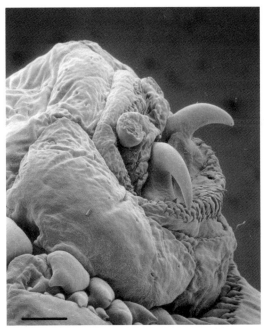

Fig. 14.82. Scanning electron micrograph of the cephalic and first thoracic segments of third-instar *Gyrostigma rhinocerontis* (bar = 0.5 mm).

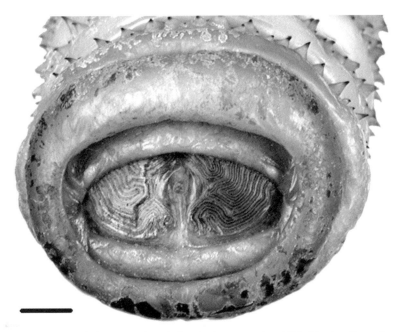

Fig. 14.83. View of the posterior abdominal segments of third-instar *Gyrostigma rhinocerontis* showing the spiracular plates with the convoluted spiracular openings (bar = 2 mm).

Fig. 14.84.   Lateral view of puparium of *Gyrostigma rhinocerontis* (bar = 5 mm).

Cobboldia Brauer

Biology

Larvae are parasites in the stomach of elephants. Females attach eggs firmly to the base of the tusk. Details of hatch, entry into the host and developmental site of the first instars are not known although all three instars have been recovered from the stomach. Second and third instars are found between the food bolus and the gastric mucosa. Unlike *Gasterophilus* and *Gyrostigma* the larvae are not attached to the mucosa. Mature third instars migrate upwards to the mouth where they escape the host and pupate in the soil. Pupal development takes 2–3 weeks. There may be several generations per year as adults were seen during all months in southern Africa.

Both species or subspecies (see Barriel *et al.*, 1999; Roca *et al.*, 2001) of African elephant are infested. Both *Cobboldia loxodontis* and *Cobboldia roeveri* have been reported from the two species/subspecies (Condy, 1974; Kinsella *et al.*, 2004) but it is unclear whether there are differences as most of the earlier work did not differentiate the hosts.

Disease, pathogenesis and significance

There have been no descriptions of disease associated with larvae of these species.

Management

The absence of significant disease or pathology suggests that treatment is not required in the management of wild host populations. However, it may be presumed that treatment with macrocyclic lactone anti-parasitics will give effective control (see Chapter 13). There has been no published work to verify these suggestions.

Hosts and distribution

Table 14.24. Hosts and distribution (*Cobboldia*).

Species	Hosts	Distribution	Common names
elephantis	*Elephas maximus*	Indian subcontinent	Black elephant stomach bot fly
loxodontis (= *Platycobboldia loxodontis*)	*Loxodonta africana cyclotis*, *Loxodonta africana africana*	Sub-Saharan Africa	Blue elephant stomach bot fly
roveri (= *Rodhainomyia roveri*)	*Loxodonta africana cyclotis*, *Loxodonta africana africana*	Central Africa (Congo)	Green elephant stomach bot fly

Figures

Fig. 14.85. Dorsal view of adult female *Cobboldia loxodontis* (bar = 2 mm).

Fig. 14.86. Lateral view of adult female *Cobboldia loxodontis* (bar = 2 mm).

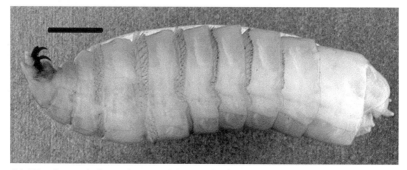

Fig. 14.87. Lateral view of second-instar *Cobboldia loxodontis*. Note the strong well-developed mouth hooks, the few number of spines on the abdominal segments and the large papillae on the terminal abdominal segment (bar = 1 cm).

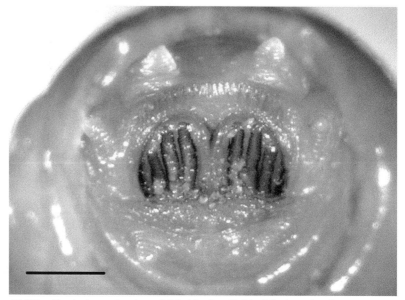

Fig. 14.88. Posterior spiracles of third-instar *Cobboldia loxodontis*.

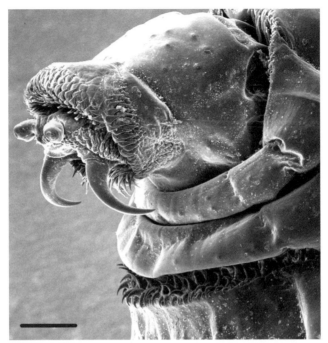

Fig. 14.89. Scanning electron micrograph of the cephalic and thoracic segments of third-instar *Cobboldia loxodontis*. Note the strong mouth hooks (bar = 0.43 mm).

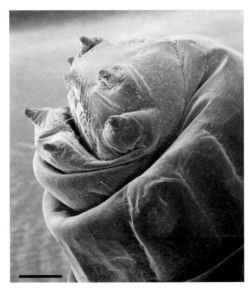

Fig. 14.90. Scanning electron micrograph of the posterior abdominal segments of third-instar *Cobboldia loxodontis*. Note the several large papillae, the relative absence of spines and the fringe of hairs on the dorsal rim of the spiracular pit (bar = 0.6 mm).

Ruttenia Rodhain

This genus is monospecific. Larvae live in cutaneous cysts on the upper parts, flanks and thighs of the host body.

Biology

No data are available on oviposition or other adult behaviours. Third instars and presumably second instars (see Stehlik, 1980) are found within cutaneous cysts that are distributed widely over the body of affected elephants. The cysts are visible for approximately 4 months (i.e. October through January) of the dry season. Pupal development takes approximately 25 days.

Disease, pathogenesis and significance

Infestations appear as discoloured patches on the skin; no swelling has been noticed and no apparent discomfort is associated. After a period of approximately 3 months the patches become slightly swollen and appear to rupture as the larvae reach maturity (see Stehlik, 1980). There is little indication of the impact the infestation has on the host.

Management

None has been shown to be necessary as the impact appears limited and there are no other known hosts that are affected. The absence of significant disease or pathology suggests that treatment is not required in the management of wild host populations. However, it may be presumed that treatment with avermectin antiparasitics will give effective control (see Chapter 13). There has been no published work to verify these suggestions.

Hosts and distribution

Table 14.25. Hosts and distribution (*Ruttenia*).

Species	Hosts	Distribution
loxodontis	*Loxodonta africanus*	Central Africa, Congo

Figures

Fig. 14.91. Ventral view of a third instar (bar = 2 mm).

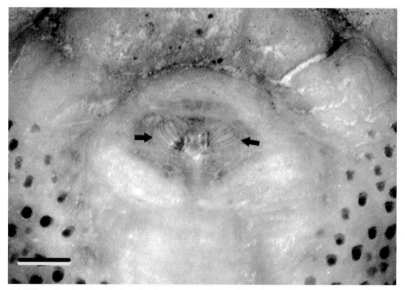

Fig. 14.92. Posterior abdominal segment showing the spiracular plates (arrows) (bar = 1 mm).

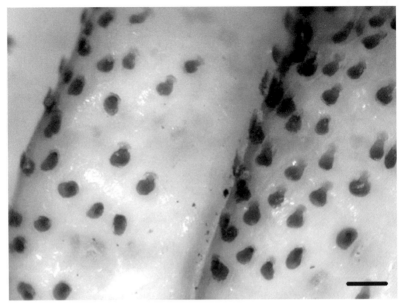

Fig. 14.93.   Detail of abdominal segment showing the spine shape (bar = 1 mm).

Neocuterebra Grünberg

This is a monospecific genus whose larvae are subcutaneous parasites, living in cysts usually located on the lower surface of the host's feet.

Biology

There is little information available on the biology of this species. Zumpt (1965) suggests that first instars may penetrate directly into the foot, although no evidence supports this claim. Second and third instars are found in cysts located on the ventral surface of the foot. The openings of the cysts are narrow and usually located in crevices of the foot.

There are no data on the prevalence or intensity and reports are few. Distribution within the range of the only host is also not clearly delimited.

Disease, pathogenesis and significance

Larvae appear to cause only slight local inflammation and there seems to be little exudate. There are no other data on the impact of these maggots on elephants.

Management

None has been shown to be necessary as the impact appears limited and there are no other known hosts that are affected. The absence of significant disease or pathology suggests that treatment is not required in the management of wild host populations. However, it may be presumed that treatment with macrocyclic lactone-based anti-parasitics will give effective control (see Chapter 13). There has been no published work to verify these suggestions.

Hosts and distribution

Table 14.26. Hosts and distribution (*Neocuterebra*).

Species	Hosts	Distribution
squamosa	*Loxodonta africanus*	Central Africa, Congo

Figures

Fig. 14.94. Section through the sole of the foot of an African elephant showing a third-instar *Neocuterebra squamosa*. Note the absence of exudates and relatively little inflammation (bar = 1 cm) (image used with permission: Spinnage, 1994).

References

Adamczyk, J.J., Silvain, J.F. and Pashley-Prowell, D. (1996) Intra- and interspecific DNA variation in a sodium channel intron in *Spodoptera* (Lepidoptera: Noctuidae). *Annals of the Entomological Society of America* 89, 812–821.

Agneessens, J., Engelen, S., Debever, P. and Vercruysse, J. (1998) *Gasterophilus intestinalis* infections in horses in Belgium. *Veterinary Parasitology* 77, 199–204.

Ahmed, M.J. and Miller, A. (1969) Pulmonary coin lesion containing a horse bot, *Gasterophilus*. *American Journal of Clinical Pathology* 52, 414–419.

Akchurin, B.S. (1945) Rhinoestrus of horses in the Bashkir ASSR. *Veterinariya* 22, 21. [In Russian]

Akre, R.D., Greene, A., MacDonald, J.F., Landolt, P.J. and Davis, H.G. (1981) *Yellowjackets of America North of Mexico*. Agriculture Handbook No. 552, United States Department of Agriculture, Washington, DC, 102 pp.

Al-Ani, F.K., Khamas, W.A., Zenad, K.H. and Al-Shareef, M.R. (1991) Camel nasal myiasis: clinical, epidemiological and pathological studies in Iraq. *Indian Journal of Animal Sciences* 61, 576–578.

Alcock, J. (1987) Leks and hilltopping in insects. *Journal of Natural History* 21, 319–328.

Alcock, J. and Schaefer, J.E. (1983) Hilltop territoriality in a Sonoran Desert bot fly (Diptera: Cuterebridae). *Animal Behaviour* 31, 518–525.

Aldrovandi, U. (1602) De Animalibus Insectis, Libri Septum, Cum Singolorum Iconibus as Vivum Experssis. *Apud. Ioan Bapt Bellagambam, Bononiae*.

Alicata, J.E. (1964) Parasitic infections of man and animals in Hawaii. *Hawaii Agricultural Experiment Station Technical Bulletin* No 61, Honolulu, Hawaii, 138 pp.

Allan, S., Day, J.F. and Edman, J.D. (1987) Visual ecology of biting flies. *Annual Review of Entomology* 32, 297–316.

Allen, J.R., Doube, B.M. and Kemp, D.H. (1997) Histology of bovine skin reactions to *Ixodes holocyclus* Neumann. *Canadian Journal of Comparative Medicine* 41, 26–35.

Amorim, D.S. and Silva, V.C. (2002) How far advanced was Diptera evolution in the Pangaea? *Annales de la Société entomologique de France (nouvelle serie)* 38, 177–200.

Anderson, J.L. (1997) *Che Guevara – a revolutionary life*. Grove Press, New York, 814 pp.

Anderson, J.R. (1974) Symposium on reproduction of arthropods of medical and veterinary importance. II. Meeting of the sexes. *Journal of Medical Entomology* 11, 7–19.

Anderson, J.R. (1975) The behavior of nose bot flies (*Cephenemyia apicata* and *C. jellisoni*) when attacking black-tailed deer (*Odocoileus hemionus columbianus*) and the resulting reactions of the deer. *Canadian Journal of Zoology* 53, 977–992.

Anderson, J.R. (1989) Use of deer models to study larviposition by wild nasopharyngeal bot flies (Diptera: Oestridae). *Journal of Medical Entomology* 26, 234–236.

Anderson, J.R. (2001) Larviposition by nasopharyngeal bot fly parasites of Columbian black-tailed deer: a correction. *Medical and Veterinary Entomology* 15, 438–442.

Anderson, J.R. and Hoy, J.B. (1972) Relationships between host attack rates and CO_2-baited insect flight trap catches of certain *Symphoromyia* species. *Journal of Medical Entomology* 9, 373–393.

Anderson, J.R. and Nilssen, A.C. (1990) The method by which *Cephenemyia trompe* (Modeer) larvae invade reindeer (*Rangifer tarandus*). *Rangifer Special Issue* 3, 291–297.

Anderson, J.R. and Nilssen, A.C. (1996a) Trapping oestrid parasites of reindeer: the response of *Cephenemyia trompe* and *Hypoderma tarandi* to baited traps. *Medical and Veterinary Entomology* 10, 337–346.

Anderson, J.R. and Nilssen, A.C. (1996b) Trapping oestrid parasites of reindeer: the relative age, fat body content and gonotrophic conditions of *Cephenemyia trompe* and *Hypoderma tarandi* females caught in baited traps. *Medical and Veterinary Entomology* 10, 347–353.

Anderson, J.R. and Olkowski, W. (1968) Carbon dioxide as an attractant for host-seeking *Cephenemyia* females (Diptera: Oestridae). *Nature* 220, 190–191.

Anderson, J.R. and Yee, W.L. (1995) Trapping black flies (Diptera: Simuliidae) in northern California. I. Species compositions and seasonal abundance on horses, models and in insect flight traps. *Bulletin of Vector Ecology* 20, 7–25.

Anderson, J.R., Nilssen, A.C. and Folstad, I. (1994) Mating behavior and thermoregulation of the reindeer warble fly, *Hypoderma tarandi* L. (Diptera: Oestridae). *Journal of Insect Behavior* 7, 679–706.

Anderson, W.B. (1935) Ophthalmomyiasis. *American Journal of Ophthalmology* 18, 699–705.

Araujo-Chaveron, N., Charbon, J.L. and Pfister, K. (1994) Influence of hypodermosis on the incidence of other diseases in cattle. In:

Pfister, K., Charbon, J.L., Tarry, D.W. and Pithan, K. (eds) *Improvements in the Control Methods for Warble-fly in Cattle and Goats.* Proceedings, 23–25 September 1993, Thun, Switzerland. Commission of the European Communities, Brussels, pp. 121–126.

Arnheim, N. (1983) Concerted evolution of multigene families. In: Nei, M. and Koehn R.K. (eds) *Evolution of Genes and Proteins.* Sinauer, Sunderland, Massachussetts, pp. 38–61.

Badia, L. and Lund, Y.J. (1994) Vile bodies: an endoscopic approach to nasal myiasis. *Journal of Laryngology and Otology* 12, 1083–1085.

Baird, C.R. (1971) Development of *Cuterebra jellisoni* (Diptera: Cuterebridae) in six species of rabbits and rodents. *Journal of Medical Entomology* 8, 615–622.

Baird, C.R. (1972a) Development of *Cuterebra ruficrus* (Diptera: Cuterebridae) in six species of rabbits and rodents with a morphological comparison of *C. ruficrus* and *C. jellisoni* third instars. *Journal of Medical Entomology* 9, 81–85.

Baird, C.R. (1972b) Termination of pupal diapause in *Cuterebra tenebrosa* (Diptera: Cuterebridae) with injections of ecdysterone. *Journal of Medical Entomology* 9, 77–80.

Baird, C.R. (1972c) Development of *Cuterebra ruficrus* (Diptera: Cuterebridae) in six species of rabbits and rodents with a morphological comparison of *C. ruficrus* and *C. jellisoni* third instars. *Journal of Medical Entomology* 9, 81–85.

Baird, C.R. (1974) Field behavior and seasonal activity of the rodent bot fly, *Cuterebra tenebrosa*, in central Washington (Diptera: Cuterebridae). *The Great Basin Naturalist* 34, 247–253.

Baird, C.R. (1975) Larval development of the rodent botfly, *Cuterebra tenebrosa*, in bushy-tailed wood rats and its relationship to pupal diapause. *Canadian Journal of Zoology* 53, 1788–1798.

Baird, C.R. (1983) Biology of *Cuterebra lepusculi* Townsend (Diptera: Cuterebridae) in cottontail rabbits in Idaho. *Journal of Wildlife Diseases* 19, 214–218.

Baird, C.R. (1997) Bionomics of *Cuterebra austeni* (Diptera: Cuterebridae) and its association with *Neotoma albigula* (Rodentia: Cricetidae) in the southwestern United States. *Journal of Medical Entomology* 34, 690–695.

Baird, C.R. and Smith, D.H. (1979) Case reports of bot fly myiasis in pikas (*Ochotona princeps*). *Journal of Wildlife Diseases* 15, 553–555.

Baird, C.R., Podgore, J.K. and Sabrosky, C.W. (1982) *Cuterebra* myiasis in humans; six new case reports from the United States with a summary of known cases (Diptera: Cuterebridae). *Journal of Medical Entomology* 19, 263–267.

Baird, J.K., Baird, C.R. and Sabrosky, C.W. (1989) North American cuterebrid myiasis. *Journal of the American Academy of Dermatology* 21, 763–772.

Baker, G.T. (1986) Morphological aspects of the egg and cephalic region of the first instar larvae of *Cuterebra horripilum* (Diptera: Cuterebridae). *Wasmann Journal of Biology* 44, 66–72.

Bakos, L. and Zanini, S. (1979) Botfly infestation of the tongue. *British Journal of Dermatology* 100, 223–224.

Bal, S. and Amitava, A.K. (2000) Palpebral myiasis. *Journal of Pediatric Ophthalmology and Strabismus* 37, 309–310.

Baldelli, B., Polidori, G.A., Grelloni, V., Principato, M., Moretti, A. and Fioretti, D.P. (1981) The ELISA method for the detection of antibodies in bovines affected by *Hypoderma* infestation, preliminary results. *Parassitologia* 23, 115–119.

Balic, A., Bowles, V.M. and Meeusen, E.N. (2000) The immunobiology of gastrointestinal nematode infections in ruminants. *Advances in Parasitology* 45, 181–241.

Banegas, A.D. and Mourier, H. (1967) Laboratory observation on the life history and habits of *Dermatobia hominis* (Diptera: Cuterebridae). I. Mating behavior. *Annals of the Entomological Society of America* 60, 878–881.

Banegas, A.D., Mourier, H. and Graham, O.H. (1967) Laboratory colonization of *Dermatobia hominis* (Diptera: Cuterebridae). *Annals of the Entomological Society of America* 60, 511–514.

Barker, H.M. (1964) *Camels and the Outback.* Sir Isaac Pitman & Sons, Melbourne, Australia, p. ix (255 pp).

Baron, R.W. and Colwell, D.D. (1991a) Mammalian immune responses to myiasis. *Parasitology Today* 7, 353–355.

Baron, R.W. and Colwell, D.D. (1991b) Enhanced resistance to cattle grub infestation (*Hypoderma lineatum* de Vill.) in calves immunized with purified hypodermin A, B and C plus monophosphoryl lipid A (MPL). *Veterinary Parasitology* 38, 185–197.

Baron, R.W. and Weintraub, J. (1986) Immunization of cattle against hypodermatosis *Hypoderma lineatum* (De Vill.) and *H. bovis* (L.) using *H. lineatum* antigens. *Veterinary Parasitology* 21, 43–50.

Baron, R.W. and Weintraub, J. (1987) Immunological responses to parasitic arthropods. *Parasitology Today* 3, 77–82.

Barquet, P., Chabaudie, N., Nicolas, I. and Boulard, C. (1993) Action des hypodermines, enzymes sécrétées par *Hypoderma lineatum* (insecte, oestridae) sur la migration et l'expression des neutrophiles bovins. *Veterinary Research* 24, 364 (in French).

Barrett, C.C. (1981) A new technique for collecting cattle grub larvae. *Southwestern Entomologist* 6, 144–146.

Barriel, V., Thuet, E., Tassy, P. (1999) Molecular phylogeny of Elephantidae. Extreme divergence of the extant forest African elephant. *Comptes Rendus de l'Academie des Sciences Serie III* 322, 447–454.

Basson, P.A. (1962a) Studies on specific oculovascular myiasis of domestic animals (uitpeuloog). II. Experimental transmission. *Onderstepoort Journal of Veterinary Research* 29, 203–210.

Basson, P.A. (1962b) Studies on specific oculovascular myiasis (uitpeuloog). III. Symptomatology, pathology, aetiology and epizootiology. *Onderstepoort Journal of Veterinary Research* 29, 211–240.

Basson, P.A. (1966) Gedoelstial myiasis in antelopes of southern Africa. *Onderstepoort Journal of Veterinary Research* 33, 77–91.

Bates, M. (1943) Mosquitoes as vectors of *Dermatobia* in eastern Columbia. *Annals of the Entomological Society of America* 36, 21–24.

Battisti, A., Panfil, G., DiGuardo, G. and Principato, M. (1997) *Gasterophilus pecorum* larvae in an aged lion. *Veterinary Record* 140, 664.

Bauer, M.S. (1978) Warble-infested cattle attacked by starlings. *Modern Veterinary Practice* 59, 536–538.

Bauer, C. (1994) Does the use of microdosed ivermectin to control bovine hypodermosis increase the risk of selecting drug resistant nematodes? In: Pfister, K., Charbon, J.L., Tarry, D.W. and Pithan, K. (eds) *Improvements in the Control Methods for Warble-fly in Cattle and Goats*. Proceedings, 23–25 September 1993, Thun, Switzerland. COST 811. Commission of the European Communities, Brussels, pp. 145–147.

Bautista-Garfias, C.R., Angulo-Contreras, R.M. and Garay-Garzon, E. (1988) Serologic diagnosis of *Oestrus ovis* (Diptera: Oestridae) in naturally infested sheep. *Medical and Veterinary Entomology* 2, 351–355.

Beamer, R.H. (1950) An observation on the egg-laying of *Cuterebra buccata* Fabr. in nature. *Journal of the Kansas Entomological Society* 23, 16.

Beaucournu, J.C. (1965) Contribution à l'étude des hypodermoses humaines dans l'ouest de la France. Thesis, Faculté Mixte de Médecine et de Pharmacie, Rennes, France, 171 pp. (in French).

Beesley, W.N. (1966) The use in Britain of systemic insecticides for the control of *Hypoderma* (Diptera: Oestridae). *Veterinary Medicine Review* 1, 37–48.

Beesley, W.N. (1967) Observations on the biology of the ox warble-fly (*Hypoderma*: Diptera, Oestridae). I. *In vitro* culture of the first-instar larvae. *Annals of Tropical Medicine and Parasitology* 61, 175–181.

Beesley, W.N. (1968) Observations on the biology of the ox warble-fly (*Hypoderma*: Diptera, Oestridae). II. Bacteriostatic properties of larval extracts. *Annals of Tropical Medicine and Parasitology* 62, 8–12.

Beesley, W.N. (1969) Observations on the biology of the ox warble-fly (*Hypoderma*: Diptera, Oestridae). III. Dermolytic properties of larval extracts. *Annals of Tropical Medicine and Parasitology* 63, 157–159.

Beesley, W.N. (1970) Observations on the biology of the ox warble-fly (*Hypoderma*: Diptera, Oestridae). IV. Some serological reactions against the parasite in natural and artificial hosts. *Annals of Tropical Medicine and Parasitology* 64, 277–281.

Beesley, W.N. (1971) Observations on the biology of the ox warble-fly (*Hypoderma*: Diptera, Oestridae). V. Anaphylactoid shock in laboratory animals and calves following exposure to extracts of larvae of *Hypoderma*. *Annals of Tropical Medicine and Parasitology* 65, 567–572.

Bekenov, A.B., Grachev, I.A. and Milner-Gulland, E.J. (1998) The ecology and management of the saiga antelope in Kazakhstan. *Mammal Review* 28, 1–52.

Belem, A.M.G. and Rouille, D. (1988) Strose des petits ruminants au Burkina Faso. *Revue Elevage et Médecine Vétérinaire des Pays Tropicaux* 41, 59–64 (in French).

Bell, G. (1980) The costs of reproduction and their consequences. *American Naturalist* 116, 45–76.

Bello, T.R. (1989) Efficacy of ivermectin against experimental and natural infections of *Gasterophilus* spp. in ponies. *American Journal of Veterinary Research* 50, 2120–2123.

Benakhla, A., Lonneux, J.F., Mekroud, A., Losson, B. and Boulard, C. (1999) Hypodermose bovine dans le nord-est algérien: prévalence et intensité d'infestation. *Veterinary Research* 30, 539–545 (in French).

Bennett, G.F. (1955) Studies on *Cuterebra emasculator* Fitch 1856 (Diptera: Cuterebridae) and a discussion of the status of the genus *Cephenemyia* Ltr. *Canadian Journal of Zoology* 33, 75–98.

Bennett, G.F. (1962) On the biology of *Cephenemyia phobifera* (Diptera: Oestridae), the pharyngeal bot of the white-tailed deer, *Odocoileus virginianus*. *Canadian Journal of Zoology* 40, 1195–1210.

Bennett, G.F. (1972) Observations on the pupal and adult stages of *Cuterebra emasculator* Fitch (Diptera: Cuterebridae). *Canadian Journal of Zoology* 50, 1367–1372.

Bensasson, D., Zhang, D.X., Hartl, D.L. and Hewitt, G.M. (2001) Mitochondrial pseudogenes: evolution's misplaced witnesses. *Trends in Ecology and Evolution* 16, 314–321.

Benz, G.W. (1985) Animal health applications of ivermectin. *Southwestern Entomologist* Suppl. 7, 43–50.

Bergeaud, J.P., Duranton, C. and Dorchies, Ph. (1994) L'œstrose ovine en Aveyron: Résultat d'une enquête sur 1036 têtes à l'abattoir de Rodez. *Revue de Médecine Vétérinaire* 145, 863–866 (in French).

Bergman, A.M. (1917) Om renens oestrider. *Entomologisk Tidskrift* 38, 1–32, 113–146 (in Swedish with English translation).

Bergman, A.M. (1919) Über die Oestriden des Renntieres. *Zeitschrift für Infektionskrankheiten, Haustiere* 20, 65–116, 179–201.

Bergman, A.M. (1932) *Skin and Nasal Insects in Reindeer*. Vlast Sovietov, Moscow, Russia, pp. 234–257 (304 pp).

Berkenkamp, S.D. and Drummond, R.O. (1990) Hypodermosis. Part II. *Compendium of Continuing Education Practicing Veterinarian* 12, 881–888.

Biggs, H.C., McClain, E., Muller, G.L., Anthonissen, M. and Hare, K.M. (1998) A prediction model for strike in the sheep nasal fly, *Oestrus ovis*, in Namibia. *Preventive Veterinary Medicine* 33, 267–282.

Bishopp, F.C., Laake, E.W., Brundrett, H.M. and Wells, R.W. (1926) The cattle grubs or ox warbles, their biologies and suggestions for control. *United States Department of Agriculture Bulletin* No. 1369, Washington, DC, 119 pp.

Bishopp, F.C., Laake, E.W. and Wells, R.W. (1929) Cattle grubs or heel flies, with suggestions for their control. *United States Department of Agriculture Bulletin* No. 1596, Washington, DC, 22 pp.

Bisley, G.C. (1972) A case of intraocular myiasis in man due to the first stage larva of the oestrid fly *Gedoelstia* spp. *East African Medical Journal* 49, 768–771.

Bisseru, B. (1967) *Diseases of Man Acquired from His Pets*. Heinemann, London, 400 pp.

Blagoveshchensky, D.I. (1970) The discovery of *Trichopria* sp. (Insecta: Hymenoptera) in the puparia of *Hypoderma bovis* De Geer. *Parazitology* 4, 265–266 (in Russian with English summary). (Abstract: *Review of Applied Entomology* B 60, 1745, 1972.)

Bodenheimer, F.S. (1928) *Material en sur Geschichte der Entomologie bis Linné*, vols I and II. W. Junk, Berlin, Germany.

Boiko, G.P., Belozerskaya, N.I. and Pirozhok, V.S. (1978) Larvae of the bot fly in the conjunctival sac of man. *Meditsinskaya Parazitologiya i Parazitarnye Bolezni* 47, 116–117.

Boulard, C. (1968) Differenciation et developpement des gonades males et femelles chez les larves d'*Hypoderma bovis* et d'*Hypoderma lineatum* (Dipt., groupe biologique des Oestriformes). *Annales de la Société Entomologique de France* 4, 349–364 (in French).

Boulard, C. (1969) Anatomie et histologie du tube digestif de la larve d'*Hypoderma bovis* (Diptères Oestriformes). *Annales de la Société Entomologique de France* 5, 371–387 (in French).

Boulard, C. (1970) Etude préliminaire d'une collagénase brute extraite de la larve de premier stade d'*Hypoderma lineatum* (De Villers). *Comptes Rendus de l'Académie des Sciences, Paris* 270, 1349–1351 (in French).

Boulard, C. (1975a) Evolution des anticorps circulants chez les bovins traités contre l'hypodermose. *Annales de Recherches Vétérinaires* 6, 143–154 (in French).

Boulard, C. (1975b) Modifications histologiques de la sous-muqueuse oesophagienne de bovins parasités par *Hypoderma lineatum* (De Vill.) (Diptera Oestriforme). *Annales de Recherches Vétérinaires* 6, 131–142 (in French).

Boulard, C. (1979) Circulating antibodies and blood histamine in cattle after treatment against hypodermosis. *Veterinary Parasitology* 18, 379–387.

Boulard, C. (1989) Degradation of bovine C3 by serine proteases from parasites *Hypoderma lineatum* (Diptera: Oestridae). *Veterinary Immunology and Immunopathology* 20, 387–398.

Boulard, C. (2002) Durably controlling hypodermosis. *Veterinary Research* 33, 455–464.

Boulard, C. and Bencharif, F. (1984) Changes in the haemolytic activity of bovine serum complement by *Hypoderma lineatum* larval proteinases in naive and immune cattle. *Parasite Immunology* 6, 459–467.

Boulard C. and Garrone, R. (1978) Characterization of a collagenolytic enzyme from larvae of *Hypoderma lineatum* (Insecta: Diptera: Oestriform). *Comparative Biochemistry and Physiology* 59, 251–255.

Boulard, C. and Petithory, J. (1977) Serological diagnosis of human hypodermosis: a preliminary report. *Veterinary Parasitology* 3, 259–263.

Boulard, C. and Moiré, N. (1998) Immuno-epi demiology in low prevalence conditions of bovine hypodermosis. In: Boulard, C., O'Brien, D., Pithan, K., Sampimon, O., Sol, J. and Webster. K. (eds) *Improvements in the Control Methods for Warblefly in Cattle and Goats*. Proceedings, COST 811. Commission of the European Communities, Brussels.

Boulard, C. and Troccon, J.L. (1984) Attempts to reduce adverse reactions following hypodermosis chemotherapy. In: Boulard, C. and Thornberry, H. (eds) *Warble Fly Control in Europe: a Symposium in the EC Programme of Coordination of Research on Animal Pathology*, 16–17 September 1982, Brussels. A.A. Balkema, Rotterdam, The Netherlands, pp. 111–124.

Boulard, C. and Villejoubert, C. (1991) Use of pooled serum or milk samples for the epidemiological surveillance of bovine hypodermosis. *Veterinary Parasitology* 39, 171–183.

Boulard, C. and Weintraub, J. (1973) Immunological responses of rabbits artificially infested with the cattle grubs *Hypoderma bovis* and *H. lineatum*. *International Journal of Parasitology* 3, 379–386.

Boulard, C., Soria, J. and Soria, C. (1970) Possibilité d'emploi de la rédaction d'hémagglutination passive pour le diagnostic de l'hypodermose, en utilisant comme antigène une collagénase brute extraite des larves de 1er stade d'*Hypoderma lineatum*. *Comptes Rendus de l'Académie des Sciences, Paris* 270, 1965–1968 (in French).

Boulard, C., Villejoubert, C. and Moiré, N. (1996a) Cross-reactive, stage-specific antigens in the Oestridae family. *Veterinary Research* 27, 535–544.

Boulard, C., Villejoubert, C., Moire, N., Losson, B. and Lonneux, J.F. (1996b) Sero-surveillance of hypodermosis in a herd under therapeutic control. Effect of a low level of infestation. *Veterinary Parasitology* 66, 109–117.

Boulard, C., Banting, A., de L. and Cardinaud, B. (1998) Activity of moxidectin 1% injectable solution against first instar *Hypoderma* spp. in cattle and effects of antibody kinetics. *Veterinary Parasitolology* 77, 205–210.

Bowles V., Carnegie, P. and Sandeman, R. (1988) Characterization of proteolytic and collagenolytic enzymes from the larvae of *Lucilia cuprina*, the sheep blowfly. *Australian Journal of Biological Research* 41, 269–278.

Bowles, V.M., Grey, S.T. and Brandon, M.R. (1992) Cellular immune responses in the skin of sheep infected with larvae of *Lucilia cuprina*, the sheep blowfly. *Veterinary Parasitology* 44, 151–167.

Bradbury, J.W. and Gibson, R. (1983) Leks and mate choice. In: Bateson, P. (ed.) *Mate Choice*. Cambridge University Press, Cambridge, UK, pp. 109–138.

Bradley, W.H. (1931) Origin and microfossils of the oil shale of the Green River formation of Colorado and Utah. *US Geological Survey Professional Paper* No. 168, Washington, DC, 58 pp.

Brain, V., Van Der Merwe, H.E. and Horak, I.G. (1983) *Strobiloestrus* sp. larvae in Merino sheep. *Journal of the South African Veterinary Association* 54, 185–186.

Bram, R.A. (1994) Integated control of ectoparasites. *Revue Scientifique et Technique Office International des Epizooties* 13, 1357–1365.

Brasavola Musa, A. (1541) In octos libros aphorismorum Hippocratis et Galeni commentaria et annotations. In: *officina frobeniana, Balilae*.

Brauer, F.M. (1863) *Monographie der Oestriden*. Carl Veberrentez, Wien, Austria, 292 pp. (in German).

Breev, K.A. (1950) The behavior of blood-sucking Diptera and warble-flies during their attack on reindeer and the resulting reactions in reindeer. *Parazitologicheskii sbornik* 12, 167–198 (in Russian with English translation). (*Review of Applied Entomology* B 42, 67–68, 1954.)

Breev, K.A. (1956) Die Stärke des Befalls der Renntiere durch die hautbremse (*Oedemagena tarandi* L.) und die Nasenbremse (*Cephenomya* {sic} *trompe* L.) und der Einfluss der sie regulierenden Faktoren. *Parazitologicheskii sbornik* 16, 155–183 (in German).

Breev, K.A. (1967) New data on the migration of instar-I larvae of *Hypoderma bovis* De Geer in the host organism. *Parazitologicheskii sbornik* 23, 191–221.

Breev, K.A. and Minar, J. (1976) On the regularities of distribution of *Hypoderma bovis* De Geer larvae parasitizing cattle herds in different parts of the range of this warble fly. *Folia Parasitologica (Praha)* 23, 343–356.

Breev, K.A. and Minar, J. (1979) Statistical characteristics of the host parasite relationship of the northern cattle grub (*Hypoderma bovis*) (Hypodermatidae) in various parts and its distribution in bovin. *Parazitologiya* 13, 93–102.

Breev, K.A. and Obrezha, G.N. (1969) On the antigenic properties of material from 1st and 3rd instar larvae of reindeer warble-fly. *Parasitologia* 3, 489–492.

Breev, K.A. and Savel'ev, D.V. (1958) *Skin Insects of Reindeer and Its Control*. Zoological Institute of Academy of Sciences, St Petersburg, Russia, 96 pp.

Breev, K.A., Zagretdinov, R.G. and Minár, J. (1980) Influence of constant and variable temperatures on pupal development of the sheep bot fly (*Oestrus ovis* L.). *Folia Parasitologica (Praha)* 27, 359–365.

Brett, C.H. (1946) Insecticidal properties of the indigobush (*Amorpha fruticosa*). *Journal of Agricultural Research* 73, 81–96.

Brewer, T.F., Wilson, M.E., Gonzalez, E. and Felsenstein, D. (1993) Bacon therapy and furuncular myiasis. *Journal of the American Medical Association* 270, 2087–2088.

Breyev, K.A. (1973) Some general principles of the control of parasitic arthropods as exemplified by the control of warble flies. *Entomological Review* 52, 142–151.

Breyev, K.A. and Breyeva, Z.F. (1946) Data on the biology of the reindeer warble-fly (*Oedemagena tarandi*). *Review of Applied Entomology* 34, 111.

Breyev, K.A. and Karazeeva, Z.F. (1957) Materials relating to the biology of the warble fly (*Oedemagena tarandi* L). III. Observations on the pupa and adult flies. *Parazitologiceskij Sbornik Akademija Nauk SSSR Moskva Zoologiceskij muzei* 17, 199–228 (in Russian).

Briere, J.-F., Pracros, P., Le Roux, A.-Y. and Pierre, J.-S. (1999) A novel rate model of temperature-dependent development for arthropods. *Environmental Entomology* 28, 22–29.

Broutin, I., Arnoux, B., Riche, C., Lecroisey, A., Keil, B., Pascard, C. and Ducruix, A. (1996) Structure of *Hypoderma lineatum* collagenase: a member of the serine proteinase family. *Acta Crystalographia* D 52, 380–392.

Brown, J.M., Pellmyr, O., Thompson, J.N. and Harrison, R.G. (1994) Phylogeny of *Greya* (Lepidoptera: Prodoxidae) based on nucleotide sequence variation in mitochondrial cytochrome oxidase I and II: consequence with morphological data. *Molecular Biology and Evolution* 11, 128–141.

Bruce, W.G. (1938) Soil moisture and its relation to the mortality of *Hypoderma* pupae. *Journal of Economic Entomology* 31, 639–642.

Bruel, H., Guiguen, C., Chevrier, S., Geilh, B. and le Gall, E. (1995) Hypodermose humaine. *Medecine et Maladies Infectieuses* 25, 965–969.

Brugnone, G. (1781) Trattato delle razze di cavalla. *Appresso i fratelli Reicends, Torino.*

Brum, J.G.W., Valente, A., Costa, P.R.P., Giacomini, C. and Bohrer, J.L. (1996) *Gyrostigma pavesii* (Diptera: Gasterophilidae) em rinoceronte (*Ceratotherium simum*) proveniente da Africa do Sul. *Revista Brasileira de Parasiologia Veterinaria* 5, 57–58.

Bukshtynov, V.I. (1978) Predicting periods of development of bot fly (*Oestris ovis*) in sheep. *Veterinariia* 9, 60–62 (in Russian with English translation).

Cameron, A.E. (1932) The nasal bot fly, *Cephenomyia auribarbis* Meigen (Diptera, Tachinidae) of the Red Deer, *Cervus elaphus* L. *Parasitology* 24, 185–195.

Campbell, W.C., Fisher, M.H., Stapley, E.O., Albers-Schonberg, G. and Jacob, T.A. (1983) Ivermectin: a potent new antiparasitic agent. *Science* 221, 823–828.

Capelle, K.J. (1970) Studies on the life history and development of *Cuterebra polita* (Diptera: Cuterebridae) in four species of rodents. *Journal of Medical Entomology* 7, 320–327.

Capelle, K.J. (1971) Myiasis. In: Davis, J.W. and Anderson, R.C. (eds) *Parasitic Diseases of Wild Mammals*. Iowa State University Press, Ames, Iowa, pp. 279–305.

Caracappa, S., Loria G.R., Donn, A., Manfredi, M.T. and Guarda, F. (1996) *Hypoderma bovis* neuropathology in veal calves: case report. *European Jouranl of Veterinary Pathology* 2, 35–38.

Carpenter, G.H., Hewitt, T.R. and Reddin, T.K. (1914) *The Warble-Flies. Fourth Report on Experiments and Observations as to Life History and Treatment*. Department of Agriculture and Technical Instruction for Ireland XV (n1), Dublin, pp. 105–132.

Carter, G.T., Nietsche, J.A. and Borders, D.B. (1987) Structure determination of LL-F28249α, β, γ, and λ, potent antiparasitic macrolides from *Streptomyces cyaneogriseus* sp. noncyanogenus. *Journal of the Chemical Society, Chemistry Communications* Part 1, 402–404.

Carvalho, C.J.B. de (1999) Revision, cladistics and biogeography of the neotropical genus *Souzalopesmyia* Albuquerque (Diptera: Muscidae). *Proceedings of the Entomological Society of Washington* 101, 123–137.

Carvalho, C.J.B. de Bortolanza, M., Cardoso da Silva, M.C. and Soares, E.D.G. (2003) Distributional patterns of the neotropical Muscidae (Diptera). In: Morrone, J.J. and Llorente, B.J. (eds) *Una perspectiva Latinoamericana de la Biogeografía*. Las Prensas de Ciencia, Facultad de Ciencias, Universidad Autónoma de México, México, pp. 263–274.

Caterino, M.S., Cho, S. and Sperling, F.A.H. (2000) The current state of insect molecular systematics. *Annual Review of Entomology* 45, 1–54.

Catts, E.P. (1964a) Laboratory colonization of rodent bot flies (Diptera: Cuterebridae). *Journal of Medical Entomology* 1, 195–196.

Catts, E.P. (1964b) Field behavior of adult *Cephenemyia* (Diptera: Oestridae). *Canadian Entomologist* 96, 579–585.

Catts, E.P. (1967) Biology of a California rodent bot fly, *Cuterebra latifrons* Coq. *Journal of Medical Entomology* 4, 87–101.

Catts, E.P. (1979) Hilltop aggregation and mating behavior by *Gasterophilus intestinalis* (Diptera: Gasterophilidae). *Journal of Medical Entomology* 16, 461–464.

Catts, E.P. (1982) Biology of New World bot flies: Cuterebridae. *Annual Review of Entomology* 27, 313–338.

Catts, E.P. and Garcia, R. (1963) Drinking by adult *Cephenemyia* (Diptera: Oestridae). *Annals of the Entomological Society of America* 56, 660–663.

Catts, E.P., Garcia, R. and Poorbaugh, J.H. (1965) Aggregation sites of males of the common cattle grub, *Hypoderma lineatum* (De Villers) (Diptera: Oestridae). *Journal of Medical Entomology* 1, 357–358.

Cencek, T. (1995) The dependence of kinetic development of serum antibody to *H. bovis* proteins in infested climate on the climate of northeastern and central Poland. In: Tarry, D.W., Pithan, K. and Webster, K. (eds) *Improvements in the Control Methods for Warble-fly in Cattle and Goats*. Proceedings, 8–10 September 1994, Guidford, UK. COST 811. Commission of the European Communities, Brussels, pp. 69–73.

Cepeda-Palacios, R. and Scholl, P.J. (1999) Gonotrophic development in *Oestrus Ovis* (Diptera:Oestridae). *Journal of Medical Entomology* 36, 435–440.

Cepeda-Palacios, R. and Scholl, P.J. (2000a) Factors affecting the larvipositional activity of *Oestrus ovis* gravid females (Diptera: Oestridae). *Veterinary Parasitology* 91, 93–105.

Cepeda-Palacios, R. and Scholl, P.J. (2000b) Intra-puparial development in *Oestrus ovis* (Diptera: Oestridae). *Journal of Medical Entomology* 37, 239–245.

Cepeda-Palacios, R., Avila, A., Ramirez-Orduna, R. and Dorchies, Ph. (1999) Estimation of the growth patterns of *Oestrus ovis* L. larvae hosted by goats in Baja California Sur, Mexico. *Veterinary Parasitology* 86, 119–126.

Cepeda-Palacios, R., Monroy, A., Mendoza, M.A. and Scholl, P.J. (2001) Testicular maturation in the sheep bot fly *Oestrus ovis*. *Medical and Veterinary Entomology* 15, 275–280.

Cepelák, J., Bucek, G. and Mandelik, D. (1972) Further observations on the biology of warble flies in the area of Kovácovo hills. *Pol'nokospoda'rstvo* 18, 812–823.

Céspedes, F., Arguedas, R.J., Guillén, E. and Hevla, U. (1962) Dermatobiosis mortal. *Acta medica Costarricence* 5, 175–182.

Chabaudie, N. and Boulard, C. (1992) Effect of hypodermin A, an enzyme secreted by *Hypoderma lineatum* (insect Oestridae), on the bovine immune system. *Veterinary Immunology and Immunopathology* 31, 167–177.

Chabaudie, N., Villejoubert, C. and Boulard, C. (1991) The response of cattle vaccinated with hypodermin A to a natural infestation of *Hypoderma bovis* and *Hypoderma lineatum*. *International Journal for Parasitology* 21, 859–862.

Chabert, Ph. (1782) Traité des malaides vermineuse dans les animaux. *Imprimerie Royale*, Paris.

Chamberlain, W.F. (1964a) Survival of first-instar common cattle grubs under anaerobic conditions *in vitro*. *Annals of the Entomological Society of America* 57, 799–800.

Chamberlain, W.F. (1964b) Effect of various gases on survival of first-instar cattle grubs (*Hypoderma lineatum*) (de Villers). *Folia Entomologia Mexico* 7/8, 66.

Chamberlain, W.F. (1970) *In vitro* culture of the common cattle grub for six weeks after hatching. *Annals of the Entomological Society of America* 63, 1465–1466.

Chamberlain, W.F. (1989) Survival of second and third instars of the cattle grubs, *Hypoderma lineatum* (Villers) and *Hypoderma bovis* (L.), in artificial media. *Southwestern Entomologist* 14, 233–239.

Chamberlain, W.F. and Scholl, P.J. (1991) New procedures to enhance survival of third instar *Hypoderma lineatum* (Villers) (Diptera: Oestridae) in artificial media. *Journal of Medical Entomology* 28, 266–269.

Chapman, R.F. (1998) *The Insects: Structure and Function.* Cambridge University Press, Cambridge, 770 pp.

Charbon, J.L., Tieche, M.A., Villejoubert, C., Boulard, C. and Pfister, K. (1995) Monitoring the infestation of *Hypoderma* spp. in the canton of Vaud: a comparison between bulk milk serology and visual inspection of cattle. In: Tarry, D.W., Pithan, K. and Webster, K. (eds) *Improvements in the Control Methods for Warble-fly in Cattle and Goats.* Proceedings, 8–10 September 1994, Guildford, UK. COST 811. Commission of the European Communities, Brussels, pp. 147–151 (181 pp.).

Cheema, A.H. (1977) Observations on the histopathology of warble infestation in goats by the larvae of *Przhevalskiana silenus. Zentralblatt fur Veterinarmedizin Reihe B* 24, 648–655.

Chen, X.N. (2001) A case with skin myiasis caused by *Gasterophilus nigricornis. Chinese Journal of Parasitology and Parasitic Diseases* 19, 60.

Chereshnev, N.A. (1951) Biological peculiarities of the botfly *Gasterophilus pecorum* Fabr. (Diptera: Gasterophilidae). *Doklady Akademii Nauk SSSR* 77, 765–768 (in Russian).

Chirico, J., Stenkula, S., Eriksson, B., Gjotterberg, M., Ingemansson, S.O., Pehrson- Palmqvist, G. and Stenkula, E. (1987) Larvae of the reindeer warble fly causing three cases of human myiasis. *Lakartidningen* 84, 2207–2208.

Chiszar, D., Melcer, T., Lee, R., Radcliffe, C.W. and Duvall, D. (1990) Chemical cues used by prairie rattlesnakes (*Crotalus viridis*) as they follow the trails of rodent prey. *Journal of Chemical Ecology* 16, 79–86.

Chu-Wang, I.-W. and Axtell, R.C. (1971) Fine structure of the dorsal organ of the house fly larvae, *Musca domestica* L. *Zeitschrift fur Zellforchung mikroskopie und Anatomie* 117, 17–34.

Chu-Wang, I.-W. and Axtell, R.C. (1972a) Fine structure of the terminal organ of the house fly larvae, *Musca domestica* L. *Zeitschrift fur Zellforchung mikroskopie und Anatomie* 127, 287–305.

Chu-Wang, I.-W. and Axtell, R.C. (1972b) Fine structure of the ventral organ of the house fly larvae, *Musca domestica* L. *Zeitschrift fur Zellforchung mikroskopie und Anatomie* 130, 489–495.

Claerebout, E. and Vercruysse, J. (2000) The immune response and the evaluation of acquired immunity against gastrointestinal nematodes in cattle: a review. *Parasitology* 120, 25–42.

Clark, B. (1815) *An Essay on the Bots of Horses and Other Animals.* W. Flint, Old Bailey, London, 94 pp.

Clements, A.N. (1992) *The Biology of Mosquitoes,* vol. I. *Development, Nutrition and Reproduction.* Chapman & Hall, New York, 509 pp.

Cobb, M. (1999) What and how do maggots smell? *Biological Reviews,* 74, 425–459.

Cobbett, N.G. and Mitchell, W.C. (1941) Further observations on the life cycle and incidence of the sheep bot, *Oestrus ovis,* in New Mexico and Texas. *American Journal of Veterinary Research* 2, 358–366.

Cogley, T.P. (1989) Effects of migrating *Gasterophilus intestinalis* larvae (Diptera: Gasterophilidae) on the mouth of the horse. *Veterinary Parasitology* 31, 317–331.

Cogley, T.P. (1990) Morphology of the eggs of the rhinoceros bot flies *Gyrostigma conjugens* and *G. pavesii* (Diptera: Gasterophilidae). *International Journal of Insect Morphology and Embryology* 19, 323–326.

Cogley, T.P. (1991) Warble development by the rodent bot *Cuterebra fontinella* (Diptera: Cuterebridae) in the deer mouse. *Veterinary Parasitology* 38, 275–288.

Cogley, T.P. (1999) Morphology of a newly discovered sensory array on the mouthhooks of *Gasterophilus* larvae. *Medical and Veterinary Entomology* 13, 439–446.

Cogley, T.P. and Anderson, J.R. (1981) Invasion of black-tailed deer by nose bot fly larvae

(Diptera: Oestridae: Oestrinae). *International Journal of Parasitology* 11, 281–286.

Cogley, T.P. and Anderson, J.R. (1983) Ultrastructure and function of the attachment organ of *Gasterophilus* eggs (Diptera: Gasterophilidae). *International Journal of Insect Morphology and Embryology* 12, 13–23.

Cogley, T.P. and Cogley, M.C. (1999) Inter-relationship between *Gasterophilus* larvae and the horse's gastric and duodenal wall with special reference to penetration. *Veterinary Parasitology* 86, 127–142.

Cogley, T.P. and Cogley, M.C. (2000) Field observations of the host–parasite relationship associated with the common horse bot fly, *Gasterophilus intestinals*. *Veterinary Parasitology* 88, 93–105.

Cogley, T.P., Anderson, J.R. and Weintraub, J. (1981) Ultrastructure and function of the attachment organ of warble fly eggs (Diptera: Oestridae: Hypodermatinae). *International Journal of Insect Morphology and Embryology* 10, 7–18.

Cogley, T.P., Anderson, J.R. and Cogley, L.J. (1982) Migration of *Gasterophilus intestinalis* larvae (Diptera: Gasterophilidae) in the equine oral cavity. *International Journal for Parasitology* 12, 473–480.

Colebrook, E. and Wall, R. (2004) Ectoparasites of livestock in Europe and the Mediterranean region. *Veterinary Parasitology* 120, 251–274.

Coles, G.C. and Pearson, G.R. (2000) *Gasterophilus nasalis* infection: prevalence and pathological changes in equids in south-west England. *The Veterinary Record* 146, 222–223.

Colman, J.E., Pedersen, C., Hjermann, D.Ø., Holand, Ø., Moe, S.R. and Reimers, E. (2003) Do wild reindeer exhibit grazing compensation during insect harassment? *Journal of Wildlife Management* 67, 11–19.

Columbre, A. (1518) Della natura deçavalli et del modo di medicare le loro infermita. Libri III. *Appresso Francesco Fagiani, Vinetia*.

Colwell, D.D. (1986) Cuticular sensilla on newly hatched larvae of *Cuterebra fontinella* Clark (Diptera: Cuterebridae) and *Hypoderma* spp. (Diptera: Oestridae). *International Journal for Insect Morphology and Embryology* 15, 385–392.

Colwell, D.D. (1989a) Host–parasite relationships and adaptations to parasitism of first instar cattle grubs, *Hypoderma bovis* (L.) and *H. lineatum* (Vill.). PhD thesis, University of Guelph, Guelph, Ontario, Canada, 257 pp.

Colwell, D.D. (1989b) Scanning electron microscopy of the posterior spiracles of cattle grubs. *Medical and Veterinary Entomology* 3, 391–398.

Colwell, D.D. (1991) Ultrastructure of the integument of first-instar *Hypoderma lineatum* and *H. bovis* (Diptera: Oestridae). *Journal of Medical Entomology* 28, 86–94.

Colwell, D.D. (1992) *Cattle Grubs: Biology and Control*. Agriculture Canada Publication 1880/E, Ottawa, 17 pp.

Colwell, D.D. (2000) Persistence of the cattle grubs (Diptera: Oestridae) on a Canadian ranch with long-term, continuous therapeutic control. *Veterinary Parasitology* 94, 127–132.

Colwell, D.D. (2001a) Bot flies and warble flies (Order Diptera: Family Oestridae). In: Samuel, W.M., Pybus, M. and Kocan, A. (eds) *Parasitic Diseases of Wild Mammals*, 2nd edn. Iowa State University Press, Ames, Iowa, pp. 46–71.

Colwell, D.D. (2001b) Stage specific mortality and humoral immune responses during pulse and trickle infestations of the common cattle grub, *Hypoderma lineatum* (Diptera: Oestridae). *Veterinary Parasitology* 99, 231–239.

Colwell, D.D. (2002) Persistence of hypodermosis in North America and prospects for the development of vaccines. In: Good, M., Hall, M., Losson, B., O'Brien, D., Pithan, K. and Sol, J. (eds) *Mange and Myiasis of Livestock*. Proceedings, 3–6 October 2001, Toulouse, France. COST 833. Commission of the European Communities, Luxembourg, pp. 7–15.

Colwell, D.D. and Baron, R.W. (1990) Early detection of cattle grub (*Hypoderma lineatum* and *H. bovis*) (Diptera, Oestridae) using ELISA. *Medical and Veterinary Entomology* 4, 35–42.

Colwell, D.D. and Berry, N.M. (1993) Tarsal sensilla of the warble flies *Hypoderma bovis* and *H. lineatum* (Diptera: Oestridae). *Annals of the Entomological Society of America* 86, 756–765.

Colwell, D.D. and Jacobsen, J.A. (2002) Persistent activity of topical invermectin against *Hypoderma lineatum* (Diptera: Oestridae). *Veterinary Parasitology* 105, 247–256.

Colwell, D.D. and Kokko, E.G. (1986) Preparation of dipteran larvae for scanning electron microscopy using a freeze substitution technique. *Canadian Journal of Zoology* 64, 797–799.

Colwell, D.D. and Leggett, F.L. (1998) Ultrastructure of *in utero* first instar *Oestrus ovis* (L.) (Oestridae). In: *Proceedings of the 4th International Congress of Dipterology*. Oxford, UK, p. 38 (150 pp).

Colwell, D.D. and Leggett, F.L. (2003) Uptake of bovine IgG by first instars of the common cattle grub, *Hypoderma lineatum* (Diptera: Oestridae). *International Journal for Parasitology* 34, 219–223.

Colwell, D.D. and Milton, K. (1998) Development of *Alouattamyia baeri* (Diptera: Oestridae) from howler monkeys (Primates: Cebidae) on Barro Colorado Island, Panama. *Journal of Medical Entomology* 35, 674–680.

Colwell, D.D. and Milton, K. (2002) *Cuterebra* (= *Alouattamyia*) *baeri* from howler monkeys on Barro Colorado Island, Panama. In: Good, M., Hall, M., Losson, B., O'Brien, D., Pithan, K. and Sol, J. (eds) *Mange and Myiasis of Livestock*. Proceedings, 3–6 October 2001, Toulouse, France. COST 833. Commission of the European Communities, Luxembourg, pp. 206–210.

Colwell, D.D. and Scholl, P.J. (1995) Cuticular sensilla on newly hatched larvae of *Gasterophilus intestinalis* (De Geer) and *Oestrus ovis* (L.). *Medical and Veterinary Entomology* 9, 85–93.

Colwell, D.D., Martinez-Moreno, F.J., Martinez-Moreno, A., Hernandez-Rodriguez, S., de la Fuente-Lopez, C., Alunda, J.M. and Hall, J.R. (1998) Comparative scanning electron microscopy of third-instar *Hypoderma* spp. (Diptera: Oestridae). *Medical and Veterinary Entomology* 12, 181–186.

Colwell, D.D., Baird, C.R., Lee, B. and Milton, K. (1999) Scanning electron microscopy and comparative morphometrics of eggs from six bot fly species (Diptera: Oestridae: Cuterebrinae). *Journal of Medical Entomology* 36, 803–810.

Condamine, C.M. de la (1745) Relation abregee d'un voyage fait dans l'interieur de l'Amerique Meridionale. Depuis la cote de la Mer du Sud, Jusq'aux cotes du Bresil et de la Guiane, en descendant la riviere des Amazones; lue a l'assemblee publique de l'Academie des Sciences Chez la Vueve Pissot, 28 avril 1745, Paris.

Condy, J.B. (1974) Observations on internal parasites in Rhodesian elephant, *Loxodonta Africana* Blumenbach 1797. *Proceedings and Transactions of the Rhodesian Scientific Association* 55, 67–99.

Cooper, J.E. and Houston, D.C. (1972) Lesions in the crop of vultures associated with bot fly larvae. *Transactions of the Royal Society of Tropical Medicine and Hygiene* 66, 515–516.

Cope, S.E. and Catts, E.P. (1991) Parahost behavior of adult *Gasterophilus intestinalis* (Diptera: Gasterophilidae) in Delaware. *Journal of Medical Entomology* 28, 67–73.

Court, N. and Hartenberger, J.L. (1992) A new species of the hyracoid mammal *Titanohyrax* from the Eocene of Tunisia. *Palaeontology* 35, 309–317.

Crosskey, R.W. (1990) *The Natural History of Black Flies*. John Wiley & Sons, New York, 711 pp.

Cruz, J.B., Benitez-Usher, C., Crame, L.G., Gross, S.J. and Kohn, A.B. (1993) Efficacy of abamectin injection against *Dermatobia hominis* in cattle. *Parasitology Research* 79, 183–185.

Curtice, C. (1890) *The Animal Parasites of Sheep*. United States Government Printing Office, Washington, DC, 222 pp.

Curtice, C. (1891) The ox warble of the United States. *Journal of Comparative Medicine and Veterinary Archives* 6, 265–274.

Dame, D.A. (1984) Control of insects of veterinary importance by genetic techniques. *Preventative Veterinary Medicine* 2, 515–522.

Danilov, L.N. (1973) A case of multiple linear myiasis in a man. *Meditsinskaya Parazitologiya i Parazitarnye Bolezni* 42, 361.

Daoud, M.S., Abid, T.A. and Al-Amary, A.M. (1989) Rectal prolapse due to *Gasterophilus intestinalis* larvae in a horse: a case report. *Iraqi Journal of Veterinary Sciences* 2, 1–2.

Dar, M.S., Ben-Amer, M., Dar, F.K. and Papazotos, V. (1980) Ophthalmomyiasis caused by the sheep nasal bot, *Oestrus ovis* (Oestridae) larvae, in the Benghazi area of eastern Libya. *Transactions of the Royal Society of Tropical Medicine and Hygiene* 74, 303–306.

Dart, A.J., Hutchins, D.R. and Begg, A.P. (1987) Suppurative spenlitis and peritonitis in a horse after gastric ulceration caused by larvae of *Gasterophilus intestinalis*. *Australian Veterinary Journal* 64, 155–158.

Daubenton, C. (1753) In: Buffon, L.L. (ed.) *Histoire Generale et Particuliere Avec la Description du Cabinet du Roi*, vol. IV. Imprimerie Royal, Paris.

Davis, C.C., Bell, C.D., Mathews, S. and Donoghue, M.J. (2002) Laurasian migration explains Gondwanan disjunctions: evidence from Malpighiaceae. *Proceedings of the National Academy of Sciences USA* 99, 6833–6837.

Dawson, M. (1987) Pathogenesis of *maedi-visna*. *The Veterinary Record* 120, 451–454.

De Barro, P.J., Driver, F., Trueman, J.W.H. and Curran, J. (2000) Phylogenetic relationships of world populations of *Bemisia tabaci* (Gennadius) using ribosomal ITS1. *Molecular Phylogenetics and Evolution* 16, 29–36.

DeBry, R.W. (2003) Identifying conflicting signal in a multigene analysis reveals a highly resolved tree: the phylogeny of Rodentia (Mammals). *Systematic Biology* 52, 604–617.

Dechant, W., Pamba, H.O. and Awan, A.M. (1981) A case of internal ophthalmomyiasis in Kenya. *Klinische Monatsblätter für Augenheilkundi* 179, 368–369.

D'Elía, G. (2000) Comments on recent advances in understanding sigmodontine phylogeny and evolution. *Journal of Neotropical Mammals* 7, 47–54.

D'Erchia, A.M., Gissi, C., Pesole, G., Saccone, C. and Arnason, U. (1996) The guinea-pig is not a rodent. *Nature* 381, 597–600.

De Gray, T. (1684) *The Compleat Horse-Man and Expert Ferrier*. Samuel Lowndes, London, 124 pp.

Dempsey, J. (1983) Warbles in sheep. *Meat Hygienist* 38, 15.

Denlinger, D.L. and Zdárek, J. (1994) Metamorphosis behavior of flies. *Annual Review of Entomology* 39, 243–266.

Dewey, J.F., Pitman, W.C., Ryan, W.B.F. and Bonin, J. (1973) Plate tectonics and the evolution of the Alpine system. *Bulletin of the Geological Society of America* 84, 3137–3180.

Dill, R. (1934) Grub in the head in sheep in northeastern Nevada. Methods of herding which favored injury and methods of range management which practically eliminate losses. *University of Nevada Agriculture Experiment Station Bulletin* No. 135, 12 pp.

Disney, R.H.L. (1974) Speculations regarding the mode of evolution of some remarkable associations between Diptera (Cuterebridae, Simuliidae and Sphaeroceridae) and other

arthropods. *Entomologist's Monthly Magazine* 110, 67–74.

Doby, J.M. and Deunff, J. (1982) Considerations sur la frequence respective des especes dhypodermes (Insecta: Diptera: Oestroidea) a l'origine des cas humains d'hypodermose en France. *Anales de Parasitologie Humaine et Comparee* 57, 497–505 (in French with English summary).

Dolbois, L. and Dorchies, Ph. (1993) *Oestrus ovis* of sheep: pituitary eosinophils and masts cells. *Veterinary Research* 24, 362–363.

Dorchies, P. and Yilma, J.M. (1996) Current knowledge in immunology of *Oestrus ovis* infection. *Acta parasitologica Turcica* Suppl. 1, 563–580.

Dorchies, P., Yilma, J.M. and Savey, J. (1993) Lung involvement in ovine oestrosis: prevalence of lung abscesses and interstitial pneumonia. *The Veterinary Record* 133, 325.

Dorchies, P., Cardinaud, B. and Fournier, R. (1996) Efficacy of moxidectin as a 1% injectable solution and a 1% oral drench against nasal bots, pulmonary and gastrointestinal nematodes in sheep. *Veterinary Parasitology* 65, 163–168.

Dorchies, P., Jacquiet, P. and Duranton, C. (1998) Pathophysiology of *Oestrus ovis* infection in sheep and goats: review. *The Veterinary Record* 142, 487–489.

Dorchies, P., Bergeaud, J.P., Tabouret, G., Duranton, C., Prevot, F. and Jacquiet, Ph. (2000) Prevalence and larval burden of *Oestrus ovis* (Linne, 1761) in sheep and goats in northern Mediterranean region of France. *Veterinary Parasitology* 88, 269–273.

Dorchies, P., Jacquiet, P., Bergeaud, J.P., Duranton, C., Prevot, F., Alzieu, J.P. and Gosselin, J. (2001) Efficacy of doramectin injectable against *Oestrus ovis* and gastrointestinal nematodes in sheep in the southwestern region of France. *Veterinary Parasitology* 96, 147–154.

Dorst, J. and Dandelot, P. (1972) *A Field Guide to the Larger Mammals of Africa*, 2nd edn. Collins, London, 287 pp.

Dove, W.E. (1918) Some biological and control studies of *Gasterophilus haemorrhoidalis* and other bots of horses. *United States Department of Agriculture Bulletin* No. 597, 51 pp.

Dove, W.E. (1937) Myiasis of man. *Journal of Economic Entomology* 30, 29–39.

Downes, C.M., Smith, S.M., Theberge, J.B. and Dewar, H.J. (1985) Hilltop aggregation sites and behavior of male *Cephenemyia trompe* (Diptera: Oestridae). *Canadian Entomologist* 117, 321–326.

Downes, C.M., Theberge, J.B. and Smith, S.M. (1986) The influence of insects on the distribution, microhabitat choice, and behaviour of the Burwash caribou herd. *Canadian Journal of Zoology* 64, 622–629.

Draber-Mońko, A. (1974) *Horse Bot Flies – Gasterophilidae* (translated from Polish). Published for United States Department of Agriculture by Foreign Scientific Publication Department, National Center for Scientific Technology and Economic Information. Warsaw, Poland. From: *Polskie Towarz. Entomol., Klucze Do Oznaczania Owadow Polske*, Ser. 61, part 28, 1–57.

Drummond, R.O. (1985) Effectiveness of ivermectin for control of arthropod pests of livestock. *Southwestern Entomologist* Suppl. 7, 34–42.

Drummond, R.O., Lambert, G., Smalley, H.E. Jr and Terrill, C.E. (1981) Estimated losses of livestock to pests. In: Pimentel, D. (ed.) *CRC Handbook of Pest Management in Agriculture*. CRC Press, Boca Raton, Florida, pp. 111–127 (597 pp).

Drummond, R.O., George, J.E. and Kunz, S.E. (1988) *Control of Arthropod Pests of Livestock: A Review of Technology*. CRC Press, Boca Raton, Florida, 245 pp.

Dudzinski, W. (1970) Studies on *Cephenemyia stimulator* (Clark) (Diptera: Oestridae), the parasite of the European roe deer, *Capreolus capreolus* (L.). I. Biology. *Acta Parasitologica Polonica* 18, 555–572.

Dunn, L.H. (1930) Rearing the larvae of *Dermatobia hominis* Linn. in man. *Psyche* 37, 327–342.

Dunn, L.H. (1934) Prevalence and importance of the tropical warble fly, *Dermatobia hominis* Linn. (sic) in Panama. *Journal of Parasitology* 20, 219–226.

Duque-Araujo, A.M., Reina, D., Hernandez-Rodriquez, S., Santiago, V. and Navarrete, I. (1995) Preliminary approach to the prevalence of bovine hypodermosis in Portugal. In: Tarry, D.W., Pithan, K. and Webster, K. (eds) *Improvements in the Control Methods for Warble-fly in Cattle and Goats*. Proceedings, 8–10 September 1994, Guildford, UK. COST 811. Commission of the European Communities, Brussels, pp. 37–41 (181 pp.).

Duranton C. and Dorchies, P. (1997) *In vitro* culture of *Oestrus ovis* (Linné, 1761) first instar larvae: its application to antiparasitic drug screening. *International Journal for Parasitolology* 27, 125–128.

Duranton, C., Bergeaud, J.P. and Dorchies, Ph. (1995) Le Dot Enzyme-Linked immunosorbent Assay (Dot-ELISA): méthode de dépistage rapide de l'oestrose ovine. *Revue de Médecine Vétérinaire* 146, 283–286 (in French).

Duranton, C., Bergeaud, J.P. and Dorchies, Ph. (1996) Infestations expérimentales du chevreau par des larves L1 d'*Oestrus ovis*. *Veterinary Research* 27, 473–477.

Duranton, C., Dorchies, Ph., Grand, S., Lesure, C. and Oswald, I.P. (1999) Changing reactivity of caprine and ovine mononuclear phagocytes throughout part of the life cycle of *Oestrus ovis*: assessment through spontaneous and inductible NO production. *Veterinary Research* 30, 371–376.

Echevarria, F.A.M., Armour, J., Bairden, K. and Duncan, J.L. (1993) Laboratory selection for ivermectin resistance in *Haemonchus contortus*. *Veterinary Parasitology* 49, 265–270.

Edwards, K.M., Meredith, T.A., Hagler, W.S. and Healy, G.R. (1984) Ophthalmomyiasis interna causing visual loss. *American Journal of Ophthalmology* 97, 605–610.

Efron, B., Halloran, E. and Holmes, S. (1996) Bootstrap confidence levels for phylogenetic trees. *Proceedings of National Academy of Science USA* 93, 13429–13434.

Emborsky, M.E. and Faden, H. (2002) Ophthalmomyiasis in a child. *Pediatric Infectious Disease Journal* 21, 82–83.

Emmons, L.H. (1997) *Neotropical Rainforest Mammals, A Field Guide*, 2nd edn. The University of Chicago Press, Chicago, Illinois, 307 pp.

Enigk, G. and Pfaff, W. (1954) Bau und Zusammensetzung der larvencuticula von *Hypoderma bovis* (Oestridae). *Zeitschrift für Morphologie und Ökologi der Tierre* 43, 124–153 (in German).

Enileeva, N.Kh. (1987) Ecological characteristics of horse botflies in Uzbeckistan. *Parazitologiya*

21, 577–579. (*Review of Applied Entomology* B 77, 303, 1989).

Erzinclioglu, Y.Z. (1990) The means of attachment of the larvae of horse, zebra and rhinoceros bot-flies (Diptera: Gasterophilidae). *Medical and Veterinary Entomology* 4, 57–59.

Escartin-Peña, M. and Bautista-Garfias, C.R. (1993) Comparison of five tests for the serologic diagnosis of myiasis by *Gasterophilus* spp. larvae (Diptera: Gasterophilidae) in horses and donkeys: a preliminary study. *Medical and Veterinary Entomology* 7, 233–237.

Espmark, Y. (1967) Observations of defence reactions to oestrid flies by semidomestic forest reindeer (*Rangifer tarandus* L.) in Swedish Lapland. *Zoologische Beiträge* 14, 155–167.

Espmark, Y. (1972) Skator som 'predator' på ren. *Fauna och Flora* 67, 250–253 (in Swedish).

Evangelista, L.G. and Leite, A.C.R. (2003) Midgut ultrastructure of the third instar of *Dermatobia hominis* (Diptera: Cuterebridae) based on transmission electron microscopy. *Journal of Medical Entomology* 40, 133–140.

Evenhuis, N.L. (1994) *Catalogue of the Fossil Flies of the World* (*Insecta*: *Diptera*). Backhuys Publishers, Leiden, The Netherlands, 608 pp.

Everard, C.O.R. and Aitken, T.H.G. (1972) Cuterebrid flies from small mammals in Trinidad. *Journal of Parasitology* 58, 189–190.

Evstafjev, M.N. (1980) The role of immunity during hypodermatosis of cattle. *Parazitolgiya* 14, 197–205.

Evstafjev, M.N. (1982) The effect of acquired immunity on warble flies of *Hypoderma bovis* and *H. lineatum* (Hypodermatidae) during hypodematosis of cattle. *Parasitologia* 16 (6), 476–482.

Eyre, P., Boulard, C. and Deline, T.R. (1980) Reaginic (type 1 anaphylactic) antibodies produced by calves in response to *Hypoderma* larvae. *The Veterinary Record* 107, 280–281.

Eyre, P., Boulard, C. and Deline, T.R. (1981) Local and systemic reactions in cattle to *Hypoderma lineatum* larval toxin: protection by phenylbutazone. *American Journal of Veterinary Research* 42, 25–28.

Fain, A., Devos, E., Depoortere, G., Demuynck, G. and Vandepitte, J. (1977) A case of wandering subcutaneous myiasis due to *Hypoderma lineatum* in Belgium. *Louvain Medical* 94, 479–483 (in French with English summary).

Fan, Z. (1992) *Key to the Common Flies of China*. Shanghai Institute of Entomology, Academica Sinica. [In Chinese with English subtitle and and preface. English descriptions of all new taxa, pp. 912–927.]

Faust, E.C., Russell, P.F. and Jung, R.C. (1970) *Craig and Faust's Clinical Parasitology*, 8th edn. Lea and Febiger, Philadelphia, pp. 729–730.

Felsenstein, J. (1985) Confidence limits on phylogenies: an approach using the bootstrap. *Evolution* 39, 783–791.

Ferrar, P. (1987) A guide to the breeding habits and immature stages of Diptera Cyclorrhapha (2 vols) *Entomonograph* 8, 1–907.

Fidler, A.H. (1987) Migrierende dermale Myiasis durch *Hypoderma diana*. *Mitteilungen der Österreichischen Gesellschaft für Tropenmedizin und Parasitologie* 9, 111–119.

Figuier, L. (1867) *Les Insects*. L. Hachette et Cie, Paris.

Filippis, T. de and Leite, A.C.R. (1997) Scanning electron microscopy studies on the first-instar larva of *Dermatobia hominis*. *Medical and Veterinary Entomology* 11, 165–171.

Filippis, T. de and Leite, A.C.R. (1998) Morphology of second- and third-instar larvae of *Dermatobia hominis* by scanning electron microscopy. *Medical and Veterinary Entomology* 12, 160–168.

Fischer, M.S. (1989) Hyracoids, the sistergroup of perissodactyls. In: Prothero, D.R. and Schoch, R.M. (eds) *The Evolution of Perissodactyls*. Oxford University Press, New York and Oxford, pp. 37–56 (537 pp).

Fisher, W.F., Pruett, J.H., Howard, V.M. and Scholl, P.J. (1991) Antigen-specific lymphocyte proliferative responses in vaccinated and *Hypoderma lineatum*-infested calves. *Veterinary Parasitology* 40, 135–145.

Floate, K.D., Colwell, D.D. and Fox, A.S. (2002) Reductions of non-pest insects in dung of cattle treated with endectocides: a comparison of four products. *Bulletin of Entomological Research* 92, 471–481.

Folstad, I., Nilssen, A.C., Halvorsen, O. and Andersen, J. (1991) Parasite avoidance: the cause of post-calving migrations in Rangifer. *Canadian Journal of Zoology* 69, 2423–2429.

Forbes, A.B. (1994) The use of unlicensed dosages of ivermectin for *Hypoderma* control. In: Pfister, K., Charbon, J.L., Tarry, D.W. and Pithan, K. (eds) *Improvements in the Control Methods for Warble-fly in Cattle and Goats*. Proceedings, 23–25 September 1993, Thun, Switzerland. COST 811. Commission of the European Communities, Brussels, pp. 151–153.

Force, D.C. (1975) Succession of r and K strategies. In: Price, P.W. (ed.) *Evolutionary Strategies of Parasitic Insects and Mites*. Plenum Press, New York, pp. 112–129.

Forrest, T.G. (1987) Insect size tactics and developmental strategies. *Oecologia* (Berlin) 73, 178–184.

Fowler, M.E. and Paul-Murphy, J. (1985) *Cephenemyia* sp. infestations in the Llama. *California Veterinarian* 39, 10–12.

Fraenkel, G. and Bhaskaran, G. (1973) Pupariation and pupation in cyclorrhaphous flies (Diptera): terminology and interpretation. *Annals of the Entomological Society of America* 66, 1418–1422.

Fraiha, H., Chaves, L.C.L., Borges, I.C. and Freitas, R.B. de (1984) Miiases humanas na Amazõnia – III. Miiase pulmonary por *Alouattamyia baeri* (Shannon and Greene, 1926) (Diptera: Cuterebridae). *Seperata da Revista da Fundasão Sesp* 29, 63–68.

French, F.E. and Kline, D.L. (1989) 1-Octen-3-ol, and effective attractant for Tabanidae (Diptera). *Journal of Medical Entomology* 26, 459–461.

Frugère, S., Cota Leon, A., Prevot, F., Cepeda Palacios, R., Tabouret, G., Bergeaud, J.P., Duranton, C., Dorchies, Ph. and Jacquiet, Ph. (2000) Immunization of lambs with excretory secretory products of *Oestrus ovis* third instar larvae and subsequent experimental challenge. *Veterinary Research* 31, 527–535.

Fukuda, K., Ohnishi, Y., Inomata, H., Kirita, M. and Tsuji, M. (1987) Two cases of aveitis due to migrating visceral larva. *Rinsho-Ganka* 41, 845–849.

Gansser, A. (1951) *Dasselfliegen. Biologies, Schäden und Bekämpfung von Oestriden*. Verlag Schweizerischen Häuteschädenkommission, Basel, Switzerland, 128 pp.

Gansser, A.W.E. (1956) *Warble Flies and Other Oestridae: Biology and Control*. The Hide and Allied Trades Improvement Society, Surrey, UK, 63 pp.

Gansser, V.A. (1957) Zur Biologie der Dasselfliegen und zur Bekämpfung der Dasselphage durch Abfanger der Dasselfliegen. *Schweizer Archiv für Tierheilkunde* 99, 17–27 (in German).

Garsault, F.A. de (1770) Le nouveau parfait maréchal ou connaissance générale et universelle de cheval, divisé en sept traits. *Chez D'Houry, Paris*. 4th edn.

Garzozi, H., Lang, Y. and Barkay, S. (1989) External ophthalmomyiasis caused by *Oestris ovis*. *Israel Journal of Medical Sciences* 25, 162–163.

Gasser, R.B. (1999) PCR-based technology in veterinary parasitology. *Veterinary Parasitology* 84, 229–258.

Gaunt, M.W. and Miles, M.A. (2002) An insect molecular clock dates the origin of the insects and accords with palaeontological and biogeographic landmarks. *Molecular Biology and Evolution* 19, 748–761.

Gawor, J.J. (1995) The prevalence and abundance of internal parasites in working horses autopsied in Poland. *Veterinary Parasitology* 58, 99–108.

Gebauer, O. (1939) Das Verhalten der grossen Dasselfliege (*Hypoderma bovis* De Geer) in Tierversuch und die perkutane Invasion der Larvae des estern Stadiums. *Zeitschrift für Parasitenkunde* 11, 391–399 (in German).

Giangaspero, A., Auteri, P., Panunzio, M., Faliero, S.M., Nardella la Porta, M.C.F. and Cafiero, M.A. (1994) Two cases of human myiasis (ocular and naso-laryngeal) caused by *Oestrus ovis* L. 1758 in the Apulia region, Italy. *Giornale di Malattie Infettive e Parassitarie* 46, 50–53.

Gibson, G. and Torr, S.J. (1999) Visual and olfactory responses of haematophagous Diptera to host stimuli. *Medical and Veterinary Entomology* 13, 2–23.

Gingrich, R.E. (1980) Differentiation of resistance in cattle to larval *Hypoderma lineatum*. *Veterinary Parasitology* 7, 243–254.

Gingrich, R.E. (1981) Migratory kinetics of *Cuterebra fontinella* (Diptera: Cuterebridae) in the white-footed mouse, *Peromyscus leucopus*. *Journal of Parasitology* 67, 398–402.

Gingrich, R.E. (1982) Acquired resistance to *Hypoderma lineatum*: comparative immune response of resistant and susceptible cattle. *Veterinary Parasitology* 9, 233–242.

Gingrich, R.E. and Barrett, C.C. (1976) Natural and acquired resistance in rodent hosts to myiasis by *Cuterebra fontinella* (Diptera: Cuterebridae). *Journal of Medical Entomology* 13, 61–65.

Glaser, H. (1913) Über Dasselfliegen. In: *Mitteilungen des Ausschusses zur Bekämpfung der dasselplage*, No. 5, Berlin, pp 1–39.

Goddard, J. (1997) Human infestation with rodent botfly larvae: a new route of entry? *Southern Medical Journal* 90, 254–255.

Goddard, P., Bates, P. and Webster, K.A. (1999) Evaluation of a direct ELISA for the serodiagnosis of *Oestrus ovis* infections in sheep. *Veterinary Record* 144, 497–501.

Goldsmid, J.M. and Phelps, R.J. (1977) A review of myiasis of man in Rhodesia. *Central African Journal of Medicine* 23, 174–179.

Gomoyunova, N.P. (1973) *Hypoderma* in cattle in the Kemerovo region. *Trudy Biologicheskogo Instituta, Sibirskoe Otdelenie, Akademiya Nauk SSSR* 16, 158–170 (in Russian).

Gomoyunova, N.P. (1976) Biologiya ov odor severnykh oleney. *Novosibirsk 'Nauka', Sibirskoe Otdedenie*. Novosibirsk, Russia, 112 pp. (in Russian).

Goodman, R.L., Montalvo, M.A., Reed, J.B., Scribbick, F.W., McHugh, C.P., Beatty, R.L. and Aviles, R. (2000) Anterior orbital myiasis caused by human botfly *Dermatobia hominis*. *Archives of Ophthalmology* 7, 1002–1003.

Goodrich, A.L. (1940) Starling attacks upon warble infested cattle in the Great Plains area. *Journal of the Kansas Entomological Society* 13, 33–40.

Gorrochotegui-Escalante, N., Munoz, M.L., Fernandez-Salas, I., Beaty, B.J. and Black, W.C. (2000) Genetic isolation by distance among *Aedes aegypti* populations along the northeastern coast of Mexico. *American Journal of Tropical Medicine and Hygiene* 62, 200–209.

Goudie A.C., Evans, N.A., Gration, K.A.F., Bishop, B.F., Gibson, S.P., Holdom, B., Kaye, K.S., Wicks, S.R., Lewis, D., Weatherley, A.J., Bruce, C.I., Herbert, A. and Seymour, D.J. (1993) Doramectin – a potent novel endectocide. *Veterinary Parasitology* 49, 5–15.

Goulding, R.L. (1962) Menazon as a systemic insecticide in cattle. *Journal of Economic Entomology* 55, 577–579.

Graham, O.H. and Hourrigan, J.L. (1977) Eradication programs for the arthropod parasites of livestock. *Journal of Medical Entomology* 13, 629–658.

Grandcolas, P., Deleporte, P., Desutter-Grandcolas, L. and Daugeron, C. (2001) Phylogeny and ecology: as many characters as possible should be included in the cladistic analysis. *Cladistics* 17, 104–110.

Green, C.H. (1994) Bait methods for tsetse fly control. *Advances in Parasitology* 34, 229–291.

Gregson, J.D. (1958) Recent cattle grub life-history studies at Kamloops, British Columbia and Lethbridge, Alberta. *Proceedings of the 10th International Congress of Entomology Montreal* 3, 725–734.

Griffith, F.L. (1898) *Hieratic papyri from Kahum and Gurob*. Bernard Quaritch, London.

Grimaldi, D.A. and Cumming, J. (1999) Brachyceran Diptera in Cretaceous ambers and Mesozoic diversification of the Eremoneura. *Bulletin of the American Museum of Natural History* 239, 1–124.

Grogan, T.M., Payne, C.M., Payne, T.B., Spier, C., Cromey, D.W., Rangel, C. and Richter, L. (1987) Cutaneous myiasis, immunohistologic and ultrastructural morphometric features of a human botfly lesion. *American Journal of Dermatopathology* 9, 232–239.

Grunin, K.J. (1957) Nosoglotochnye ovoda (Oestridae) (Nasal botflies (Oestridae)). *Zool. Inst. Akad. Nauk. SSSR, Fauna SSSR, Nasekomye dvukryilye* 19(3), 1–145 (in Russian).

Grunin, K.J. (1959) Reasons for the congregation of male warble flies at high points in a locality. *Zoologichesky Zhurnal* 38, 1683–1688 (in Russian with English translation).

Grunin, K.J. (1962) *Botflies (Hypodermatidae)*. Fauna USSR, Insecta: Diptera, Leningrad, 19, No. 4, 237 pp. (in Russian).

Grunin, K.J. (1965) Hypodermatidae. In: Lindner, E. (ed.) *Die Fliegen der Paläarktischen Region*, vol. 8, pp. 1–154 (41 pls., addendum 155–160). (pp. 1–40 published 31.xii.1964; pp. 155–160 published 30.ix.1969.)

Grunin, K.J. (1966) Oestridae (Fam. 64a). In: Lindner, E. (ed.) *Die Fliegen der Paläarktischen Region*, vol. 8, pp. 1–96 (25 pls).

Grunin, K.J. (1969) Gasterophilidae. In: Lindner, E. (ed.) *Die Fliegen der Paläarktischen Region*, 8, pp. 1–61 (6 pls).

Grunin, K.J. (1973) The first discovery of larvae of the mammoth bot-fly *Cobboldia* (*Mamontia*, subgen. n.) *russanovi* sp. n. (Diptera, Gasterophilidae). *Éntomologicheskoe Obozrenie* 52, 228–233 (in Russian with English subtitles). (English translation in *Entomological Review, Washington* 52, 165–169.)

Grunin, K.J. (1975) Morphological and biological features of *Crivellia silenus shugnanica* Grunin, subsp. n. (Diptera, Hypodermatidae) from *Ovis ammon* L., in the Eastern Pamirs. *Entomologicheskoe obozrenie* 54, 186–190 (in Russian with English summary).

Guimarães, J.H. (1966) Nota sobre os habitos dos machos da *Dermatobia hominis* (Linneaus Jr.) (Diptera, Cuterebridae). *Papeis avulsos Zoologia, Sao Paulo* 18, 277–279.

Guimarães, J.H. and Coimbra, C.E.A. Jr (1982) Miiase humane per *Alouattamyia baeri* (Shannon and Greene) (Diptera: Cuterebridae*)*. Communicacão de dois ca sos na região Norte do Brasi. *Revista Brasileira de Zoologia* 1, 35–39.

Guimarães, J.H. and Papavero, N. (1966) A tentative annotated bibliography of *Dermatobia hominis* (Linnaeus Jr., 1781) (Diptera: Cuterebridae). *Arquivo Zoologico Estado* São *Paulo* 14, 233–294.

Guimarães, J.H. and Papavero, N. (1999) *Myiasis in Man and Animals in the Neotropical Region: Bibliographic Database.* Editora Plêiade, São Paulo, Brazil, 308 pp.

Guitton, C., Dorchies, P. and Morand, S. (1996) Scanning electron microscopy of larval instars and imago of *Rhinoestrus usbekistanicus* Gan, 1947 (Oestridae). *Parasite* 3, 155–159.

Gummer, D.L., Forbes, M.R., Bender, D. J. and Barclay, R.M.R. (1997) Botfly (Diptera: Oestridae) parasitism of Ord's kangaroo rats (*Dipodomys ordii*) at Suffield National Wildlife Area, Alberta, Canada. *Journal of Parasitology* 83, 601–604.

Haas, G.E. and Dicke, R.J. (1958) On *Cuterebra horripilum* Clark (Diptera: Cuterebridae) parasitizing cottontail rabbits in Wisconsin. *Journal of Parasitology* 44, 527–540.

Hadwen, S. (1915) The seasonal prevalence of *Hypoderma bovis* in 1915, together with observations on the terrifying effect *H. bovis* has upon cattle, and lesions produced by the larva. *Entomological Society of Ontario Annual Report* 36, 108–119.

Hadwen, S. (1917a) Anaphylaxis in cattle and sheep, produced by the larvae of *Hypoderma bovis*, *H. lineatum* and *Oestrus ovis*. *Journal of the American Veterinary Medical Association* 4, 15–44.

Hadwen, S. (1917b) The life history of *Hypoderma bovis* and *H. lineatum*. *Journal of the American Veterinary Medical Association* 4, 541–544.

Hadwen, S. (1919) *Warble Flies*: Hypoderma lineatum, *Villers and* Hypoderma bovis, *De Geer*. Department of Agriculture Canada, Health of Animals Branch, Canada, Scientific Series 27, pp. 1–24.

Hadwen, S. (1926) Notes on the life history of *Oedemagena tarandi* L. and *Cephenomyia trompe* Modeer. *Journal of Parasitology* 13, 56–65.

Hadwen, S. and Cameron, A.E. (1918) A contribution to the knowledge of the bot-flies, *Gasterophilus intestinalis*, DeG., *G. haemorrhoidalis*, L., and *G. nasalis*, L. *Bulletin of Entomological Research* 9, 91–106.

Hadwen, S. and Fulton, J.S. (1929) On the migration of *Hypoderma lineatum* from the skin to the gullet. *Parasitology* 16, 98–106.

Haeselbarth, E., Segerman, J. and Zumpt, F. (1966) The arthropod parasites of vertebrates in Africa south of the Sahara (Ethiopian region), vol. III (Insecta excl. Phthiraptera). *Publications of the South African Institute for Medical Research* 13, 1–283.

Hagan, H.R. (1951) *Embryology of the Viviparous Insects.* The Ronald Press, New York, 472 pp.

Hall, M.C. (1917) Notes in regard to bots, *Gasterophilus* spp. *Journal of the American Veterinary Medical Association* 52, 178–184.

Hall, M.J.R. (1995) Trapping the flies that cause myiasis: their responses to host-stimuli. *Annals of Tropical Medicine and Parasitology* 89, 333–357.

Hall, M.J.R. and Smith, K.G.V. (1993) Diptera causing myiasis in man. In: Lane, R.P. and Crosskey, R.W. (eds) *Medical Insects and*

Arachnids. Chapman & Hall, London, pp. 429–469 (688 pp).

Hall, M.J.R. and Wall, R. (1995) Myiasis of humans and domestic animals. *Advances in Parasitology* 35, 257–334.

Hanson, J. and Howell, J. (1981) Possible fenthion toxicity in magpies (*Pica pica*). *Canadian Veterinary Journal* 22, 18–19.

Haralampidis, S.T. (1987) ELISA in the sero-epidemiology of sheep and goats. *Bulletin of the Hellenic Veterinary Medical Society* 38, 215–223.

Harrison, R.G., Rand, D.M. and Wheeler, W.C. (1987) Mitochondrial DNA variation in field crickets across a narrow hybrid zone. *Molecular Biology and Evolution* 4, 144–158.

Harvey, J.T. (1986) Sheep botfly: ophthalmomyiasis extena. *Canadian Journal of Opthalmology* 21, 92–95.

Harwood, R.F. and James, M.T. (1979) *Entomology in Human and Animal Health*, 7th edn. Macmillan, New York, 548 pp.

Hearle, E. (1938) *Insects and Allied Parasites Injurious to Livestock and Poultry in Canada*. Canadian Department of Agriculture, Canada, Publication No 604, 108 pp.

Heath, A.C., Elliott, D.C. and Dreadon, R.G. (1968) *Gasterophilus intestinalis*, the horse botfly as a cause of cutaneous myiasis in a man. *New Zealand Medical Journal* 68, 31–32.

Hendrickx, M.O., Anderson, L., Boulard, C., Smith, D.G. and Weatherley, A.J. (1993) Efficacy of doramectin against warble fly larvae (*Hypoderma bovis*). *Veterinary Parasitology* 49, 75–84.

Hendrickx, W.M.L., Jansen, J., Vries, T.J. and DeVries, T.J. (1989) A *Hypoderma diana* (Diptera: Hypodermatidae) infection in a horse. *Veterinary Quarterly* 11, 56–57.

Henny, C.J., Blus, L.J., Kolbe, E.J. and Fitzner, R.E. (1985) Organophosphate insecticide (famphur) topically applied to cattle kills magpies and hawks. *Journal of Wildlife Management* 49, 648–658.

Herd, R.P. (1986) Pasture hygiene: a nonchemical approach to equine endoparasite control. *Modern Veterinary Practice*, 36–38.

Hightower, B.G., Adams, A.L. and Alley, D.A. (1965) Dispersal of released irradiated laboratory-reared screw-worm flies. *Journal of Economic Entomology* 58, 373–374.

Hillis, D.M. and Bull, J.J. (1993) An empirical test of bootstrapping as a method for assessing confidence in phylogenetic analysis. *Systematic Biology* 42, 182–192.

Hillis, D.M. and Dixon, M.T. (1991) Ribosomal DNA: molecular evolution and phylogenetic inference. *Quarterly Review of Biology* 66, 411–453.

Hogue, C.L. (1993) *Latin American Insects and Entomology*. University of California Press, Berkeley, California, 536 pp.

Holste, J.E., Colwell, D.D., Kumar, R., Lloyd, J.E., Pinkall, N.P.M., Sierra, M.A., Waggoner, J.W., Langholff, W.K., Barrick, R.A. and Eagleson, J.S. (1998) Efficacy of eprinomectin against *Hypoderma* spp. in cattle. *American Journal of Veterinary Research* 59, 56–58.

Hope, F.W. (1840) On insects and their larvae occasionally found in the human body. *Transactions of the Royal Entomological Society of London* 2, 256–271.

Horak, I.G. (1977) Parasites of domestic and wild animals in South Africa. I. *Oestrus ovis* in sheep. *Onderstepoort Journal of Veterinary Research* 44, 55–64.

Horak, I.G. (1987) Arthropod parasites of some wild animals in South Africa and Namibia. *Journal of the South African Veterinary Association* 58, 207–211.

Horak, I.G. and Boomker, J. (1981) *Strobiloestrus* sp. larvae in cattle. *Journal of the South African Veterinary Association* 52, 211–212.

Horak, I.G., De Vos, V. and De Klerk, B.D. (1984) Parasites of domestic and wild animals in South Africa. XVII. Arthropod parasites of Burchell's zebra, *Equus burchelli*, in the eastern Transvaal Lowveld. *Onderstepoort Journal of Veterinary Research* 51, 145–154.

Hoste, H., Leveque, H. and Dorchies, Ph. (2001) Comparison of nematode infections of gastro-intestinal tract in Angora and dairy goats in a rangeland environment: relations with the feeding behaviour. *Veterinary Parasitology* 101, 127–135.

Howard, G.W. (1980) Immature stages and affinities of the southern lechwe warblefly, *Strobiloestrus vanzyli* Zumpt (Diptera: Oestridae). *Journal of Natural History* 14, 669–683.

Howard, G.W. (1981) Stomach bots (Diptera: Gasterophilidae) infesting zebra in Zambia. In: Fowler, M.E. (ed.) *Wildlife Diseases of the Pacific Basin and Other Countries.* Wildlife Disease Association, Ames, Iowa, 262 pp.

Howard, G.W. and Conant, R.A. (1983) Nasal botflies of migrating wildebeest from western Zambia. *Journal of Natural History* 17, 619–626.

Howland, D.E. and Hewitt, G.M. (1995) Phylogeny of the Coleoptera based on mitochondrial cytochrome oxidase I sequence data. *Insect Molecular Biology* 4, 203–215.

Hu, Q. and Wang, Y. (2000) A case of cutaneous myiasis caused by *Hypoderma bovis* larvae. *Chinese Journal of Parasitology and Parasitic Diseases* 18, 268.

Huchon, D., Madsen, O., Sibbald, M.J.J.B., Ament, K., Stanhope, M.J., Catzeflis, F., de Jong, W.W. and Douzery, E.J.P. (2002) Rodent phylogeny and a timescale for the evolution of glires: evidence from an extensive taxon sampling using three nuclear genes. *Molecular Biology and Evolution* 19, 1053–1065.

Huelsenbeck, J.P. (1995) The performance of phylogenetic methods in simulation. *Systematic Biology* 44, 17–48.

Huffman, B. (2004) The Ultimate Ungulate Page. Available at: www.ultimateungulate.com

Humphreys, W.F. and Reynolds, S.E. (1980) Sound production and endothermy in the horse bot-fly, *Gasterophilus intestinalis.* *Physiological Entomology* 5, 235–242.

Hunter, D.M. and Webster, J.M. (1973a) Aggregation behavior of adult *Cuterebra grisea* and *C. tenebrosa* (Diptera: Cuterebridae). *Canadian Entomologist* 105, 1301–1307.

Hunter, D.M. and Webster, J.M. (1973b) Determination of the migratory route of botfly larvae, *Cuterebra grisea* (Diptera: Cuterebridae) in deermice. *International Journal for Parasitology* 3, 311–316.

Hussein, M.F., Hassan, H.A.R., Bilal, H.K., Basmae'il, S.M., Younis, T.M., Al-Motlaq, A.A.R. and Al-Sheikh, M.A. (1983) *Cephalopina titillator* (Clark 1797) infection in Saudi Arabian camels. *Zentralblatt fur Veterinaria und Medicina B* 30, 553–558.

Ilchmann, G. and Hiepe, T. (1985) Immunological studies on the diagnostis of *Oestrus ovis* infestation. *Monatshefte für Veterinärmedizin* 40(9), 304–307.

Ilchmann, G., Betke, P., Gräfe, D. and Gossing, S. (1986) Untersuchungen zur oestrose und ihre bekäkampfung in der Mongolischen Volksrepublik. *Monatshefte fur Veterinarmedizin* 41, 128–132.

Infante-Malachias, M.E., Yotoko, K.S.C. and Azeredo-Espin, A.M.L. (1999) Random amplified polymorphic DNA of screwworm fly populations (Diptera: Calliphoridae) from southeastern Brazil and northern Argentina. *Genome* 42, 772–779.

Innocenti, L., Masetti, M., Macchioni, G. and Giorgi, F. (1995) Larval salivary gland proteins of the sheep nasal bot fly (*Oestrus ovis* L.) are major immunogens in infested sheep. *Veterinary Parasitology* 60, 273–282.

Innocenti, L., Lucchesi, P. and Giorgi, F. (1997) Integument ultrastructure of *Oestrus ovis* (L.) (Diptera: Oestridae) larvae: host immune response to various cuticular components. *International Journal for Parasitology* 27, 495–506.

Ivey, M.C., Hoffman, R.A. and Claborn, H.V. (1968) Residues of Gardona in the body tissues of cattle sprayed to control *Hypoderma* spp. *Journal of Economic Entomology* 61, 1647–1648.

Jackson, H.C. (1989) Ivernectin as a systemic insecticide. *Parasitology Today* 5, 146–156.

Jagannah, M.S., Cozab, N. and Vijayasarathi, S.K. (1989a) Histopathological changes in the nasal passage of sheep and goats infested with *Oestrus ovis* (Diptera: Oestridae). *Indian Journal of Animal Science* 59, 87–91.

Jagannah, M.S., Cozab, N., Rahman, S.A. and Honnappa, T.G. (1989b) Serodiagnosis of *Oestrus ovis* infestation in sheep and goats. *Indian Journal of Animal Science* 59, 1220–1224.

James, M.T. (1947) *The Flies that Cause Myiasis in Man.* United States Department of Agriculture, Miscellaneous Publication No. 631, Washington, D.C., 175 pp.

Janzen, D.H. (1976) The occurrence of the human warble fly (*Dermatobia hominis*) in the dry deciduous forest lowlands of Costa Rica. *Biotropica* 8, 210.

Janzen, D.H. (1983) *Tapirus bairdii* (Danto, Danta, Baird's Tapir). In: Janzen, D.H. (ed.) *Costa Rican Natural History.* The University of Chicago Press, Chicago (Illinois) and London, pp. 496–497 (816 pp).

Jellison, Wm.L. (1935) *Cephenomyia pratti* (Diptera: Oestridae) reared from blacktailed deer. *Proceedings of the Helminthological Society of Washington* 2, 69.

Jettmar, H.M. (1953) Uber bakteriostatische Stoffe im Darm der Hypodermalarve. *Zeitschrift für Hygiene* 137, 61–66 (in German).

Jones, S.R. (2000) Hair suitability and selection during oviposition by *Hypoderma lineatum* (Diptera: Oestridae). *Annals of the Entomological Society of America* 93, 525–528.

Jordan, P.A., Botkin, D.B. and Wolfe, M.L. (1971) Biomass dynamics in a moose population. *Ecology* 52, 147–152.

Juzarte, T.J. (1999) Diario da navegacao do Rio Tiete, Rio Grande e Rio Gatemi. *UNICAMP*, Campinas, SP. 1769–1770 (in Portuguese).

Kaboret, Y., Deconinck, P., Panguy, L.J., Akakpo, J. and Dorchies, Ph. (1997) Lésions de la rhinoestrose spontanée à *Rhinoestrus usbekistanicus* (Gan 1947) chez l'âne (*Equus asinus*) au Sénégal. *Revue de Médecine Vétérinaire* 148, 123–126 (in French).

Kalelioglu, M., Akturk, G., Akturk, F., Komsuoglu, S.S., Kuzeyli, K., Tigin, K., Karaer, Z. and Bingol, R. (1989) Intracerebral myiasis from *Hypoderma bovis* larva in a child: case report. *Journal of Neurosurgery* 71, 929–931.

Kal'vish, T.K. (1990) Morphophysiological features of the entomopathogenic fungus *Tolypocladium niveum* (O. Rostr.) Bissett (Deuteromycotina: Moniliales) new to the flora of the USSR. *Mikologiya i Fitopatologiya* 24, 210–215 (in Russian).

Karter, A.J., Folstad, I. and Anderson, J.R. (1992) Abiotic factors influencing embryonic development, egg hatching, and larval orientation in the reindeer warble fly, *Hypoderma tarandi*. *Medical and Veterinary Entomology* 6, 355–362.

Kearney, M.S., Nilssen, A.C., Lyslo, A., Syrdalen, P. and Dannevig, L. (1991) Ophthalmomyiasis caused by the reindeer warble fly larva. *Journal of Clinical Pathology* 44, 276–284.

Keilin, D. (1994) Respiratory systems and respiratory adaptations in larvae and pupae of Diptera. *Parasitology* 36, 1–66.

Kennaugh, J.H. (1972) Some observations on the cuticle of the larvae of *Hypoderma bovis* and *Hypoderma diana* (Oestridae). *Parasitology* 65, 121–130.

Kettle, D.S. (1990) *Medical and Veterinary Entomology*. CAB International, Wallingford, UK, 658 pp.

Kettle, D.D. and Utsi, M.N.P. (1955) *Hypoderma diana* (Diptera, Oestridae) and *Lipotena cervi* (Diptera, Hippoboscidae) as parasites of reindeer (*Rangifer tarandus*) in Scotland with notes on the second-stage larva of *Hypoderma diana*. *Parasitology* 45, 116–122.

Khan, M.A. (1969) Systemic pesticides for use on animals. *Annual Review of Entomology* 14, 369–386.

Khan, M.A., Connell, R. and Darcel, C. (1960) Immunization and parenteral chemotherapy for the control of cattle grub, *H. lineatum* and *H. bovis*. *Canadian Journal of Comparative Medicine and Veterinary Science* 24, 177–180.

Khan, M.A., Scholl, P.J. and Weintraub, J. (1985) Ivomec for control of hypodermal larvae. In: *Pesticide Research Report – 1984*. Expert Committee for Pesticide Use in Agriculture, Ottawa, Canada, p. 193.

Kilani, M., Kacem, H., Dorchies, Ph. and Franc, M. (1986) Observations sur le cycle annuel d'*Oestrus ovis* en Tunisie. *Revue de Médecine Vétérinaire* 137, 451–457.

Kilgore, W.W. and Doutt, R.L. (eds) (1967) *Pest Control: Biological, Physical and Selected Chemical Methods*. Academic Press, New York and London, 477 pp.

Kinsella, J.M., Deem, S.L., Blake, S. and Freeman, A.S. (2004) Endoparasites of African forest elephants (*Loxodonta africana cyclotis*) from the Republic of Congo and Central African Republic. *Comparative Parasitology* 71, 104–110.

Klei, T.R., Rehbein, S., Visser, M., Langholff, A.K., Chapman, M.R., French, D.D. and Hanson, P. (2001) Re-evaluation of ivermectin efficacy against equine intestinal parasites. *Veterinary Parasitology* 98, 315–320.

Klein, K.K., Fleming, C.S., Colwell, D.D. and Scholl, P.J. (1990) Economic analysis of an integrated approach to cattle grub (*Hypoderma* spp.) control. *Canadian Journal of Agricultural Economics* 38, 59–173.

Knapp, F.W., Brethour, J.R., Harvey, T.L. and Roan, C.C. (1959) Field observations of increasing resistance of cattle to cattle grubs. *Journal of Economic Entomology* 52, 1022–1023.

Knipling, E.F. (1955) Possibilities of insect control or eradication through the use of sexually sterile males. *Journal of Economic Entomology* 48, 459–462.

Knipling, E.F. and Wells, R.W. (1935) Factors stimulating hatching of eggs of *Gasterophilus intestinalis* (De Geer) and the application of warm water as a practical method of destroying these eggs on the host. *Journal of Economic Entomology* 28, 1065–1072.

Kocher, T.D. and Xiong, B. (1991) Comparison of mitochondrial DNA sequences of seven morphospecies of black flies (Diptera: Simulidae). *Genome* 34, 306–311.

Koekemoer, L.L., Lochouarn, L., Hunt, R.H. and Coetzee, M. (1999) Single-strand conformation polymorphism analysis for identification of four members of the *Anopheles funestus* (Diptera: Culicidae) group. *Journal of Medical Entomology* 36, 125–130.

Koone, H.D. and Banegas, A.D. (1959) Biology and control of *Dermatobia hominis* in Honduras (Diptera: Cuterebridae). *Journal of the Kansas Entomological Society* 32, 100–108.

Krafsur, E.S., Bryant, N.L., Marquez, J.G. and Griffiths, N.T. (2000) Genetic distances among North American, British, and West African house fly populations (*Musca domestica* L.). *Biochemical Genetics* 38, 275–284.

Krümmel, H. and Brauns, A. (1956) Myiasis des Auges: medizinische und entomologische Grundlagen. *Zeitschrift für angewandte Zoologie* 43, 129–190.

Kubie, J.L. and Halpern, M. (1979) Chemical senses involved in garter snake prey trailing. *Journal of Physiology and Psychology* 93, 648–667.

Kühl, R. (1949) Beiträge zur Biologie und Bekämpfung der Rinderdasselfliegen *Hypoderma bovis* und *Hypoderma lineatum*. *Anzeiger für Schädlingskunde* 22, 74–78 (in German).

Kunz, S.E., Scholl, P.J., Colwell, D.D. and Weintraub, J. (1990) Use of the sterile insect technique for control of cattle grubs (*Hypoderma lineatum* and *H. bovis*) (Diptera: Oestridae). *Journal of Medical Entomology* 27, 523–529.

Kutzer, E. (2000) Die bekampfung der Oestinose und Hypodermose bei Rothirsch (*Cervus elaphus hippelaphus*) und Reh (*Capreolus c. capreolus*) mittels Ivermectin (IvomecR). *Berliner und Munchener Tierarztliche Wochenschrift* 113, 149–151.

Lafosse, P.-E. (1772) *Cours d'hippiatrique ou Traité Complet de la Médecine des Chevaux*. Chez Edemé, Paris.

Landi, S. (1960) Bacteriostatic effect of haemolymph of larvae of various botflies. *Canadian Journal of Microbiology* 6, 115–119.

Lane, R.P., Lovell, C.R., Griffiths, W.A.D. and Sonnex, T.S. (1987) Human cutaneous myiasis – a review and report of two cases due to *Dermatobia hominis*. *Clinical and Experimental Dermatology* 12, 40–45.

Languillat, G., Garin, Y., Beauvais, B., Lariviere, M. and Schaison, G. (1976) Meningitis caused by *Hypoderma*. *Nouvelle Presse Medicale* 5, 984–986.

Lanham, U.N. (1975) A mountain-top swarm of the Hemipteran *Nysius raphanus* in New Mexico, with notes on other insects. *Pan-Pacific Entomologist* 51, 166–167.

Lapointe, J.M., Céleste, C. and Villeneuve, A. (2003) Septic peritonitis due to colonic perforation associated with aberrant migration of a *Gasterophilus intestinalis* larva in a horse. *Veterinary Pathology* 40, 338–339.

Leaning, W.H.D. (1984) Ivermectin as an antiparasitic agent in cattle. *Modern Veterinary Practice* 65, 669–672.

Lecroisey, A., Boulard, C. and Keil, B. (1979) Chemical and enzymatic characterization of the collagenase from the insect *Hypoderma lineatum*. *European Journal of Biochemistry* 101, 385–393.

Lecroisey, A., De Wolf, A. and Keil, B. (1980) Crystallization and preliminary X-ray diffraction data for the collagenase of *Hypoderma lineatum* larvae. *Biochemical and Biophysics Research Communication* 94, 1261–1265.

Lecroisey, A., Tong, N.T. and Keil, B. (1983) Hypodermin B, a trypsin-related enzyme from the insect *Hypoderma lineatum*. Comparison with hypodermin A and *Hypoderma* collagenase, two serine proteinases from the same source. *European Journal of Biochemistry* 134, 261–267.

Lecroisey, A., Gilles, A.M., De Wolf, A. and Keil, B. (1987) Complete amino acid sequence of the collagenase from the insect

Hypoderma lineatum. Journal of Biological Chemisty 262, 7546–7551.

Leebens-Mack, J., Pellmyr, O. and Brock, M. (1998) Host specificity and the generic structure of two yucca moth species in a yucca hybrid zone. *Evolution* 52, 1376–1382.

Leinati, L., Oberosler, R. and Beber, L. (1971) The disinfestation of mountain pastures for control of bovine hypodermosis: II. The biological method. *25th Atti Della Societa Italiana Delle Scienze Veterinarie* 25, 450–452 (in Italian with English summary).

Leite, A.C.R. (1988) Scanning electron microscope of the egg and first instar larva of *Dermatobia hominis* (Diptera: Cuterebridae). *Memorias de Instituto Oswaldo Cruz* 83, 253–257.

Leite, A.C.R. and Scott, F.B. (1999) Scanning electron microscopy of the second-instar larva of *Gasterophilus nasalis*. *Medical and Veterinary Entomology* 13, 288–294.

Leite, A.C.R. and Williams, P. (1989) Morphological observation on the egg and first instar larva of *Metacuterebra apicalis* (Diptera: Cuterebridae). *Memorias do Instituto Oswaldo Cruz* 84, 123–130.

Leite, A.C.R. and Williams P. (1999) Experimental infection of rodents with larvae of *Metacuterebra apicalis*, a neotropical cuterebrid. *Revista Brasileira de Parasitologia Veterinaria* 8, 35–39.

Leite, A.C.R., Scott, F.B. and Evangelista, L.G. (1999) Scanning electron microscope observations on third-instar *Gasterophilus nasalis* (Diptera: Oestridae). *Journal of Medical Entomology* 36, 643–648.

Lello, E. and de Boulard, C. (1990) Rabbit antibody responses to experimental infestation with *Dermatobia hominis*. *Medical and Veterinary Entomology* 4, 303–309.

Lello, E. and Peraçoli, M.T.S. (1993) Cell-mediated and humoral immune responses in immunized and/or *Dermatobia hominis* infested rabbits. *Veterinary Parasitology* 47, 129–138.

Lello, E. and de Rosis, A.M.B. (2003) Inflammatory reaction to the human botfly, *Dermatobia homini*, in infested and re-infested mice. *Medical and Veterinary Entomology* 17, 55–60.

Lello, E.D., Gregório, E.A. and Toledo, L.A. (1985) Desenvolvimento das gônadas de *Dermatobia hominis* (Diptera: Cuterebridae). *Memorias do Instituto Oswaldo Cruz (Rio de Janeiro)* 80, 159–170 (in Portuguese).

Lello, E., Oliveira-Sequeira, T.C.G. and Peraçoli, M.T.S. (1999) Inflammatory response in *Dermatobia hominis* infested rabbits. *Revista Brasileira de Parasitologia Veterinaria* 8, 87–91.

Lelouarn, H. and Boulard, C. (1974) Observations sur *Oestromyia leporina*. *Annales de la Société française d'entomologie*, 10, 79–83 (in French).

Lessinger, A.C., Martins Junqueira, A.C., Lemos, T.A., Kemper, E.L., da Silva, F.R., Vettore, A.L., Arruda, P. and Azeredo-Espin, A.M.L. (2000) The mitochondrial genome of the primary screwworm fly *Cochliomyia hominivorax* (Diptera: Calliphoridae). *Insect Molecular Biology* 9, 521–529.

Li, C.S. (1989) A case of *Hypoderma lineatum* (Diptera: Oestridae) of the human scalp. *Chinese Journal of Parasitic Disease Control* 2, 164.

Liebisch, A. (1992) Changing patterns in the epidemiology and control of hypodermosis in cattle in Germany. In: Gasca, A., Hernandez, S., Martinez, J. and Pithan, K. (eds) *Improvements in the Control Methods for Warble-fly in Cattle and Goats*. Proceedings, 8–10 May 1991, Cordoba, Spain. COST 811. Commission of the European Communities, Brussels, pp.1–4.

Linnaeus, C. (1758) *Systema Naturae Per Regna Tria Naturae, Secundum Classes, Ordines, Genera, Species, cum Characteribus, Differentiis, Synonimis, Locis*, 10th edn. Laurentis Salvius, Stockholm, Sweden, 823 pp.

Little, V.A. (1931a) A preliminary report on the insecticidal properties of devil's shoestring, *Cracca virginiana* Linn. *Journal of Economic Entomology* 24, 743–754.

Little, V.A. (1931b) Devil's shoe-string as an insecticide. *Science* 73, 315–316.

Lloyd, J.E., Kumar, R., Waggoner, J.W. and Phillips, F.E. (1996) Doramectin systemic activity against cattle grubs, *Hypoderma lineatum* and *H. bovis* (Diptera: Oestridae), and cattle lice, *Bovicola bovis* (Mallophaga: Trichodectidae), *Linognathus vituli* and *Solenopotes capillatus* (Anoplura: Linognathidae), and *Haematopinus eurysternus* (Anoplura: Haematopinidae), in Wyoming. *Veterinary Parasitology* 63, 307–317.

Losson, B. and Lonneux, J.F. (1993) Activity of moxidectin 0.5% pour-on against *Hypoderma bovis* in naturally infested cattle. In: Losson, B., Lonneux, J.F. and Pithan, K. (eds) *Improvements in the Control Methods for Warble-fly in Cattle and Goats*. Proceedings, 16–18 September 1992, Liege, Belgium. Commission of the European Communities, Brussels, pp. 39–47.

Lucet, M.A. (1914) Experiments on the life-history of *Hypoderma bovis* and means of destroying it. *Comptes Rendus de l'Academie des Sciences* 158, 812–814 (in French). (Abstract: *Review of Applied Entomology* B 2, 90, 1914.)

Lucientes, J., Clavel, A., Ferrer-Dufol, M., Valles, H., Peribanez, M.A., Gracia-Salinas, M.J. and Castillo, J.A. (1997) One case of nasal human myiasis caused by third stage instar larvae of *Oestrus ovis*. *American Journal of Tropical Medicine and Hygiene* 56, 608–609.

Luckett, W.P. and Hartenberger, J.-L. (1993) Monophyly or polyphyly of the order Rodentia: possible conflict between morphological and molecular interpretations. *Journal of Mammalian Evolution* 1, 127–147.

Lujan, L., Vazquez, J., Lucientes, J., Panero, J.A. and Varea, R. (1998) Nasal myiasis due to *Oestrus ovis* infestation in a dog. *Veterinary Record* 11, 282–283.

Lunt, D.H., Zhang, D.X., Szymura, J.M. and Hewitt, G.M. (1996) The insect cytochrome oxidase I gene: evolutionary patterns and conserved primers for phylogenetic studies. *Insect Molecular Biology* 5, 153–165.

Lyon, S., Stebbings, H.C. and Coles, G.C. (2000) Prevalence of tapeworms, bots and nematodes in abattoir horses in south-west England. *The Veterinary Record* 147, 456–457.

Lysyk, T.J., Colwell, D.D. and Baron, R.W. (1991) A model for estimating abundance of cattle grub (Diptera: Oestridae) from the proportion of uninfested cattle as determined by serology. *Medical and Veterinary Entomology* 5, 53–258.

Ma, ZhiHua. (1999) A case of myiasis caused by *Gasterophilus haemorrhoidalis*. *Chinese Journal of Parasitology and Parasitic Diseases* 17, 57.

MacDiarmid, J.A., Clarke, R., McClure, S.J., Bowen, F.R. and Burrel, D.H. (1995) Use of a monoclonal antibody to ovine IgE for fly-strike studies in sheep. *International Journal for Parasitology* 25, 1505–1507.

Maddison, W.P. and Maddison, D.R. (1992) *MacClade. Analysis of Phylogeny and Character Evolution*, ver. 3.0.1. Computer software and documentation, Sinauer Association, Sunderland, Massachusetts.

Maes S., and Boulard C. (2000) Deer hypodermosis. Workshop on hypodermosis: Hypodermosis in Central Asia. Lanzhou (CHI) 16-20 October 2000.

Maes, S. and Boulard, C. (2001) Deer myiasis in France. In: Good, M., Hall, M., Losson, B., O'Brien, D., Pithan, K. and Sol, J. (eds) *Proceedings of COST Action 833, Mange and myiasis of livestock*, 28–30 September 2000, Ceské Budejovice, Czech Republic. Commission of the European Communities, Brussels, pp. 181–186.

Magat, A. and Boulard, C. (1970) Essais de vaccination contre l'hypodermose bovine avec un vaccin contenant une collagénase brute extraite des larves de 1er stade *d'Hypoderma lineatum*. *Comptes Rendus de l'Academie des Sciences (Paris)* 270, 728–730.

Maia, A.A.M. and Guimarães, M.P. (1986) Uso da ivermectina no controle de larvas de *Dermatobia hominis* (Linnaeus, 1781) (Diptera: Cuterebridae) em bovinos de corte. *Arquivo Brasileiro Medicina e Veterinaria Zootaxonomia* 38, 57–64 (in Portuguese).

Marchenko, V.A. and Marchenko, V.P. (1989) Survival of *Oestrus ovis* larvae depends on the state of the immune system of sheep. *Parazitologiya* 2, 129–133.

Marcilla, A., Bargues, M.D., Ramsey, J.M., Magallon-Gastelum, E., Salazar-Schettino, P.M., Abad-Franch, F., Dujardin, J.P., Schofield, C.J. and Mas-Coma, S. (2001). The ITS-2 of the nuclear rDNA as a molecular marker for populations, species, and phylogenetic relationships in Triatominae (Hemiptera: Reduviidae), vector of Chagas disease. *Molecular Phylogenetics and Evolution* 18, 136–142.

Marley, S.E. and Conder, G.A. (2002) The use of macrocyclic lactones to control parasites of domesticated wild ruminants. In: Vercruysse, J. and Rew, R. (eds) *Macrocyclic Lactones in Antiparasitic Therapy*. CABI Publishing, Wallingford, UK, 432 pp.

Marquez, J.G. and Krafsur, E.S. (2002) Gene flow among geographically diverse housefly populations (*Musca domestica* L.): a worldwide survey of mitochondrial diversity. *Journal of Heredity* 93, 254–259.

Marquez, J.G. and Krafsur, E.S. (2003) Mitochondrial diversity evaluated by the single strand conformation polymorphism method in African and North American house flies (*Musca domestica* L.) *Insect Molecular Biology* 12, 99–106.

Martinez-Gomez, F., Gasca-Arroyo, A., De Juan, F. and Hernanez-Rodriquez, S. (1991) Maternal anti-*Hypoderma* antibodies in calves. *Medical and Veterinary Entomology* 5, 381–383.

Martinez-Moreno, J., Reina, D., Navarrete, I., Jimenez, V., Martinez-Moreno, A. and Hernandez, S. (1996) Epidemiological survey of hypodermosis in western Spain. *Veterinary Record* 139, 340–343.

Matton, J.S., Gerros, T.C., Parker, J.E., Carter, C.A. and LaMarche, R.M. (1997) Upper airway obstruction in a Llama caused by aberrant nasopharyngeal bots (*Cephenemyia* sp.). *Veterinary Radiology and Ultrasound* 38, 384–386.

McAlpine, J.F. (1970) First record of calypterate flies in the mesozoic era (Diptera: Calliphoridae). *Canadian Entomologist* 102, 342–346.

McAlpine, J.F. (1981) Morphology and terminology – adults. In: J.F. McAlpine, B.V. Peterson, G.E. Shewell, H.J. Teskey, J.R. Vockeroth and D.M. Wood (eds). *Manual of Nearctic Diptera*, Vol. 1 Agriculture Canada Research Branch Monograph 27, 674 pp.

McMahon, D.C. and Bunch, T.D. (1989) Bot fly larvae (*Cephenemyia* spp., Oestridae) in mule deer (*Odocoileus hemionus*) from Utah. *Journal of Wildlife Diseases* 25, 636–638.

McMullin, P.F., Cramer, L.G., Benz, G., Jeromel, P.C. and Gross, S.J. (1989) Control of *Dermatobia hominis* infestation in cattle using an ivermectin slow-release bolus. *Veterinary Record* 124, 465.

Medownick, M., Lazarus, M., Finkelstein, E. and Weiner, J.M. (1985) Human external ophthalmomyiasis caused by the horse bot fly larva *Gasterophilus* spp. *Australian and New Zealand Journal of Ophthalmology* 13, 387–390.

Meeusen, E.N. and Balic, A. (2000) Do eosinophils have a role in the killing of helminth parasites? *Parasitology Today* 16, 95–101.

Meleney, W.P., Cobbett, N.G. and Peterson, H.O. (1962) The natural occurrence of *Oestrus ovis* in sheep from the southwestern United States. *American Journal of Veterinary Research* 23, 1246–1251.

Meyer, R.P. and Bock, M.E. (1980) Aggregation and territoriality of *Cuterebra lepivora* (Diptera: Cuterebridae). *Journal of Medical Entomology* 17, 489–493.

Michelsen, V. (1991) Revision of the aberrant New World genus *Coenosopsia* (Diptera: Anthomyiidae), with a discussion of anthomyiid relationships. *Systematic Entomology* 16, 85–104.

Michelsen, V. (2000) Oldest authentic record of a fossil calyptrate fly (Diptera): a species of Anthomyiidae from early Coenozoic Baltic amber. *Studia dipterologica* 7, 11–18.

Mikkola, K., Selvennoinen, J. and Hackman, W. (1982) Ophthalmomyiasis caused by the elk throat botfly *Cephenemyia ulrichii* in man. *Duodecim* 98, 1022–1225.

Miller, H.R.P. (1996) Mucosal mast cells and the allergic response against nematode parasites. *Veterinary Immunology and Immunopathology* 54, 331–336.

Milner-Gulland, E.J., Bukreeva, O.M., Coulson, T., Lushchekina, A.A., Kholodova, M.V., Bekenov, A.B. and Grachev, I.A. (2003) Reproductive collapse in saiga antelope harems. *Nature* 422, 135.

Milton, K. (1996) Effects of bot fly (*Alouattamyia baeri*) parasitism on a free-ranging howler monkey (*Alouatta palliata*) population in Panama. *Journal of Zoology* 239, 39–63.

Minár, J. (1974) Experimental laboratory rearing of warble-flies, bot-flies and gad-flies. *Folia Facultatis Scientiarum Naturalium Universitatis Purkynianae Brunensis* 15, 105–108.

Minár, J. (1987) The attacking of a horse by the deer warble fly *Hypoderma diana* Brauer, 1858 (Diptera, Hypodermatidae). *Veterinarni Medicina (Praha)* 32, 187–191.

Minár, J. and Breev, K.A. (1982) Laboratory and field rearing of the warble-fly *Hypoderma bovis* (De Geer) (Diptera, Hypodermatidae) in the research of its population ecology. *Folia Parasitologia (Praha)* 29, 351–360.

Minár, J. and Breev, K.A. (1983) Studies on the low and fundamental populations of the warble fly *Hypoderma bovis* (De Geer) (Diptera: Hypodermatidae). *Folia Parasitologia (Praha)* 30, 57–71.

Minár, J. and Samsinakova, A. (1977) Entomophagous fungus *Beauveria tenella* (Deuteromycetes), parasite of pupae of warble fly *Hypoderma bovis* (Diptera, Hypodermatidae). *Folia Parasitology* 24, 93.

Mock, D.E. (1973) Eggs of the horse bot fly *Gasterophilus intestinalis* (Diptera: Gasterophilidae) on pastured cattle. *Journal of Medical Entomology* 10, 34–37.

Moiré, N., Bigot, Y., Periquet, G. and Boulard, C. (1994) Sequencing and gene expression of hypodermins A, B and C in larval stages of *Hypoderma lineatum*. *Molecular Biochemistry Parasitology* 66, 233–240.

Moiré, N., Nicolas-Gaulard, I., Le Vern, Y. and Boulard, C. (1997) Enzymatic effect of hyodermin A, a parasite protease, on bovine lymphocyte membrane antigens. *Parasite Immunology* 19, 21–27.

Monfray, K. and Boulard, C. (1990) Preliminary evaluation of four immunological tests for the early diagnosis of *Hypoderma tarandi* causing hypodermosis in reindeer. *Medical Veterinary Entomology* 4, 297–302.

Moroz, A.A. (1979) The effect of bacterial preparations of the group *Bacillus thuringiensis* on larvae of cattle warble-flies. *Entomol. Issledov. Kirgiz.* 13, 105–106 (in Russian). (Abstract: *Review of Applied Entomology* B 70, 1322, 1982.)

Mota, N.G.S., Peracoli, M.T.S. and Lello, E. (1980) Anticorpos circulantes em coelhos imunizados obtidos de larvas de *Dermatobia hominis* Linnaeus (Diptera: Cuterebridae). *Ciencia e Cultura* 32, 453–457.

Mouffet, T. (1634) *Insectorum Sive Minimorum Animalum Theatrum*. T. Cotes, London, (326 pp).

Mourier, H. and Banegas, A.D. (1970) Observations on the oviposition and ecology of the eggs of *Dermatobia hominis* (Diptera: Cuterebridae). *Videnskabelige Meddelelser Fra Dansk Naturhistorisk Forening* 133, 59–68.

Moya-Borja, G.E., Muniz, R.A., Sanavria, A., Goncalves, L.C.B. and Rew, R.S. (1993) Therapeutic and persistent efficacy of doramectin against *Dermatobia hominis* in cattle. *Veterinary Parasitology* 49, 85–93.

Muharsini, S., Sukarsih, S., Riding, G., Partoutomo, S., Hamilton, S., Willadsen, P. and Wijfells, G. (2000) Identification and characterization of the excreted/secreted serine proteases of larvae of the Old World screwwom fly, *Chrysomya bezziana*. *International Journal for Parasitology* 30, 705–714.

Muirhead-Thompson, R.C. (1982) *Behaviour Patterns of Blood-Sucking Flies*. Pergamon Press, New York, (224 pp).

Muller, G.H., Kirk, R.W. and Scott, D.W. (1983) *Small Animal Dermatology*. W.B. Saunders Co., Philadelphia, Pennsylvania, 889 pp.

Mykytowycz, R. (1964) Occurrence of bot-fly larvae *Tracheomyia macropi* Froggatt (Diptera: Oestridae) in wild red kangaroos, *Megaleia rufa* (Desmarest). *Proceedings of the Linnean Society of New South Wales* 88, 307–312.

Nachlupin, N.G. and Pawlowsky, E.N. (1932) Zur Biologie der Hautbremse des Renntiers – *Oedemagena tarandi* – in der Tundra Bolschesemelsk. *Parazitologiceskij Sbornik Akademija Nauk SSSR. Moskva Zoologiceskij muzei* 3, 115–129 (in Russian).

Nachtigall, W.A. (1974) *Insects in Flight*. McGraw-Hill, New York, 153 pp.

Natvig, L.R. (1916) Beitrag zur Biologie der Dasselfliegen des Renntieres. *Tromsø Museums Aarshefter* 38–39, 117–133 (in German).

Natvig, L.R. (1937) Om Kubremsene og deres optreden I Norge (The cattle grubs and their occurrence in Norway). *Norsk Veterinaer Tidsskrift* 5–10, 1–183 (in Norwegian).

Natvig, L.R. (1939) Fliegen larven als fakultative Parasiten bei Menschen und Tieren im Norwegen. In: Jordan, K. and Hering, E.M. (eds) *VII Internationalen Kongresse für Entomologie*, vol. 3. pp. 1641–1655 (in German).

Navajas, A., Cardenal, I., Pinan, M.A., Ortiz, A., Astigarrga, I. and Idez-Teijeiro, A. (1998) Hypereosinophilia due to myiasis. *Acta Haematologica* 1, 27–30.

Neefs, J.M., Van der Peer, Y., Hendricks, L. and De Wachter, R. (1990) Compilation of small ribosomal subunit RNA sequences. *Nucleic Acids Research* 18, 2237–2243.

Nei, M. (1991) Relative efficiencies of different tree-making methods for molecular data. In: Miyamoto, M.M. and Cracraft,

J. (eds). *Phylogenetic Analysis of DNA Sequences*. Oxford University Press, New York, pp. 90–128 (368 pp).

Neiva, A. and Gomes, J.F. (1917) Biologia da mosca da berne (*Dermatobia hominis*) observada em todas as suas phases. *Annales Paulista de Medicina e Cirurgia* 8(9), 206 (in Portuguese).

Nelson, D.L. and Knapp, F.W. (1961) A flocculation test for the detection of antibodies resulting from the injection of an extract of *Hypoderma* larvae. *American Journal of Veterinary Research* 22, 332–334.

Nelson, W.A. (1952) A note on the presence of first-stage larvae of *Gasterophilus intestinalis* (DeGeer) (Diptera: Gasterophilidae) in the mouth of the horse. *The Canadian Entomologist* 84, 356–357.

Nelson, W.A. and Weintraub, J. (1972) *Hypoderma lineatum* (De Vill.) (Diptera: Oestridae): invasion of the bovine skin by newly hatched larvae. *Journal of Parasitology* 58, 614–624.

Nguyen Van Khanh, N.V., Bourge, N., Concordet, D. and Dorchies, Ph. (1996) Recherche des mastocytes et des éosinophiles de la muqueuse respiratoire chez le mouton infesté naturellement par *Oestrus ovis* (Linné 1761). *Parasite* 3, 217–221 (in French).

Nguyen Van Khanh, N.V, Delverdier, M., Jacquiet, Ph. and Dorchies, Ph. (1998) Expression tissulaire de l'épitope Ki-67 dans le compartiment épithélial de la muqueuse nasale du mouton et de la chèvre infestés naturellement par *Oestrus ovis* (Linné 1761). *Revue de Médecine Vétérinaire* 149, 1109–1113 (in French).

Nguyen Van Khanh, N.V, Jacquiet, Ph., Cabanie, P. and Dorchies, Ph. (1999a) Etude semi-quantitative des lésions des muqueuses respiratoires supérieures chez le mouton infesté naturellement par *Oestrus ovis* (Linné 1761). *Revue de Médecine Vétérinaire* 150, 43–46 (in French).

Nguyen Van Khanh, N.V, Jacquiet, Ph., Duranton, C., Bergeaud, J.P., Prevot, F., Dorchies, Ph. (1999b) Réactions cellulaires des muqueuses nasales et sinusales des chèvres et des moutons à l'infestation naturelle par *Oestrus ovis* (Linné 1761) (Diptera: Oestridae). *Parasite* 2, 141–149 (in French).

Nicolas, G. and Sillans, D. (1989) Immediate and latent effects of carbon dioxide on insects. *Annual Review of Entomology* 34, 97–116.

Nicolas-Gaulard, I., Moire, N. and Boulard, C. (1995) Effect of the parasite enzyme, hypodermin A on bovine lymphocyte proliferation and interleukin-2 production via the prostaglandin pathway. *Immunology* 84, 160–165.

Nigro, L., Solignac, M., Sharp, P.M. (1991) Mitochondrial DNA sequence in the *Melanogaster* and oriental species subgroups of *Drosophila*. *Journal of Molecular Evolution* 33, 156–162.

Nilssen, A.C. (1997a) Factors affecting size, longevity and fecundity in the reindeer oestrid flies *Hypoderma tarandi* (L.) and *Cephenemyia trompe* (Modeer). *Ecological Entomology* 22, 294–304.

Nilssen, A.C. (1997b) Effect of temperature on pupal development and eclosion dates in the reindeer oestrids *Hypoderma tarandi* and *Cephenemyia trompe* (Diptera: Oestridae). *Environmental Entomology* 26, 296–306.

Nilssen, A.C. (1998) Effect of 1-octen-3-ol in field trapping *Aedes* spp. (Diptera: Culicidae) and *Hybomitra* spp. (Diptera: Tabanidae) in subarctic Norway. *Journal of Applied Entomology* 122, 465–468.

Nilssen, A.C. and Anderson, J.R. (1995a) Flight capacity of the reindeer warble fly, *Hypoderma tarandi* (L.), and the reindeer nose bot fly, *Cephenemyia trompe* (Modeer) (Diptera: Oestridae). *Canadian Journal of Zoology* 73, 1228–1238.

Nilssen, A.C. and Anderson, J.R. (1995b) The mating sites of the reindeer nose bot fly: not a practical target for control. *Rangifer* 15, 55–61.

Nilssen, A.C. and Haugerud, R.E. (1994) The timing and departure rate of larvae of the reindeer warble fly *Hypoderma* (= *Oedemagena*) *tarandi* (L.) and the reindeer nose bot fly *Cephenemyia trompe* (Modeer) (Diptera: Oestridae) from the reindeer. *Rangifer* 14, 113–122.

Nilssen, A.C. and Haugerud, R.E. (1995) Epizootiology of the reindeer nose bot fly, *Cephenemyia trompe* (Modeer) (Diptera: Oestridae), in reindeer, *Rangifer tarandus* (L.), in Norway. *Canadian Journal of Zoology* 73, 1024–1036.

Nilssen, A.C., Anderson, J.R. and Bergersen, R. (2000) The reindeer oestrids, *Hypoderma tarandi* and *Cephenemyia trompe* (Diptera: Oestridae): Batesian mimics of bumblebees (Hymenoptera: Apidae: *Bombus* spp.). *Journal of Insect Behavior* 13, 307–320.

Nilssen, A.C., Hemmingsen, W. and Haugerud, R.E. (2002) Failure of two consecutive annual treatments with ivermectin to eradicate the reindeer parasites (*Hypoderma tarandi, Cephenemyia trompe* and *Linguatula arctica*) from an island in northern Norway. *Rangifer* 22, 115–122.

Nirmala, X., Hypsa, V. and Zurovec, M. (2001) Molecular phylogeny of Calyptratae (Diptera: Brachycera): the evolution of 18S and 16S ribosomal rDNAs in higher dipterans and their use in phylogenetic inference. *Insect Molecular Biology* 10, 475–485.

Nogge, G. and Staack, W. (1969) Das Flugverhalten der Dasselfliege (*Hypoderma* Latreille) (Diptera, Hypodermatidae) und das Biesen der Rinder. *Behaviour* 35, 200–211 (in German).

Nogge, G. and Werner, H. (1970) Investigations on the contents of bacteria in larvae and boils of *Hypoderma bovis* (De Geer), Diptera, Hypodermatidae. *Zentralblat für Bacteriologie, Parasitenkunde, Infektion* 125, 326–330 (in German with English summary).

Norman, J.O. and Younger, R.L. (1979) Microbial flora of *Hypoderma* (Diptera: Oestridae) larvae taken from cattle treated with juvenile hormone analogues. *Journal of Medical Entomology* 16, 43–47.

Novacek, M.J. (1992) Mammalian phylogeny: shaking the tree. *Nature* 356, 121–125.

Numan, A. (1834) *Waarnemingen Omtrent de Horsel-Maskers, Welke in de Maag van Het Paard Huisvesten.* G.G. Sulpke, Amsterdam.

Nylin, S., Wiklund, C. and Wickman, P.-O. (1993) Absence of trade-offs between sexual size dimorphism and early male emergence in a butterfly. *Ecology* 74, 1414–1427.

Obitz, K. (1937) The cattle grubs in Poland, their distribution and control. *Pam. Panstw. Inst. Naukow. Gosp. Wiejsk.* 1, 100–106 (in Polish with English summary). (Abstract: *Review of Applied Entomology* B 26, 61, 1938).

O'Brien, C.S. and Allen, J.H. (1939) Ophthalmomyiasis interna anterior. *American Journal of Ophthalmology* 22, 996–998.

Oefele, F. von. (1901) Studien uber altegyptischen Parasitologie. Erster Teil: Aussere Parasiten. *Archs. Parasit.* 4, 481–530 (in German).

Oksanen, A. (1995) Therapeutic efficacy of doramectin against *Cephenemyia trompe* (reindeer throat bot) and *Hypoderma tarandi* (reindeer warble). *Bulletin of the Scandinavian Society of Parasitologists* 5, 17.

Oksanen, A. (1999) Endectocide treatment of the reindeer. *Rangifer* Special Issue No. 11, 19, 1–49 (plus appendices).

Oksanen, A. and Nieminen, M. (1998) Moxidectin as an endectocides in reindeer. *Acta Veterinaria Scandinavia* 39, 483–489.

Oksanen, A., Nieminen, M., Soveri, T. and Kumpula, K. (1992) Oral and parenteral administration of ivermectin to reindeer. *Veterinary Parasitology* 41, 241–247.

Oliveira-Sequeira, T.C.G., Sequeira, J.L., Schmitt, F.L. and De Lello, E. (1996) Histological and immunological reaction of cattle skin to first-instar larvae of *Dermatobia hominis. Medical Veterinary Entomology* 10, 323–330.

Ono, S. (1931) Studies on 'Hypodermatotoxin,' toxin obtained from the larvae of *Hypoderma* sp. at the oesophageal stage I. Its action on the blood coagulation. *Journal of the Japanese Society of Veterinary Science* 10, 230–231.

Ono, S. (1932) Studies on 'Hypodermatotoxin,' toxin obtained from the larvae of *Hypoderma* sp. at the oesophageal stage II. Biologic significance of Hypodermatotoxin from point of view of its hemorrhagenic and dermolytic actions. *Journal of the Japanese Society of Veterinary Science* 11, 53–54.

Ono, S. (1933) Morphologic studies on larvae of *Hypoderma lineatum* at the esophageal and hypodermic stages, with special reference to the organ containing 'Hypodermatotoxin.' *Journal of the Japanese Society of Veterinary Science* 11, 220–223.

Onyabe, D.Y. and Conn, J.E. (1999) Intragenomic heterogeneity of a ribosomal DNA spacer (ITS2) varies regionally in the neotropical malaria vector *Anopheles nuneztovari* (Diptera: Culicidae). *Insect Molecular Biology* 8, 435–442.

Oppliger, F.Y., Guerin, P.M. and Vlimant, M. (2000) Neurophysiological and behavioural evidence for an olfactory function for the dorsal organ and a gustatory one for the terminal organ in *Drosophila melanogaster* larvae. *Journal of Insect Physiology*, 46, 135–144.

Ostlind, D.A., Cifelli, S. and Lang, R. (1979) Insecticidal activity of the anti-parasitic avermectins. *Veterinary Record* 5, 168.

Otranto, D. (2001) The immunology of myiasis: parasite survival and host defense strategies. *Trends in Parasitology* 17, 176–182.

Otranto, D. and Puccini, V. (2000a) Further evidence on the internal life cycle of *Przhevalskiana silenus* (Diptera: Oestridae). *Veterinary Parasitology* 88, 321–328.

Otranto, D. and Puccini, V. (2000b) Cytochrome oxidase I (COI) gene of some obligate myiasis-causing larvae (Diptera: Oestridae): which perspective for phylogenetic studies? Preliminary data. *Proceedings of COST Action 833*. Ceskè Budejovice (Czech Republic), September 28-30, 2000, pp. 195–201.

Otranto, D. and Puccini, V. (2001) Cytochrome oxidase I (COI) gene of some obligate myiasis-causing larvae (Diptera: Oestridae): Which perspective for phylogenetic studies? Preliminary data. *Proceedings of COST Action 833*, 28–30 September 2000. Ceské Budejovice (Czech Republic), Bohemia, pp. 195–201.

Otranto, D. and Stevens, J.R. (2002) Molecular approaches to the study of myiasis-causing larvae. *International Journal for Parasitology* 32, 1345–1360.

Otranto, D. and Traversa, D. (2004) Molecular evidence indicating that *Przhevalskiana silenus*, *Przhevalskiana aegagri* and *Przhevalskiana crossii* (Diptera, Oestridae) are one species. *Acta Parsitologica* 49, 173–176.

Otranto, D., Boulard, C., Giangaspero, A., Caringella, M.P., Rimmele, D. and Puccini, V. (1999) Serodiagnosis of goat warble fly infestation by *Przhevalskiana silenus* with a commercial ELISA kit. *The Veterinary Record* 144, 726–729.

Otranto, D., Tarsitano, E., Giangaspero, A. and Puccini, V. (2000) Differentiation by polymerase chain reaction-restriction fragment length polymorphism of some Oestridae lar-

vae causing myiasis. *Veterinary Parasitology* 90, 305–313.

Otranto, D., Testini, G., Sottili, R., Capelli, G. and Puccini, V. (2001) Screening of commercial milk samples using ELISA for immuno-epidemiological evidence of infection by the cattle grub (Diptera: Oestridae). *Veterinary Parasitology* 99, 241–248.

Otranto, D., Colwell, D.D., Traversa, D. and Stevens, J.R. (2003a) Species identification of *Hypoderma* affecting domestic and wild ruminants by morphological and molecular characterization. *Medical and Veterinary Entomology* 17, 1–10.

Otranto, D., Traversa, D., Tarsitano, E. and Stevens, J.R. (2003b) Molecular differentiation of *Hypoderma bovis* and *Hypoderma lineatum* (Ditpera, Oestridae) by polymerase chain reaction description fragment length polymorphism (PCR-RFLP). *Veterinary Parasitology* 112, 197–201.

Otranto, D., Colwell, D.D., Traversa, D. and Stevens, J.R. (2003c) Species identification of *Hypoderma* affecting domestic and wild ruminants by morphological and molecular characterization. *Medical and Veterinary Entomology* 17, 316–325.

Otranto, D., Traversa, D., Colwell, D.D., Giangaspero, A., Boulard, C. and Yin, H. (2004) *Hypoderma sinense* as a proper species of *Hypoderma* (Diptera, Oestridae): molecular and morphological evidence. *Journal of Parasitology* 90, 958–965.

Pampiglione, S. (1958) Indagine epidemiologica sulla miasi conginntivale umana da *Oestrus ovis* in Italia. Nota I: Inchiesta tra i medici italiani. *Nuori Annali d'Igiene e Microbiologia* 9, 242–263.

Pampiglione, S., Giannetto, S. and Virga, A. (1997) Persistence of human myiasis by *Oestrus ovis* L. (Diptera: Oestridae) among shepherds of the Etnean area (Sicily) for over 150 years. *Parassitologia Roma* 39, 415–418.

Panadero Fontan, R., Lopez Sandez, C., Morrondo Pelayo, P. and Diez Banos, P. (1997) Advances in the diagnosis of bovine hypodermosis. *Medicina Veterinaria* 14, 647–655.

Panciera, R.J., Ewing, S.A., Johnson, E.M., Johnson, B.J. and Whitenack, D.L. (1993) Eosinophilic mediastinitis, myositis, pleuritis, and pneumonia of cattle associated with migration of first-instar larvae of *Hypoderma lineatum*. *Journal of Veterinary Diagnostic Investigation* 5, 226–231.

Pandey, V.S. and Ouhelli, H. (1984) Epidemiology of *Oestrus ovis* infection of sheep in Morocco. *Tropical Animal Health and Production* 16, 246–252.

Pandey, V.S., Ouhell, H. and Verhulst, A. (1992) Epidemiological observations on *Gasterophilus intestinalis* and *G. nasalis* in donkeys from Morocco. *Veterinary Parasitology* 41, 285–292.

Pangui, L.J., Dorchies, Ph. and Belot, J. (1988) Contribution à l'étude epidémiologique de l'œstrose ovine au Sénégal. *Revue de Médecine Vétérinaire* 139, 701–704 (in French).

Papadopoulos, E., Himonas, C. and Boulard, C. (1997) The prevalence of goat hypodermosis in Greece. *Parasitologia* 39, 427–429.

Papavero, N. (1977) *The World Oestridae (Diptera), Mammals and Continental Drift.* Series Entomologica, vol. 14, W. Junk, The Hague, The Netherlands, 240 pp.

Pape, T. (1992) Phylogeny of the Tachinidae family-group (Diptera: Calyptratae). *Tijdschrift voor Entomologie* 135, 43–86.

Pape, T. (2001a) Phylogeny of Oestridae (Insecta: Diptera). *Systematic Entomology* 26, 133–171.

Pape, T. (2001b) *Bezzimyia* – a genus of native New World Rhinophoridae (Insecta: Diptera). *Zoologica Scripta* 30, 257–297.

Pape, T. and Arnaud, P.H. Jr (2001) *Bezzimyia* – a genus of native New World Rhinophoridae (Insecta: Diptera). *Zoologica Scripta* 30, 257–297.

Patton, W. S. (1920) Some notes on the arthropods of medical and veterinary importance in Mesopotamia, and on their relation to disease. III. The botflies of Mesopotamia. *Indian Journal of Medical Research* 8, 1–16.

Patton, W.S. (1922a) Note on the egg-laying habits of *Cobboldia elephantis* Cobbold. *The Indian Journal of Medical Research* 10, 63–65.

Patton, W.S. (1922b) Notes on the myiasis-producing Diptera of man and animals. *Bulletin of Entomological Research* 12, 239–261.

Patton, W.S. (1937) Studies on the higher Diptera of medical and veterinary importance. Illustrations of the terminalia of *Cobboldia elephantis* Steel, *C. chrysidiformis* Rodhain and Bequaert and *Ruttenia loxodon-tis* Rodhain. *Annals of Tropical Medicine and Parasitology* 31, 341–349.

Payne, J.A. and Cosgrove, G.E. (1966) Tissue changes following *Cuterebra* infestations in rodents. *American Midland Naturalist* 75, 205–213.

Penso, G. (1973) *La Conquista del Mondo Invisibile.* Feltrinelli, Milan, Italy, 349 pp.

Peracoli, M.T.S., Lello, E. and Mota, N.G.S. (1980) Comportomento da resposta immune-humoral em coelhos imunizados com antigenos de *Dermatobia hominis* Linnaeus, frente as larvas desse parasita (Diptera: Cuterebridae). *Ciencia e Cultura* 32, 1537–1541.

Pereira, M.C.T., Leite, V.H.R. and Leite, A.C.R. (2001) Experimental skin lesions from larvae of the bot fly *Dermatobia hominis*. *Medical and Veterinary Entomology* 15, 22–27.

Pérez, J.M., Granados, J.E. and Ruiz-Mart'inez, I. (1995) Studies on the hypodermosis affecting red deer in central and southern Spain. *Journal of Wildlife Diseases* 31, 486–490.

Pérez, J.M., Granados, J.E., Soriguer, R.C. and Ruiz-Mart'inez, I. (1996) Prevalence and seasonality of *Oestrus caucasicus* Grunin, 1948 (Diptera: Oestridae) parasitizing the Spanish ibex, *Capra pyrenaica* (Mammalia: Artiodactyla). *Journal of Parasitology* 82, 233–236.

Peter, B. (1929) Immunization of cattle against the larvae of the ox warble fly. A theoretical consideration. *Berl. Tieraerztl. Wochenschr.* 1929, 209 (in German).

Peter, G. and Gaehtgens, H. (1934) Ueber den serologischen Nachweis der Dassellarveninfektion beim Rind un beim Menschen. (On the serological proof of ox warble infestation in cattle and man.) *Zbl. Bakt.* 132, 81–90 (in German). (*Review of Applied Entomology* B 22, 255–256, 1934.)

Pfadt, R.E. (1947) Effects of temperature and humidity on larval and pupal stages of the common cattle grub. *Journal of Economic Entomology* 40, 293–303.

Pfadt, R.E., Lloyd, J.E. and Sharafi, G. (1975) Pupal development of cattle grubs at constant and alternating temperatures. *Journal of Economic Entomology* 68, 325–328.

Philippe, H. (1997) Rodent monophyly: pitfalls of molecular phylogenies. *Journal of Molecular Evolution* 45, 712–715.

Plinius, S.C. (1507) Historia Naturalis libri XXXVII ab Alexandro Benedicto Ve. In: *Physico Emendations Rediti*. Venice.

Pollock, J.N. (1999) A study of the male abdomen of *Gyrostigma rhinocerontis* Hope (Diptera: Gasterophilidae), the stomach bot of the African rhinoceroses, with notes on the ground plan and affinities of Gasterophilidae. *Journal of Natural History* 33, 777–788.

Polyakova, P.E. and Gomoyunova, N.P. (1973) Ecological features of some species of blood-sucking mosquitoes of Central Chukotka. *Parazitologiya* 7, 327–330 (in Russian with English summary).

Pont, A.C. (1977a) Family Oestridae. In: Delfinado, M.D. and Hardy, D.E. (eds) *A Catalogue of the Diptera of the Oriental Region*, vol. 3. Suborder Cyclorrhapha (excluding Division Aschiza). The University Press of Hawaii, Honolulu, Hawaii, pp. 700–702 (854 pp).

Pont, A.C. (1977b) Family Gasterophilidae. In: Delfinado, M.D. and Hardy, D.E. (eds) *A Catalogue of the Diptera of the Oriental Region*, vol. 3. Suborder Cyclorrhapha (excluding Division Aschiza). The University Press of Hawaii, Honolulu, Hawaii, pp. 698–699 (854 pp).

Pont, A.C. (1980a) Family Gasterophilidae. In: Crosskey, R.W. (ed.) *Catalogue of the Diptera of the Afrotropical Region*. British Museum of Natural History, London, pp. 883–885.

Pont, A.C. (1980b) Family Oestridae. In: Crosskey, R.W. (ed.) *Catalogue of the Diptera of the Afrotropical Region*. British Museum of Natural History, London, pp. 886–888.

Portchinskii, I.A. (1913) The sheep bot (*Oestrus ovis*), its biology, characteristics, methods of control and relation to man. *Trudy Byuro Entomologii Department Zemi Saint Petersburg* 10, 1–65 (translated from Russian).

Portchinskii, I.A. (1915) *Rhinoestrus purpurens* Br., a parasite of the horse injecting its larvae into the eyes of man. *Trudy Byuro Entomologii*, 3rd edn. 6, 1–47 (in Russian with English translation).

Povolný, D., Holišova, U. and Zapletal, M. (1960) Kritischer Beitrag zur systematisch – biono-muschen Erforschung der Dasselfliege *Oestromyia leporina* (Pallas 1778) mit berücksichtigung der Gattung *Oestromyia* Brauer 1863.

Práce, Bruenské základny Ceskosloovenské akademie ved Acta Academie Scientiarum Cechoslovenicae Basis'brunensis 32, 33–79 (in German).

Preston, J.M. (1984) The avermectins: new molecules for use in warble fly control. In: Boulard, C. and Thornberry, H. (eds) *Warble Fly Control in Europe: a symposium in the EC Programme of Coordination of Research in Animal Pathology*, 16-17 September 1982, Brussels. A.A. Balkema, Rotterdam, The Netherlands, pp. 17–20.

Principato, M. (1987) Étude comparative sur les stigmates prothoraciques et post-abdomiaux des troisièmes stades larvaires des espèces italiennes de *Gasterophilus* (Diptera: Gasterophilidae). *Annals de Parasitologie Humaine et Comparee* 62, 175–180.

Principato, M. (1988) Classification of the main macroscopic lesions produced by larvae of *Gasterophilus* spp. (Diptera: Gasterophilidae) in free-ranging horses in Umbria. *Cornell Veterinarian* 78, 43–52.

Principato, M. (1989) Observations on the occurrence of five species of *Gasterophilus* larvae in free-ranging horses in Umbria, central Italy. *Veterinary Parasitology* 31, 173–177.

Principato, M. and Tosti, M. (1988) Scanning electron microscope observations on the anterior thoracic and post-abdominal spiracles of *Gasterophilus* larvae (Diptera: Gasterophilidae). *International Journal for Parasitology* 18, 191–196.

Prothero, D.R. and Schoch, R.M. (eds) (1989) Classification of Perissodactyla, the evolution of perissodactyls. *Oxford Monographs on Geology and Geophysics* 15, 1–537.

Pruett, J.H. (1993a) Proteolytic cleavage of bovine IgG by hypodermin A, serine protease of *Hypoderma lineatum* (Diptera: Oestridae). *Journal of Parasitology* 79, 829–833.

Pruett, J.H. (1993b) A summary of progress toward the development of a recombinant vaccine against *Hypoderma lineatum*, an insect parasite of cattle. *Proceedings of the 38th Annual Meeting of the American Association of Veterinary Parasitologists*, 45.

Pruett, J.H. (1999) Immunological control of arthropod ectoparasites – a review. *International Journal of Parasitology* 29, 25–32.

Pruett, J.H. and Barrett, C.C. (1983) Development by the laboratory rodent host of humoral antibody activity to *Cuterebra fontinella* (Diptera: Cuterebridae) larval antigens. *Journal of Medical Entomology* 20, 113–119.

Pruett, J.H. and Barrett, C.C. (1985) Kinetic development of humoral anti-*Hypoderma lineatum* antibody activity in the serum of vaccinated and infested cattle. *Southwestern Entomologist* 10, 39–47.

Pruett, J.H. and Kunz, S.E. (1996) Thermal requirements for *Hypoderma lineatum* (Diptera: Oestridae) egg development. *Journal of Medical Entomology* 33, 976–978.

Pruett, J.H. and Stromberg, P. (1995) Effects of adjuvants on bovine humoral and cellular responses to hypodermin A. *Veterinary Parasitology* 58, 143–153.

Pruett, J.H. and Temeyer, K.B. (1989) Colostral transfer of antibodies specific for *Hypoderma lineatum* proteins. *Veterinary Research Communications* 13, 213–223.

Pruett, J.H., Fisher, W.F. and Temeyer, K.B. (1989) Evaluation of purified proteins of *Hypoderma lineatum* as candidate immunogens for a vaccine against bovine hypodermiasis. *Southwestern Entomologist* 14, 363–373.

Pruett, J.H., Scholl, P.J. and Temeyer, K.B. (1990) Shared epitopes between the soluble proteins of *Hypoderma lineatum* and *Hypoderma bovis* first instars. *Journal of Parasitology* 76, 881–888.

Puccini, V. and Otranto, D. (2000) Goat warble fly infestation by *Przhevalskiana silenus* (Diptera, Oestridae). In: Tempesta, M. (ed.) *Recent Advances in Goat Diseases*. International Veterinary Information Service, Ithaca, New York, 247 pp.

Puccini, V., Tassi, P. and Arru, E. (1983) Efficacia dell' ivermectina iniettable contro le larvae di *Oestrus ovis* (L. 1761) in ovini infettati naturalmente. *Atti della Societa Italiano Scienze Veterinario* 37, 732–734 (in Italian).

Puccini, V., Giangaspero, A. and Fasanella, A. (1994) Efficacy of moxidectin against *Oestrus ovis* larvae in naturally infested sheep. *Veterinary Record* 135, 600–601.

Puccini, V., Giangaspero, A. and Caringella, M.P. (1995) Immunological response against *Hypoderma* spp. in naturally infested cattle. In: Tarry, D.W., Pithan, K. and Webster, K. (eds). *Improvements in the Control Methods for Warble-fly in Cattle and Goats*. Proceedings,

8–10 September 1994, Guildford, UK. Commission of the European Communities, Brussels, pp. 157–170 (181 pp.).

Puccini, V., Otranto, D., Caringella, M.P. and Giangaspero, A. (1997) Study of individual kinetics of antibodies in naturally infested goats by *Przhevalskiana silenus* using *Hypoderma lineatum* antigen. In: Puccini, V. and Giangaspero, A. (eds). *Improvement of means of control of warble fly in cattle and goats.* COST 811. Parma, 5–6 September 1996, pp. 31–36.

Queiroz, K. de (1996) Including the characters of interest during tree reconstruction and the problems of circularity and bias in studies of character evolution. *American Naturalist* 148, 700–708.

Quesada, P., Navarrete, M.L. and Maeso, J. (1990) Nasal myiasis due to *Oestrus ovis* larvae. *European Archives of Otorhinolaryngology* 247, 131–132.

Rakusin, W. (1970) Ocular myiasis intera caused by the sheep nasal bot fly (*Oestrus ovis* L.). *South African Medical Journal* 44, 1155–1157.

Rankin, M.A. (1989) Hormonal control of flight. In: Goldsworthy, G.J. and Wheeler, C.H. (eds) *Insect Flight*. CRC Press, Boca Raton, Florida, pp. 139–163 (371 pp).

Ransom, B.H. (1912) Grub eradication. *Proceedings of the Meeting of the National Asociation of Tanners* 1912, 8–33.

Raschi, S. (1928) Kommentar zum Talmud. In: Bodenheimer, F.S. (ed.) *Materialen zur Geschichte der Entomologie bis Linne*. W. Junk, Berlin, Germany, pp. 121–123 (428 pp.).

Rastegaev, Yu.M. (1972) On parasitism by *Hypoderma bovis* (De Geer) of sheep and goats in Western Siberia. *Parasitologyia* 6, 531–533.

Rastegaev, Yu.M. (1978) Subcutaneous myiasis in man caused by larvae of the horse bot fly. *Meditsinskaya Parazitologiya i Parazitarnye Bolezni* 47, 72–73.

Rastegaev, Yu.M. (1980) Myiasis in man caused by larvae of *Rhinoestrus purpureus* Br. *Meditsinkaya Parazitologiya i Parazitarnye. Bolezni* 49, 86–88.

Rastegaev, Yu.M. (1982) Ophthalmomyiasis in parasitism by larvae of horse warble flies. *Oftal'mologicheskii Zhurnal* 34, 444–445.

Rastegaev, Yu.M. (1984) Ecological characteristics of the warble and bot flies attacking horses (Diptera: Oestridae, Gasterophilidae)

in the desert zone of the Caspian Region. *Entomologicheskoye Obozreniye* 63, 455–459. (*Review of Applied Entomology* 64, 35–39, 1985.)

Rastegaev, Yu.M. (1988) Efficacy of ivermectin against *Rhinoestrus* and *Gasterophilus* in horses. *Izvetiya Akad. Nauk Turkmen. SSR, Ser. Biol. Nauk* 6, 67–69.

Raven, P.H. and Axelrod, D.I. (1972) Plate tectonics and Australasian paleobiogeography. *Science* 176, 1379–1386.

Raven, P.H. and Axelrod, D.I. (1974) Angiosperm biogeography and past continental movements. *Annals of the Missouri Botanical Garden* 61, 539–673.

Reaumur, R.A.F. de. (1734) Memoires pour servir a l'histoire des Insects. *L'imprimerie royale, Paris*, vol. I.

Reaumur, R.A.F. de. (1738) Memoires pour servir a l'histoire des Insects. *L'imprimerie royale, Paris*, vol. IV.

Reaumur, R.A.F. de. (1740) Memoires pour servir a l'histoire des Insects. *L'imprimerie royale, Paris*, vol. V.

Redi, F. (1668) Esperienze intorno alla generazione degli insetti. *Per Piero Matini, all'insegna della Stella, Firenza.*

Rehbinder, C. (1970) Observations of 1st instar larva of nostril fly *Cephenemyia trompe* in the eye of reindeer and their relation to keratitis in this animal. *Acta Veterinaria Scandinavia* 11, 338–339.

Rehbinder, C. (1977) Clinical and epizootiological studies on keratitis in Reindeer. *Acta Veterinaria Scandinavia* Suppl. 66, 1–27.

Reina, D., Martinez-Moreno, F.J., Hurtado, F., Hernandez-Rodriquez, S. and Navarrete, I. (1995) Humoral immune response in natural and experimental bovine hypodermosis. In: Tarry, D.W., Pithan, K. and Webster, K. (eds) *Improvements in the Control Methods for Warble-fly in Cattle and Goats.* Proceedings, 8–10 September 1994, Guildford, UK. Commission of the European Communities, Brussels, pp. 47–51, 61–67.

Reinemeyer, C.R., Scholl, P.J., Andrews, F.M. and Rock, D.W. (2000) Efficacy of moxidectin equine oral gel against endoscopically-confirmed *Gasterophilus nasalis* and *Gasterophilus intestinalis* (Diptera: Oestridae) infections in horses. *Veterinary Parasitology* 88, 287–291.

Ribbeck, R., Schusser, G.F., Ilchmann, G., Schneider, A. and Schwarzer, U. (1998) Occurrence, diagnosis and control of gasterophilosis in horses. *Wiener Tierartzliche Monatsschrift* 85, 418–423.

Ribeiro, G.A., Carter, G.R., Frederiksen, W. and Riet-Correa, F. (1989) *Pasteurella haemolytica*-like bacterium from a progressive granuloma of cattle in Brazil. *Journal of Clinical Microbiology* 27, 1401–1402.

Rich, G.B. (1958) Some factors affecting populations of *Hypoderma* species. *Proceedings of the Tenth International Congress of Entomology, Montreal* 3, 735.

Riet-Correa, F., Mendez, M.C., Schild, A.L., Ribeiro, G.A. and Almeida, S.M. (1992) Bovine focal proliferative fibrogranulomatous panniculitis (Lechiguana) associated with *Pasteurella granulomatis*. *Veterinary Pathology* 29, 93–103.

Rietschel, G. (1975a) The larval development of the warble-fly *Oestromyia leporine* Pall (Diptera: Hypodermatidae). *Zeitschrift für Parasitenkunde* 47, 283–297.

Rietschel, G. (1975b) Die Laborzucht der Dasselfliege *Oestromyia leporina* Pall. (Diptera, Hypodermatidae) und ihre biologischen Voraussetzungen. (Laboratory rearing of the warble fly *Oestromyia leporina* Pall. (Diptera, Hypodermatidae) and its biological prerequisites.) *Zeitschrift für Parasitenkunde* 47, 299–306.

Rietschel, G. (1979) Histologische reaktionen des wirtes auf den befall durch die dasselfliege *Oestromyia leporina* Pall. (Diptera, Hypodermatidae). *Zeitschrift für Parasitenkunde* 60, 277–289.

Rietschel, G. (1980) Zur Biologie und Ökologie der Entwicklungsstadien der Dasselfliege *Oestromyia leporina* (Pallas, 1778) (Diptera, Hypodermatidae). *Zoologische Jahrbücher, Abteilung für Systematik, Ökologie und Geographie der Tiere* 107, 265–285.

Rietschel, G. (1981) Structure and function of the adult mouthparts of the warble-fly *Oestromyia leporina* (Diptera: Hypodermatidae). *Entomologia Generalis* 7, 161–165.

Robertson, R.H. (1980) Antibody production in cattle infected with *Hypoderma* spp. *Canadian Journal of Zoology* 58, 245–251.

Roca, A.L., Georgiadis, N., Pecon-Slattery, J. and O'Brien, S.J. (2001) Genetic evidence for two species of elephants in Africa. *Science* 293, 1473–1477.

Rocha, U.F. and Mendes, J. (1996) Pupation of *Dermatobia hominis* (L. Jr., 1781) (Diptera: Cuterebridae) associated with *Sarcodexia lambens* (Wiedmann, 1830) (Diptera: Sarcophagidae). *Memorias Do Instituto Oswaldo Cruz, Rio de Janeiro* 91, 299–300.

Rogers, C.E. (1967) An ecological study of *Oestrus ovis* Linne. PhD thesis, Department of Entomology, Graduate School, University of Kentucky, Lexington, Kentucky, 82 pp.

Rogers, C.E. and Knapp, F.W. (1973) Bionomics of the sheep bot fly, *Oestrus ovis*. *Environmental Entomology* 2, 11–23.

Rognes, K. (1997) The Calliphoridae (Blowflies) (Diptera: Oestroidea) are not a monophyletic group. *Cladistic* 13, 27–66.

Rogoff, W.M. (1961) Chemical control of insect pests of domestic animals. In: Metcalf, R.L. (ed.) *Advances in Pest Control Research*, vol. IV. Interscience Publishers, New York, pp. 153–181.

Romer, A.S. (1966) *Vertebrate Paleontology*, 3rd edn. The University of Chicago Press, Chicago, Illinois, 468 pp.

Roncalli, R.A. (1984a) The biology and control of *Dermatobia hominis*, the tropical warble-fly of Latin America. *Preventative Veterinary Medicine* 2, 569–578.

Roncalli, R.A. (1984b) Efficacy of ivermectin against *Oestrus ovis* in sheep. *Veterinary Medicine* 79, 1095–1097.

Roncalli, R.A. (2001) The history of Italian parasitology. *Veterinary Parasitology* 98, 3–39.

Ronquist, F. (1994) Ancestral areas and parsimony. *Systematic Biology* 43, 267–274.

Ronquist, F. (1995) Reconstructing the history of host–parasite associations using generalised parsimony. *Cladistics* 11, 73–89.

Ronquist, F. (1996) DIVA version 1.1. Computer program and manual available by anonymous FTP from Uppsala University (ftp.uu.se or ft.syst-bot.uu.se).

Ronquist, F. (1997) Dispersal-vicariance analysis: a new approach to the quantification of historical biogeography. *Systematic Biology* 46,195–203.

Rooney, J.R. (1964) Gastric ulceration in foals. *Path. Vet.* 1, 497–503.

Ross, H.M. (1983) Warble infestation of sheep. *Veterinary Record* 112, 135.

Rossi, M.A. and Zucoloto, S. (1973) Fatal cerebral myiasis caused by the tropical warble fly, *Dermatobia hominis*. *American Journal of Tropical Medicine and Hygiene* 22, 267–269.

Roubaud, E. and Perard, C. (1924) Studies of *Hypoderma* or cattle grubs: extracts of oestrid larvae and immunization. *Bulletin de la Société de Pathologie Exotique* 17, 259–272 (in French).

Royce, L.A., Rossignol, P.A., Kubitz, M.L. and Burton, F.R. (1999) Recovery of a second instar *Gasterophilus* larva in a human infant: a case report. *American Journal of Tropical Medicine and Hygiene* 60, 403–404.

Rubtzov, J.A. (1939) The evolution of bot-flies (Gasterophilidae) in connection with the history of their hosts. *Zoologicheskii zhurnal* 18, 669–684 (in Russian with English summary).

Rugg, D., Gogolewski, R.P., Barrick, R.A. and Eagleson, J.S. (1997) Efficacy of ivermectin controlled-release capsules for the control and prevention of nasal bot infestations in sheep. *Australian Veterinary Journal* 75, 36–38.

Ruini, C. (1598) Anatomia del cavallo, infermita et suoi rimedii. *Heredi di Giovanni Rossi, Bologna.*

Ruíz-Martínes, I. and Palomares, F. (1993) Occurrence and overlapping of pharyngeal bot flies *Pharyngomyia picta* and *Cephenemyia auribarbis* (Oestridae) in red deer of southern Spain. *Veterinary Parasitology* 47, 119–127.

Sabrosky, C.W. (1986) North American species of *Cuterebra*, the rabbit and rodent bot flies (Diptera: Cuterebridae). *Thomas Say Foundation* 11, 1–240.

Sahibi, H., Berrag, B. and Rhalem, A. (1995) Seroepidemiology of bovine hypodermosis and its control in the Gharb. In: Tarry, D.W., Pithan, K. and Webster, K. (eds) *Improvements in the Control Methods for Warble-fly in Cattle and Goats*. Proceedings, 8–10 September 1994, Guildford, UK. Commission of the European Communities, Brussels, pp. 5–7, 84–89 (181 pp.).

Saiki, R.K., Scharf, S.J., Faloona, F., Mullis, K.B., Horn, G.T. Erlich, H. and Arnheim, N. (1985) Enzymatic amplification of β-globin genomic sequence and restriction site analysis for diagnosis of sickle anaemia. *Science* 230, 1350–1354.

Saitou, N. and Nei, M. (1987) The neighbour-joining method: a new method for reconstructing phylogenetic trees. *Molecular Biology and Evolution* 4, 406–425.

Salt, R.W. (1944) The effects of subzero temperatures on *Hypoderma lineatum* Devill. *Scientific Agriculture* 25, 156–160.

Sanavria, A. (1987) Bioecologia, patologia e alternativas de controle quimiterapico de *Dermatobia hominis* (Linnaeus Junior, 1781) (Diptera: Cuterebridae) no Rio de Janeiro. PhD thesis, Federal University of Rio de Janeiro, Rio de Janeiro, Brazil (in Portuguese with English summary).

Sancho, E. (1988) *Dermatobia*, the Neotropical warble fly. *Parasitology Today* 4, 242–246.

Sancho, E., Caballero, M. and Ruiz-Martinez, I. (1996) The associated microflora to the larvae of human bot fly *Dermatobia hominis* L. Jr. (Diptera: Cuterebridae) and its furuncular lesions in cattle. *Memorias Instituto Oswaldo Cruz (Rio de Janeiro)* 91, 293–298.

Sandeman, R.M., Collins, B.J., Carnegie, P.R. (1987) A scanning electron microscope study of *L. cuprina* larvae and the development of blowfly strike in sheep. *International Journal for Parasitology* 17, 759–765.

Sandeman, R., Chandler, R., Turner, N. and Seaton, D. (1995) Antibody degradation in wound exudates from blowfly infections on sheep. *International Journal for Parasitology* 25, 621–628.

Saraste, M. (1990) Structural features of cytochrome oxidase. *Quarterly Review of Biophysics* 23, 331–366.

Sauter, W. and Huber, P. (1988) *Pharyngomyia picta* (Meigen) (Dipt. Oestridae) als erreger einer Ophthalmomyiasis beim Menschen. *Vierteljahrsschrift der Naturforschenden Gesellshaft in Zürich* 133, 109–113 (in German).

Saveljev, D.V. (1971) Kiiliäisten biologia ja taistelu niitä vastaan. (Biology of and fight against oestrid flies.) *Poromies* 6, 6–9 (in Finnish).

Sayin, F. (1977) Incidence and seasonal activity of *Przhevalskiana silenus* (Brauer) in Angora goats in Turkey. Proceeding III, 25–28 September 1975, Gdansk, Poland. *Symposium on Medical and Veterinary Acaroentomology* 23, 157–159.

Schäefer, C.W. (1979) Feeding habits and hosts of Calyptrate flies (Diptera: Brachycera: Cyclorrhapha). *Entomologia Generalis* 5, 193–200.

Schaller, G.B. (1972) *The Serengeti Lion: a Study of Predator: Prey Relations.* University of Chicago Press, Chicago (Illinois) and London, 480 pp.

Schallig, H.D. (2000) Immunological responses of sheep to *Haemonchus contortus*. *Parasitology* 120, 263–272.

Scharff, D.K. (1950) Cattle grubs: their biologies, their distribution, and experiments in their control. *Montana State College Agricultural Experiment Station Bulletin* No 471, 74 pp.

Schiff, T.A. (1993) Furuncular cutaneous myiasis caused by *Cuterebra* larva. *Journal of American Dermatology* 28, 261–263.

Schiner, J.R. (1862) *Fauna Austriaca: Die Fliegen Diptera*, vol. I. C. Gerard's Sohn, Wien, Austria, 674 pp.

Schiner, J. R. (1864) *Fauna Austriaca: Die Fliegen Diptera*, vol. II. C. Gerard's Sohn, Wien, Austria, 658 pp.

Scholl, P.J. (1990) A review of parasites, pathogens, and predators of cattle grubs. *Southwestern Entomologist* 15, 360–365.

Scholl, P.J. (1991) Gonotrophic development in the rodent bot fly *Cuterebra fontinella* (Diptera: Oestridae). *Journal of Medical Entomology* 28, 474–476.

Scholl, P.J. (1993) Biology and control of cattle grubs. *Annual Review of Entomology* 39, 53–70.

Scholl, P.J. and Barrett, C.C. (1986) Technique to extract *Hypoderma* spp. (Diptera: Oestridae) larvae from the backs of cattle. *Journal of Economic Entomology* 79, 1125–1126.

Scholl, P.J. and Weintraub, J. (1988) Gonotrophic development in *Hypoderma lineatum* and *H. bovis* (Diptera: Oestridae), with notes on reproductive capacity. *Annals of the Entomological Society of America* 81, 318–324.

Scholl, P.J., Weintraub, J. and Khan, M.A. (1985) Late-season treatment of cattle for control of warble grubs. In: Sears, L.J.L. and Swierstra, E.E. (eds) *Research Highlights–1984.* Agriculture Canada Research Station, Lethbridge, Alberta, Canada, pp. 29–30.

Scholl, P.J., Colwell, D.D., Weintraub, J. and Kunz, S.E. (1986) Area-wide systemic insecticide treatment for control of cattle grubs, *Hypoderma* spp. (Diptera: Oestridae): two approaches. *Journal of Economic Entomology* 79, 1558–1563.

Scholl, P.J., Wedberg, J., Neher, N. and Flashinski, R. (1990) *Pest Management Principles for the Commercial Applicator: Animal Pest Control.* University of Wisconsin Press, Madison, Wisconsin, 161 pp.

Scholl, P.J., Guillott, F.S. and Wang, G.T. (1992) Moxidectin: systemic activity against common cattle grubs (*Hypoderma lineatum*) (Diptera: Oestridae) and trichostrongyle nematodes in cattle. *Veterinary Parasitology* 41, 203–209.

Scholl, P.J., Chapman, M.R., French, D.D. and Klei, T.R. (1998) Efficacy of moxidectin 2% oral gel against 2nd and 3rd instar *Gasterophilus intestinalis* (De Geer). *Journal of Parasitology* 84, 656–657.

Schöyen, W.M. (1886) Om Forekomsten af Dipterlarver under Huden hos Mennesker. *Entomologisk Tidoskrift* 7, 171–187.

Schreger, B.G. and Fischer, J.L. (1787) Observationes de oestro ovino atque bovino factae. Diputatio. *Ex officina Sobrigia, Lisiae.*

Schroeder van der Kolk, J.L.K. (1845) *Memoire Sur l'anatomie et la Physiologie du* Gastrus equi. G. G. Sulpke, Amsterdam.

Schumann, H., Schuster, R. and Ruscher, H.J. (1988) Warble infestation in a donkey. *Angewandte Parasitologie* 29, 241–243.

Schwab, L.K. (1840) Die Oestraciden-Bremsender Pferde, Rinder und Schafe. *Matthaus Possenbacher, Munich.*

Schwinghammer, K.A., Pruett, J.H. and Temeyer, K.B. (1988) Biochemical and immunochemical properties of hill peak 2, an ion exchange fraction of common cattle grub (Diptera: Oestridae). *Journal of Economic Entomology* 81, 549–554.

Sdobnikov, V.M. (1935) Relations between the reindeer (*Rangifer tarandus*) and the animal life of tundra and forest. *Transactions of the Arctic Institute, Leningrad* 24, 5–66.

Semenov, P.V. (1969) A case of the penetration of a larva of *Hypoderma lineatum* (De Villers) into the brain of a person. *Meditsinskaya Parazitologiya I Parazitarnye Bolezni* 38, 612–613.

Semenov, P.V., Gomoyunova, N.P. and Tarasenko, N.N. (1975) The flight radius of *Oestrus ovis. Veterinariia* 8, 58–59 (in Russian). (*Review of Applied Entomology* B 64, 355–356, 1976.)

Seneca, L.A. (1917) Epistle LVIII. Seneca ad Lucilium. *Epistulae Morales,* vol. I. Harvard University, Cambridge, Massachusetts, pp. 387–389.

Sequeira, J.L., Tostes, R.A., Oliveira-Sequeira, T.C.G. (2001) Prevalence and macro- and microscopic lesions produced by *Gasterophilus nasalis* (Diptera: Oestridae) in the Botucatu region, SP, Brazil. *Veterinary Parasitology* 102, 261–266.

Sergent, E. and Sergent, E. (1952) La thimni, myiase humaine d'Algerie causee par *Oestrus ovis* L. *Annals of the Pasteur Institute* 21, 392–399 (in French).

Shannon, R.C. and Greene, C.T. (1926) A bot-fly parasite in monkeys. *Zoopathology* 1, 285–290.

Sharma, L.K. (1992) Efficacy of ivermectin against nasal bots in camels. *Indian Veterinary Journal* 68, 835–836.

Sharpe, R.G., Hims, M.M., Harbach, R.E. and Butlin, R.K. (1999) PCR-based methods for identification of species of the *Anopheles minimus* group: Allele-specific amplification and single-strand conformation polymorphism. *Medical and Veterinary Entomology* 13, 265–273.

Shefstad, D.K. (1978) Scanning electron microscopy of *Gasterophilus intestinalis* lesions of the equine stomach. *Journal of the American Veterinary Medical Association* 172, 310–313.

Shewell, G.E. (1987a) Calliphoridae, chapter 106. In: McAlpine, J.F. (ed.) *Manual of Nearctic Diptera,* vol. 2. Research Branch Agriculture Canada, Monograph 28, Ottawa, Canada, pp. 1133–1145.

Shewell, G.E. (1987b) Sarcophagidae, chapter 108. In: McAlpine, J.F. (ed.) *Manual of Nearctic Diptera,* vol. 2. Research Branch Agriculture Canada, Monograph 28, Ottawa, Canada, pp. 1159–1186.

Shock, A., Rabe, K.F., Dert, G., Chambers, R.C., Gray, A.J., Chung, K.F., Barnes, P.J., Laurent, G.J. (1991) Eosinophils adhere to and stimulate replication of lung fibroblasts in vitro. *Clinical and Experimental Immunology* 86, 185–190.

Shoop, W.L. (1993) Ivermectin resistance. *Parasitology Today* 9, 154–159.

Shoshani, J. and Tassy, P. (1996) Summary, conclusions, and a glimpse into the future. In: Shoshani, J. and Tassy, P. (eds) *The Proboscidea, Evolution and Palaeoecology of Elephants and their Relatives.* Oxford University Press, Oxford, pp. 336–348 (472 pp).

Shumilov, M.F. and Nepoklonov, A.A. (1983) Ecology and control of the reindeer warble fly in the Magadan region, USSR. In: Nepoklonov, A.A. (ed.) *Veterinarnaya Entomologiya i Akarologiya.* Moscow, pp. 50–53.

Sicart, M., Bernard, S., Labatut, R. and Luffan, G. (1960) Myiasis humaines a larves d'hypodermose dans las Landes de Gascogne. *Toubuse medecine* 61, 201.

Sigalas, R. and Pautrizel, R. (1948) Surquatre cas de myiasis sons-cutanées à *Hypoderma* chez l'homine. *Bulletin de la Société de Pathologie Exotique* 41, 380–387.

Simco, J.S. and Lancaster, J.L.J. (1964) Effects of soil type, moisture level, and temperature on larval and pupal stages of the common cattle grub, *Hypoderma lineatum. Journal of the Kansas Entomological Society* 37, 11–20.

Simmons, S.W. (1937) Some histopathological changes caused by *Hypoderma* larvae in the esophagus of cattle. *Journal of Parasitology* 23, 376–381.

Simmons, S.W. (1939) Some histopathological changes in the skin of cattle infected with larvae of *Hypoderma lineatum. Journal of the American Veterinary Medicine Association* 95, 283–288.

Simon, C., Frati, F., Beckenbach, A., Crespi, B., Liu, H. and Flook, P. (1994) Evolution, weighting and phylogenetic utility of mitochondrial gene sequences and a compilation of conserved polymerase chain reaction primers. *Annals of the Entomological Society of America* 87, 651–701.

Simpson, G.G. (1980) *Splendid Isolation, the Curious History of South American Mammals.* Yale University Press, New Haven (Connecticut) and London, 266 pp.

Sinclair, B.J., Cumming, J.M. and Wood, D.M. (1994) Homology and phylogenetic implications of male genitalia in Diptera – lower Brachycera. *Entomologica scandinavica* 24, 407–432.

Sinclair, I.J. and Wassal, D.A. (1983) Enzyme-linked immunosorbent assay for the detection of antibodies to *Hypoderma bovis* in cattle. *Research Veterinary Science* 34, 251–252.

Skoda, S.R., Skoda, S.R., Pornkulat, S. and Foster, J.E. (2002) Random amplified polymorphic DNA markers for discriminating *Cochliomyia hominivorax* from *C. macellaria* (Diptera: Calliphoridae). *Bulletin of Entomological Research* 92, 89–96.

Smart, J. (1939) A case of human myiasis due to *Hypoderma. Parasitology* 31, 130–136.

Smith, D.H. (1977) The natural history and development of *Cuterebra approximata* (Diptera: Cuterebridae) in its natural host, *Peromyscus maniculatus* (Rodentia: Cricetidae), in western Montana. *Journal of Medical Entomology* 14, 137–145.

Smithcors, J.F. (1957) *Evolution of the Veterinary Art.* Veterinary Medicine Publishing Company, Kansas City, Missouri, 357 pp.

Solopov, N.V. (1989a) Ecological features of juvenile forms of reindeer flies (Oestridae, Diptera) 3. The deer fly (*Cephenemyia trompe* Modeer). *Ékologiya (Sverdlovsk)* 2, 85–87 (in Russian).

Solopov, N.V. (1989b) Features of the ecology of juvenile forms of subcutaneous gadfly in reindeer. *The Soviet Journal of Ecology* 19, 299–304.

Soni, B.N. (1942) Eggs of the goat warble-fly (*Hypoderma crossii* Patton). *Current Science* 11, 280.

Soós, Á. and Minár, J. (1986a) Family Gasterophilidae. In: Soós, Á. and Papp, L. (eds) *Catalogue of Palaearctic Diptera,* vol. 11. *Scathophagidae – Hypodermatidae.* Elsevier Science Publishers, Amsterdam, pp. 237–240 (346 pp).

Soós, Á. and Minár, J. (1986b) Family Oestridae. In: Soós, Á. and Papp, L. (eds) *Catalogue of Palaearctic Diptera,* vol. 11. *Scathophagidae – Hypodermatidae.* Elsevier Science Publishers, Amsterdam, pp. 240–244 (346 pp).

Soós, Á. and Minár, J. (1986c) Family Hypodermatidae. In: Soós, Á. and Papp, L. (eds) *Catalogue of Palaearctic Diptera,* vol. 11. *Scathophagidae – Hypodermatidae,* Elsevier Science Publishers, Amsterdam, pp. 244–251 (346 pp).

Sperling, F.A.H. and Hickey, D.A. (1994) Mitochondrial DNA sequence variation in the spruce budworm species complex (Choristoneura: Lepidoptera). *Molecular Biology and Evolution* 11, 656–665.

Sperling, F.A.H., Anderson, G.S. and Hickey, D.A. (1994) A DNA-based approach to the identification of insect species used for post-mortem interval estimation. *Journal of Forensic Science* 39, 418–427.

Spicer, G.S. (1995) Phylogenetic utility of mitochondrial cytochrome oxidase gene: molecular evolution of *Drosophila buzzatii* species complex. *Journal of Molecular Evolution* 11, 665–749.

Spinnage, C.A. (1994) *Elephants*. T & AD Poyser Natural History, London, 319 pp.

Spradbery, J.P., Mahon, R.J., Morton, R. and Tozer, R.S. (1995) Dispersal of the Old World screw-worm fly *Chrysomya bezziana*. *Medical Veterinary Entomology* 9, 161–168.

Spratt, D.M. (1984) The occurrence of *Cephalopina titillator* (Clark) (Diptera: Oestridae) in camels in Australia. *Journal of the Australian Entomological Society* 23, 229–230.

Stearns, S.C. (1976) Life-history tactics: a review of the ideas. *Quarterly Review of Biology* 51, 3–46.

Steck, T. (1932) Ein eigenartiges Vorkommen der Dasselfliege (*Hypoderma bovis* L.). *Mitteilungen Schweizer Entomologische Gesellschaft* 25, 206–207 (in German).

Stegman, P. (1920) *The Control of the Warble-Fly*. Paul Parey, Berlin, Germany, 23 pp. (in German). (*Review of Applied Entomology*, B 11, 78, 1923.)

Stehlik, M. (1980) Skin myiasis due to *Ruttenia loxodontis* Rodhain in an African elephant. *The Veterinary Record* 107, 227.

Stevens, J.R. (2003) The evolution of myiasis in blowflies (Calliphoridae). *International Journal for Parasitology* 33, 1105–1113.

Stevens, J.R. and Wall, R. (1997a) The evolution of ectoparasitism in the genus *Lucilia* (Diptera: Calliphoridae). *International Journal of Parasitology* 27, 51–59.

Stevens, J.R. and Wall, R. (1997b) Genetic variation in populations of the blowflies *Lucilia cuprina* and *Lucilia sericata*: random amplified polymorphic DNA analysis and mitochondrial DNA sequences. *Biochemical Systematics and Ecology* 25, 81–97.

Stevens, J.R. and Wall, R. (2001) Genetic relationships between blowflies (Calliphoridae) of forensic importance. *Forensic Science International* 120, 116–123.

Stevenson, L.M., Huntley, J.F., Smith, W.D. and Jones, D.G. (1994) Local eosinophils and mast cell-related responses in abomasal nematode infections of lambs. *Immunology and Medical Microbiology* 8, 167–174.

Stewart, J.R. (1932) Treatment for *Oestrus ovis*. *Journal of the American Veterinary Medical Association* 33, 108.

Strom, G. (1990) Larvae of deer bot and warble flies found in the nasal cavity of dogs. Case report. *Svensk Veterinartidning* 42, 571–572.

Sukhapesna, V., Knapp, F.W., Lyons, E.T. and Drudge, J.H. (1975) Effect of temperature on embryonic development and egg hatchability of the horse bot, *Gasterophilus intestinalis* (Diptera: Gasterophilidae). *Journal of Medical Entomology* 12, 391–392.

Sutcliffe, J.F. (1986) Black fly host location: a review. *Canadian Journal of Zoology* 64, 1041–1053.

Suter, I., Brown, G. and Hansman, D. (1972) Horse bot-fly (*Gasterophilus intestinalis*) eggs laid on human hair. *Medical Journal of Australia* 2, 609–611.

Swallow, C. (2003) Warble fly survey results, 2002/03. *The Veterinary Record* 153, 31.

Syrdalen, P., Nitter, T. and Mehl, R. (1982) Ophthalmomyiasis *interna posterior*: report of a case caused by the reindeer warble fly larva and review of previous reported cases. *British Journal of Ophthalmology* 66, 589–593.

Syrotuck, W.G. (1972) *Scent and the Scenting Dog*. Arner, Rome, New York, 102 pp.

Tabouret, G., Prevot, F., Bergeaud, J.P., Dorchies, Ph. and Jacquiet, Ph. (2001a) *Oestrus ovis* (Diptera: Oestridæ): Sheep humoral immune response to purified excreted/secreted salivary gland 28 kDa antigen complex from second and third instar larvae. *Veterinary Parasitology* 101, 53–56.

Tabouret, G., Vouldoukis, I., Duranton, C., Prevot, F., Bergeaud, J.P., Dorchies, Ph., Mazier, D. and Jacquiet, Ph. (2001b) *Oestrus ovis* (Diptera: Oestridæ): Effects of larval excretory/secretory products on nitric oxide production by murine RAW 264.7 macrophages. *Parasite Imunology* 23, 111–119.

Tabouret, G., Jacquiet, Ph., Scholl, P.J. and Dorchies, Ph. (2001c) *Oestrus ovis* in sheep: improvement in control through knowledge of relative third-instar populations. *Veterinary Research* 32, 525–531.

Tabouret, G., Bret-Bennis, L., Dorchies, Ph. and Jacquiet, Ph. (2003a) Serine protease activity in excreted-secreted products of *Oestrus ovis* (Diptera: Oestridae) larvae. *Veterinary Parasitology* 114, 305–314.

Tabouret, G., Lacroux, C, Andreoletti, O., Bergeaud, J.P., Hailu-Tolosa, Y., Hoste, H.,

Prevot, F., Duranton-Grisez, C., Dorchies, Ph. and Jacquiet, Ph. (2003b) Cellular and humoral local immune responses in sheep experimentally infected with *Oestrus ovis* (Diptera: Oestridae). *Veterinary Research* 34, 231–241.

Takken, W. and Kline, D.L. (1989) Carbon dioxide and 1-octen-3-ol as mosquito attractants. *Journal of the American Mosquito Control Association* 5, 311–316.

Tarry, D.W. (1986) Progress in warble fly eradication. *Parasitology Today* 2, 111–116.

Tarry, D.W., Sinclair, I.J. and Wassall, D.A. (1992) Progress in the British hypodermosis eradication programme: the role of serological surveillance. *The Veterinary Record* 131, 310–312.

Tashkinov, N.I. (1976) Migration of larvae of *Oedemagena tarandi* L. at the artificial infection of reindeer deer. *Parasitologiya* 10, 56–60.

Tassi, P., Puccini, V. and Giangaspero, A. (1987) Efficacy of ivermectin against goat warbles (*Przhevalskiana silenus* Brauer). *Veterinary Record* 120, 421.

Tatchell, R.J. (1958) The physiology of digestion in the larvae of the horse bot-fly *Gasterophilus intestinalis* (De Geer). *Parasitology* 48, 448–462.

Tatchell, R.J. (1960) A comparative account of the tracheal system of larvae of the horse bot-fly, *Gasterophilus intestinalis* (De Geer), and of some other dipterous larvae. *Parasitology* 50, 481–496.

Temeyer, K.B. and Pruett, J.H. (1990) Preparation of biologically active messenger RNA from larvae of the common cattle grub (Diptera: Oestridae). *Annals of the Entomological Society of America* 83, 55–58.

Terada, B. and Ono, S. (1930) The toxicological investigation in the toxic substance obtained from the larvae of *Hypoderma* sp. of Mongolian cattle at the oesophageal stage. *Journal of the Japanese Society of Veterinary Science* 9, 212–220.

Terra, W.R., Espinoza-Fuentes, F.P., Ribeiro, A.F. and Ferreira, C. (1988) The larval midgut of the housefly (*Musca domestica*): ultrastructure, fluid fluxes and ion secretion in relation to the organization of digestion. *Journal of Insect Physiology* 34, 463–472.

Terra, W.R., Ferreira, C. and Baker, J.E. (1996) Compartmentalization of digestion. In:

Lehane, M.J. and Billingsley, P.F. (eds) *Biology of The Insect Midgut*. Chapman & Hall, London, pp. 206–235 (480 pp).

Tesfaye, (1993) Prevalence of oestrosis in Ethiopian small ruminants. DVM thesis, Addis Ababa University, Ethiopia, 144 pp.

Teskey, H.J. (1981) Morphology and terminology – larvae. *Manual of Nearctic Diptera*, Vol 1 (coord. by J.F. McAlpine, B.V. Peterson, G.E. Shewell, H.J. Teskey, J.R. Vockeroth and D.M. Wood), pp. 65-88. Agriculture Canada Research Branch Monograph 27. 674 pp.

Tessier, M. (1811) Instruction sur les betes a laine, et particulierement sur la race des Merinos. *De l'imprimerie et dans la librairie de Madame Huzard (Paris)* 1811, 323 (in French).

Thomann, H. (1947) Ueber ein massenschwärm von *Cephenomyia stimulator* Clark (Diptera). *Mitteilungen Schweizer Entomologische Gesellschaft* 20, 304–305 (in German).

Thompson, J.D., Gibson, T.J., Plewniak, F., Jeanmougin, F. and Higgins, D.G. (1997) The ClustalX windows interface: flexible strategies for multiple sequence alignment aided by quality analysis tools. *Nucleic Acids Research* 24, 4876–4882.

Thornhill, R. and Alcock, J. (1983) *The Evolution of Insect Mating Systems*. Harvard University Press, Cambridge, Massachussetts, 547 pp.

Tommeräs, B.A., Wibe, A., Nilssen, A.C. and Anderson, J.R. (1993) The olfactory response of the reindeer nose bot fly, *Cephenemyia trompe* (Oestridae), to components from interdigital pheromone gland and urine from the host reindeer, *Rangifer tarandus*. *Chemoecology* 4, 115–119.

Tommeräs, B.A., Nilssen, A.C. and Wibe, A. (1996) The two reindeer parasites, *Hypoderma tarandi* and *Cephenemyia trompe* (Oestridae), have evolved similar olfactory receptor abilities to volatiles from their common host. *Chemoecology* 7, 1–7.

Tong, N.T., Imhoff, J.M., Lecroisey, A. and Keil, B. (1981) Hypodermin A, a trypsin-like neutral proteinase from the insect *Hypoderma lineatum*. *Biochemica et Biophysica Acta* 658, 209–219.

Townsend, C.H.T. (1917) *Lithohypoderma*, a new fossil genus of oestrids. *Insecutor Inscitiae Menstruus* 4, 128–130 (1916).

Townsend, C.H.T. (1938a) Five new genera of fossil *Oestromuscaria* (Diptera). *Entomological News* 49, 166–167.

Townsend, C.H.T. (1938b) *Manual of Myiology in Twelve Parts. Part VII: Oestroid Generic Diagnoses and Data (Gymnosomatini to Senostomatini).* Charles Townsend and Filhos, Itaquaquecetubo, São Paulo, Brasil, 434 pp.

Townsend, L.H., Jr, Hall, R.D. and Turner, E.C., Jr (1978) Human oral myiasis in Virginia caused by *Gasterophilus intestinalis* (Diptera: Gasterophilidae). *Proceedings of the Entomological Society of Washington* 80, 129–130.

Tunon De Lara, J.M., Berger, P. and Taytard, A. (1996) *Mastocytes, Basophileset BiologieClinique*, Biotem edns, pp. 119–129.

Ullrich, H. (1939) Zur biologie der rachenbremsen unseres einheimischen wildes genus *Cephenomyia* (sic) Latreille und genus *Pharyngomyia* Schiner. *Verhandlungen VII. International Kongress of Entomologie Weimar* 3, 2149–2171 (in German).

Urban, J.F. Jr, Fayer, R., Sullivan, C., Goldhill, J., Shea-Donohue, T., Madden, K., Morris, S.C., Katona, I., Gause, W., Ruff, M., Mansfield, L.S., Finkelman, F.D. (1996) Local TH1 and TH2 responses to parasitic infection in the intestine: regulation by IFN-gamma and IL-4. *Veterinary Immunology and Immunopathology* 54, 337–344.

Uriarte, F.J. and Ell, S.R. (1997) Doctor, there are maggots in my nose. *Journal of the Royal Society of Medicine* 11, 634–635.

Uribe, L.F., McMullin, P.F., Cramer, L.G. and Amaral, N.K. (1989) Topically applied ivermectin: efficacy against torsalo (Diptera: Cuterebridae). *Journal of Economic Entomology* 82, 847–849.

Utsi, P.M. (1979) Förekomst av svalgbroms (Cephenomyia trompe) på fjälltoppar, ansamlingsbeteende och parning. Unpublished thesis, Umeå University, Umeå, Sweden, 23 pp (in Swedish with English translation).

Uttamchandani, R.B., Trigo, L.M., Poppiti, R.J., Jr, Rosen, S. and Ratzan, K.R. (1989) Eosinophilic pleural effusion in cutaneous myiasis. *Southern Medical Journal* 82, 1288–1291.

Vaillant, J., Argente, G. and Boulard, C. (1997) Hypodermose bovine: une procédure quasi séquentielle d'observation des cheptels pour la surveillance de la recrudescence en zone éradiquée. *Veterinary Research* 28, 461–471.

Vale, G.A. (1980) Flight as a factor in the host-finding behaviour of tsetse flies (Diptera: Glossinidae). *Bulletin of Entomological Research* 70, 299–307.

Vale, G.A. (1993) Development of baits for tsetse flies (Diptera: Glossinidae) in Zimbabwe. *Journal of Medical Entomology* 30, 831–842.

Vallisnieri, A. (1696) Saggio de' dialoghi sopra la curiosa origine di molti insetti. La Galleria de Minerva. Tomo I. *Presso Girolamo Albrizzi, Venezia.*

Vallisnieri, A. (1713a) Esperienze, ed osservazioni intorno all'origine, sviluppi e constumi di varj insetti, con alter spettanti alla naturale, e medica storia. *Stamperia del Seminario, appresso Giovanni Manfre (Padoa).*

Vallisnieri, A. (1713b) Nuove osservazioni medico fisiche fatte dal Sig. Antonio Vallisnieri nella costituzione verminosa, ed epidemica seguita nelle cavalle, cavalli e puledri del mantovano. *Giornale de'letterati d'Italia* 14, 1–50.

Vallisnieri, A. (1733) Opere fisico-mediche stampate e manoscritte dal Kavalier Antonio Vallisnieri raccolte da Antonio suo figliuolo. *Appresso Sagastiano Coleti (Venezia).*

Vashkevich, R.B. (1978) Podkozhnyi ovod severnykh olenei kak nositel i perenoschik brutselleza. (The reindeer warble fly, *Oedemagena tarandi*, as a vector of *Brucella*.) Voprosy prirodnoi ochagovosti boleznei. (Contributions to the natural fatality of diseases.) *Izdatel'stvo "Nauka"; Alma Ata, Kazakhskaya SSR; USSR* 9, 119–131 (in Russian).

Vasil'eva, S.F. and Yakubovskaya, T.V. (1980) Ophthalmomyiasis. *Oftal'mologicheskii Zhurnal* 35, 443–444.

Vegetius, R.P. (1528) Scriptores rei rusticae veteres latini ex recensione Jo. Matthiae Gesneri cum Notis selectioribus. *Apud Thomam Bettinelli, Venetiis.* Tomus V, L. I, Cap. XXIV, 63.

Veraldi, S., Gorani, A., Suss, L and Tadini, G. (1998) Cutaneous myiasis caused by *Dermatobia hominis. Pediatric Dermatology* 15, 116–118.

Verger, P., Doby, J.M., Moulinier, C., Micheau, M. and Fontan, D. (1975) Infection by

Hypoderma bovis with general symptoms. *Archives Franceaisesde Pediatrie* 32, 489–490.

Viatteau, E., Nguyen Van Khanh, Jacquiet, Ph. and Dorchies, Ph. (1999) Mastocytes et osinophiles des muqueuses rhino-pharyngiennes du dromadaire infesté par *Cephalopina titillator* (Clark 1816). *Revue de Médecine Vétérinaire* 150, 4, 353–356 (in French).

Vogler, A.P. and DeSalle, R. (1994) Evolution and phylogenetic information content of the ITS-1 region in the Tiger Beetle *Cicindela dorsalis*. *Molecular Biology and Evolution* 11, 393–405.

Volf, P., Lukes, J. and Srp, V. (1990) Study on the population of the warble fly, *Oestromyia leporina* (Pallas 1778) (Diptera, Hypodermatidae) in Bohemia. *Folia Parasitologica (Praha)* 37, 187–190.

Vyslouzil, L. (1989) Occurrence of *Hypoderma diana* in an unusual host. *Veterinarstvi* 39, 222.

Wall, R. and Strong, L. (1987) Environmental consequences of treating cattle with the antiparasitic drug ivermectin. *Nature* 37, 418–421.

Wallace, R. and Ormerod, E. (1904) *Economic Entomologist. Autobiography and Correspondence.* Dutton, New York.

Wallman, J.F. and Donnellan. S.C. (2001) The utility of mitochondrial DNA sequences for the identification of forensically important blowflies (Diptera: Calliphoridae) in southeastern Australia. *Forensic Science International* 120, 60–67.

Walton, C.L. (1930) The occurrence of males of the horse bot fly. *Northwestern Naturalist* 5, 224–226.

Warnecke, V.M. and Göltenboth, R. (1977) Über das Auftreten der Magenbremse, *Cyrostigma conjugens* Enderlein, bei zwei Spitzmaulnashörnern (*Diceros bicornis* L.) des Berliner Zoologischen Gartens. *Berliner und Münchener Tierarztliche Wochenschrift* 90, 159–161.

Washburn, R.H., Klebesadel, L.J., Palmer, J.S., Luick, J.R. and Bleicher, D.P. (1980) The warble-fly problem in Alaska reindeer. *Agroborealis* 12, 23–28.

Watson, R.H. and Radford, H.M. (1960) The influence of rams on the onset of oestrus in Merino ewes in the spring. *Australian Journal of Agricultural Research* 11, 65–71.

Weber, M. (1992) Valoracion clinica del efecto de la ivermectina contra *Cephenamyia* spp. en venados cola blanca. *Veterinaria Mexico* 23, 239–242.

Webster, K.A. and Tarry, D.W. (1995) Warble fly in Great Britain. An update. In: Tarry, D.W., Pithan, K. and Webster, K. (eds) *Improvements in the Control Methods for Warble-fly in Cattle and Goats.* Proceedings, 8–10 September 1994, Guildford, UK. Commission of the European Communities, Brussels, pp. 6–7.

Webster, K.A., Dawson, C., Flowers, M. and Richards, M.S. (1997a) Serological prevalence of *Hypoderma* species in cattle in Great Britain (1995/96) and the relative value of serological surveillance over clinical observation. *Veterinary Record* 141, 261–263.

Webster, K.A., Giles, M. and Dawson, C. (1997b) A competitive ELISA for the serodiagnosis of hypodermosis. *Veterinay Parasitology* 68, 155–164.

Weintraub, J. (1961) Inducing mating and oviposition of the warble flies *Hypoderma bovis* (L.) and *H. lineatum* (De Vill.) (Diptera: Oestridae) in captivity. *Canadian Entomologist* 93, 149–156.

Weintraub, J. and Scholl, P.J. (1984) The joint Canada–U.S. pilot project on warble grub control. In: *Research Highlights – 1983.* Agriculture Canada Research Station, Lethbridge, Alberta, Canada, pp. 30–32.

Weintraub, J., McGregor, W.S. and Brundrett, H.M. (1961) Artificial infestations of the northern cattle grub, *Hypoderma bovis*, in Texas. *Journal of Economic Entomology* 54, 84–87.

Weisbroth, S.H., Wang, R. and Scher, S. (1973) *Cuterebra buccata*: Immune response in myiasis of domestic rabbits. *Experimental Parasitology* 34, 22–31.

Wells, R.W. and Knipling, E.F. (1938) A report of some recent studies on species of *Gasterophilus* occurring in horses in the United States. *Iowa State College Journal of Science* 12, 181–203.

Wells, J.D. and Sperling, F.A.H. (2001) DNA-based identification of forensically important Chrysomyinae (Diptera: Calliphoridae). *Forensic Science International* 120, 110–115.

Wells, J.D., Introna, F., Di Vella, M.D., Campobasso, C.P., Hayes, J. and Sperling, F.A.H. (2000) Human and insect mitochondrial DNA analysis from maggots. *Journal of Forensic Science* 46, 685–687.

Wells, J.D., Pape, T. and Sperling, F.A.H. (2001) DNA-based identification and molecular systematics of forensically important Sarcophagidae (Diptera). *Journal of Forensic Science* 46, 1098–1102.

Wheeler, C.H. (1989) Mobilization and transport of fuels to the flight muscles. In: *Insect Flight*. Goldsworthy, G.J. and Wheeler, C.H. (eds) CRC Press, Boca Raton, Florida, pp. 273–304.

Wiens, J.J. and Servedio, M.R. (1998) Phylogenetic analysis and intraspecific variation: performance of parsimony, likelihood and distance methods. *Systematic Biology* 47, 228–253.

Willadsen, P. (2001) The molecular revolution in the development of vaccines against ectoparasites. *Veterinary Parasitology* 101, 353–367.

Willemse, L.P.M. and Takken, W. (1994) Odor-induced host location in tsetse flies (Diptera: Glossinidae). *Journal of Medical Entomology* 31, 775–794.

Wolfe, L.S. (1959) Observations on the histopathological changes caused by the larvae of *Hypoderma bovis* (L.) and *Hypoderma lineatum* (De Vill.) (Diptera: Oestridae) in tissues of cattle. *Canadian Journal of Animal Science* 39, 145–157.

Wolstenholme, D.R. (1992) Animal mitochondrial DNA: structure and evolution. *International Review of Cytology* 141, 173–216.

Wood, D.M. (1986) Are Cuterebridae, Gasterophilidae, Hypodermatidae and Oestridae a monophyletic group? In: Darvas B. and Papp L. (eds) *Abstracts of the First International Congress of Dipterology*, 17–24 August 1986, Budapest, Hungary, 261pp.

Wood, D.M. (1987) Oestridae. In: McAlpine, J.F. (ed.) *Manual of Nearctic Diptera*, vol. 2. Research Branch Agriculture Canada, Monograph No 28, Ottawa, Canada, pp. 1147–1158 (1332 pp.).

Wratten, S.D. and Forbes, A.B. (1996) Environmental assessment of veterinary avermectins in temperate pastoral ecosystems. *Annals of Applied Biology* 128, 329–348.

Wyss, A.R., Flynn, J.J., Norell, M.A., Swisher, C.C., Charrier, R., Novacek, M.J. and McKenna, M.C. (1993) South America's earliest rodent and recognition of a new interval of mammalian evolution. *Nature* 365, 434–437.

Wyss, J.H. (2000) Screwworm eradication in the Americas. In: *Report of the 19th Conference of OIE Regional Commission for Europe*, 19–22 September 2000, Jerusalem, Israel, pp. 239–244.

Yeruham, I., Malnick, S., Bass, D. and Rosen, S. (1997) An apparently pharyngeal myiasis in a patient caused by *Oestrus ovis* (Oestridae: Diptera). *Acta Tropica* 68, 361–363.

Yilma, J.M. (1992) Contribution à L'étude de L'épidémiologie, du Diagnostic Immunologique et de la Physiopathologie de L'Oestrose ovine. Thèse Institut National Polytechnique de Toulouse, France, 223 pp. (in French).

Yilma, J.M. and Dorchies, Ph. (1991) Epidemiological study of *Oestrus ovis* in south west France. *Veterinary Parasitogy* 40, 315–323.

Yilma, J.M. and Dorchies, Ph. (1993) Essais d'infestations expérimentales de l'agneau par des larves 1 d'*Oestrus ovis*. *Bulletin de la Société Français de Parasitologie* 43, 43–47.

Yin, H., Ma, M., Yuan, G., Huang, S., Liu, Z., Luo, J. and Guan, G. (2003) Hypodermosis in China. *Journal of Animal and Veterinary Advances* 2, 179–183.

Zacharuk, R.Y. and Shields, V.D. (1991) Sensilla of immature insects. *Annual Review of Entomology* 36, 331–354.

Zayani, A., Chaabouni, M., Gouiaa, R., Ben Hadj Hamida, F. and Fki, J. (1989) Myiases conjonctivales A propos de 23 cas dans le Sahel tunisien. *Archives de l'Institut Pasteur de Tunis* 66, 289–292 (in French with English summary).

Zayed, A.A. (1992) Studies on *Rhinoestrus purpureus* (Diptera: Oestridae) larvae infesting donkeys (*Equus asinus*) in Egypt. III. Pupal duration under controlled conditions. *Veterinary Parasitology* 44, 285–290.

Zayed, A.A. (1998) Localization and migration route of *Cephalopina titillator* (Diptera: Oestridae) larvae in the head of infested camels (*Camelus dromedarius*). *Veterinary Parasitology* 80, 65–70.

Zeledon, R. and Silva, S. (1987) Attempts to culture the parasitic stage of *Dermatobia hominis* (L., Jr.) *in vitro* (Diptera: Cuterebridae). *Journal of Parasitology* 73, 907–909.

Zhang, D.X. and Hewitt, G.M. (1996a) Nuclear integrations: challenges for mitochondrial DNA markers. *Trends in Ecology and Evolution* 11, 247–251.

Zhang, D.X. and Hewitt, G.M. (1996b) Assessment of the universality and utility of a set of conserved mitochondrial COI primers in insects. *Insect Molecular Biology* 6, 143–150.

Zhang, D.X., Szymura, J.M. and Hewitt, G.M. (1995) Evolution and structural conservation of the control region of mitochondrial DNA. *Journal of Molecular Evolution* 40, 382–391.

Zrzavy, J. (1997) Phylogenetics and ecology: all characters should be included in the cladistic analysis. *Oikos* 80, 186–192.

Zumpt, F. (1957) Some remarks on the classification of the Oestridae *s.lat.* (Diptera). *Journal of the Entomological Society of Southern Africa* 20, 154–161.

Zumpt, F. (1961) The enigma of *Strobiloestrus* Brauer (Diptera: Oestridae). *Proceedings of the Royal Entomological Society of London* (B) 30, 95–102.

Zumpt, F. (1962a) The genus *Gyrostigma* Brauer (Diptera: Gasterophilidae). *Zeitschrift für Parasitenkunde* 22, 245–260.

Zumpt, F. (1962b) The oestroid flies of wild and domestic animals in the Ethiopian region, with a discussion of their medical and veterinary importance (Diptera: Oestridae and Gasterophildae). *Zeitschrift für angewandte Zoologie* 49, 393–397.

Zumpt, F. (1963) Ophthalmomyiasis in man, with special reference to the situation in southern Africa. *South African Medical Journal* 37, 425–428.

Zumpt, F. (1965) *Myiasis in Man and Animals in the Old World.* Butterworth & Co., London, 267 pp.

Index

Host Index

Akodon
    azarae 276
    molinae 276
Alcelaphus
    buselaphus 222, 226, 228
    lichtensteini 222, 226, 228
Alces
    alces 237
Alouatta
    belsebul 277
    palliata 275, 277, 279, 280, 281
Antidorcas
    marsupialis 232
Aotus
    trivirgatus 277
Apodemus
    speciosus 262
Arvicola
    terrestris 269

Bos
    grunniens 254
    indicus 282
    taurus 254, 282

Camelus
    dromedaries 247
Capra
    caucasica 228
Capra
    hirsicus 258, 272
    ibex 228

Capreolus
    capreolus 236, 237, 240, 254
Ceratotherium
    simum 294, 295
Cervus
    elaphus 236, 237, 240, 254
    nippon 240
Citellus
    undulatus 269
Connoachaetes
    gnou 222
    taurinus 222, 226, 228

Dama
    dama 237, 240
Damaliscus
    dorcas 222, 228
    korrigum 222, 226, 228
    lunatus 222, 228
Dicerorhinus
    sumatrensis 295
Diceros
    bicornis 295
Didelphis 277
Dipodomys
    ordii 275, 278

Elephas
    maximus 299
Equus 231
    burchelli 232, 289, 291
    cabalis 232
    zebra 232

Parasite Index

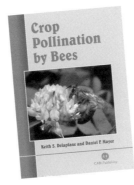
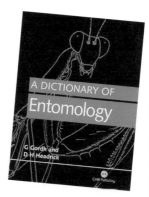
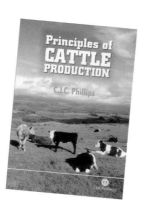